6

Topics in Heterocyclic Chemistry

Series Editor: R. R. Gupta

AF895407

Topics in Heterocyclic Chemistry
Series Editor: R. R. Gupta

Recently Published and Forthcoming Volumes

Bioactive Heterocycles I
Volume Editor: S. Eguchi
Volume 6, 2006

Marine Natural Products
Volume Editor: H. Kiyota
Volume 5, 2006

QSAR and Molecular Modeling Studies in Heterocyclic Drugs II
Volume Editor: S. P. Gupta
Volume 4, 2006

QSAR and Molecular Modeling Studies in Heterocyclic Drugs I
Volume Editor: S. P. Gupta
Volume 3, 2006

Heterocyclic Antitumor Antibiotics
Volume Editor: M. Lee
Volume 2, 2006

Microwave-Assisted Synthesis of Heterocycles
Volume Editors: E. Van der Eycken, C. O. Kappe
Volume 1, 2006

Bioactive Heterocycles I

Volume Editor: Shoji Eguchi

With contributions by
S. Eguchi · M. Kita · H. Kiyota · H. Nishino
M. Ohno · M. Somei · D. Uemura

Springer

The series *Topics in Heterocyclic Chemistry* presents critical reviews on "Heterocyclic Compounds" within topic-related volumes dealing with all aspects such as synthesis, reaction mechanisms, structure complexity, properties, reactivity, stability, fundamental and theoretical studies, biology, biomedical studies, pharmacological aspects, applications in material sciences, etc. Metabolism will be also included which will provide information useful in designing pharmacologically active agents. Pathways involving destruction of heterocyclic rings will also be dealt with so that synthesis of specifically functionalized non-heterocyclic molecules can be designed.
The overall scope is to cover topics dealing with most of the areas of current trends in heterocyclic chemistry which will suit to a larger heterocyclic community.
As a rule contributions are specially commissioned. The editors and publishers will, however, always be pleased to receive suggestions and supplementary information. Papers are accepted for *Topics in Heterocyclic Chemistry* in English.
In references *Topics in Heterocyclic Chemistry* is abbreviated *Top Heterocycl Chem* and is cited as a journal.

Springer WWW home page: springer.com
Visit the THC content at springerlink.com

ISSN 1861-9282

DOI 10.1007/11514725

ISBN 978-3-642-06998-7 e-ISBN 978-3-540-33351-7

Springer is a part of Springer Science+Business Media

springer.com

Softcover reprint of the hardcover 1st edition 2006

Cover design: *Design & Production* GmbH, Heidelberg

Printed on acid-free paper 02/3100 YL – 5 4 3 2 1 0

Topics in Heterocyclic Chemistry Also Available Electronically

For all customers who have a standing order to Topics in Heterocyclic Chemistry, we offer the electronic version via SpringerLink free of charge. Please contact your librarian who can receive a password or free access to the full articles by registering at:

springerlink.com

If you do not have a subscription, you can still view the tables of contents of the volumes and the abstract of each article by going to the SpringerLink Homepage, clicking on "Browse by Online Libraries", then "Chemical Sciences", and finally choose Topics in Heterocyclic Chemistry.

You will find information about the

- Editorial Board
- Aims and Scope
- Instructions for Authors
- Sample Contribution

at springer.com using the search function.

Preface to the Series

Topics in Heterocyclic Chemistry presents critical accounts of heterocyclic compounds (cyclic compounds containing at least one heteroatom other than carbon in the ring) ranging from three members to supramolecules. More than 50% of billions of compounds listed in *Chemical Abstracts* are heterocyclic compounds. The branch of chemistry dealing with these heterocyclic compounds is called heterocyclic chemistry, which is the largest branch of chemistry and as such the chemical literature appearing every year as research papers and review articles is vast and can not be covered in a single volume.

This series in heterocyclic chemistry is being introduced to collectively make available critically and comprehensively reviewed literature scattered in various journals as papers and review articles. All sorts of heterocyclic compounds originating from synthesis, natural products, marine products, insects, etc. will be covered. Several heterocyclic compounds play a significant role in maintaining life. Blood constituent hemoglobin and purines as well as pyrimidines, the constituents of nucleic acid (DNA and RNA) are also heterocyclic compounds. Several amino acids, carbohydrates, vitamins, alkaloids, antibiotics, etc. are also heterocyclic compounds that are essential for life. Heterocyclic compounds are widely used in clinical practice as drugs, but all applications of heterocyclic medicines can not be discussed in detail. In addition to such applications, heterocyclic compounds also find several applications in the plastics industry, in photography as sensitizers and developers, and in dye industry as dyes, etc.

Each volume will be thematic, dealing with a specific and related subject that will cover fundamental, basic aspects including synthesis, isolation, purification, physical and chemical properties, stability and reactivity, reactions involving mechanisms, intra- and intermolecular transformations, intra- and intermolecular rearrangements, applications as medicinal agents, biological and biomedical studies, pharmacological aspects, applications in material science, and industrial and structural applications.

The synthesis of heterocyclic compounds using transition metals and using heterocyclic compounds as intermediates in the synthesis of other organic compounds will be an additional feature of each volume. Pathways involving the destruction of heterocyclic rings will also be dealt with so that the synthesis of specifically functionalized non-heterocyclic molecules can be designed. Each volume in this series will provide an overall picture of heterocyclic compounds critically and comprehensively evaluated based on five to ten years of literature. Graduates, research students and scientists in the fields of chemistry, pharmaceutical chemistry, medicinal chemistry, dyestuff chemistry, agrochemistry, etc. in universities, industry, and research organizations will find this series useful.

I express my sincere thanks to the Springer staff, especially to Dr. Marion Hertel, executive editor, chemistry, and Birgit Kollmar-Thoni, desk editor, chemistry, for their excellent collaboration during the establishment of this series and preparation of the volumes. I also thank my colleague Dr. Mahendra Kumar for providing valuable suggestions. I am also thankful to my wife Mrs. Vimla Gupta for her multifaceted cooperation.

Jaipur, 31 January 2006 R.R. Gupta

Preface

In the series of Topics in Heterocyclic Chemistry, the volume of Bioactive Heterocycles aims to present comprehensive reviews on selected topics of synthetic as well as naturally occurring bioactive heterocycles.

The present volume comprises six chapters of the following specialized reviews.

The first chapter, 'Directed Synthesis of Biologically Interesting Heterocycles with Squaric Acid Based Technology' by Masatomi Ohno and Shoji Eguchi covers squaric acid and its derivatives as versatile synthons for target-oriented and diversity-oriented synthesis. The introduction of designed functional groups, followed by ring conversion induced thermally or by reactive intermediates can construct a various bioactive heterocycles including bioactive natural products.

The second chapter 'Manganese(III)-Based Peroxidation of Alkenes to Heterocycles' by Hiroshi Nishino presents a very comprehensive review on novel Mn(III)-based peroxidation chemistry, and related bioactive heterocycles based on the works of his group. The content includes synthesis of functionalized 1,2-dioxane derivatives from various 1,3-dicarbonyl compounds including nitrogen heterocycles. The spectroscopic feature, the formation mechanism of 1,2-dioxan-3-ol ring system, chemical transformations and synthetic applications are also discussed.

The third chapter 'A Frontier in Indole Chemistry: 1-Hydroxyindoles, 1-Hydroxytryptamines, and 1-Hydroxytryptophans' by Masanori Somei presents a very comprehensive review on chemistry of 1-hydroxy-indoles, -tryptamines, and -tryptophans as a frontier in indole chemistry. In fact, these new members of indole derivatives were not much known about 30 years ago in the long histry of indole alkaloids and related chemistry. Nowadays, these new families of indole compounds have been demonstrated to play their important role in life and nature by the pioneering works of Somei and his coworkers. The interesting biological and pharmaceutical activities have been found in these derivatives.

The fourth chapter 'Quinazoline Alkaloids and Related Chemistry' by Shoji Eguchi provides a perspective review focusing on developements of the synthetic methodologies and their synthetic applications. A brief historical background, aza-Wittig methodology, microwave-assisted synthesis, solid-phase synthesis, and a variety of new synthesis of quinazoline compounds by organometallic reagents, metal-catalyzed reactions, heterocyclizations, pericyclic reactions etc are briefly reviewed. Selected topics of total synthesis of various types of quinazoline alkaloids including substituted type like febrifugine and heterocycle-fused type such as pyrroloquinazolines, indolopyridoquinazolines, pyrazinoquinazolines, pyrroloquinazolinoquinolines by these methodologies are discussed.

The fifth chapter 'Bioactive Heterocyclic Alkaloids from Marine Orgin' by Masakin Kita and Daisuke Uemura presents a very comprehensive review on novel heterocyclic marine alkaloids with very intriguing structures and useful biological properties like anti-osteoprotic activity focusing on isolations, structural, synthetic, biological, and biogenetic studies mainly by Uemura group. The contents are believed to attract much attention by organic chemists, heterocyclic chemists, synthetic chemists, and workers in medicinal, pharmaceutical and bioscience fields.

The sixth chapter 'Synthetic Studies on Heterocyclic Antibiotics Containing Nitrogen Atoms' by Hiromasa Kiyota presents a very comprehensive review on a variety of heterocyclic antibiotics and phytotoxins. Early and recent examples of synthetic studies of glutarimide antibiotics, antimycins, and tabtoxins and related bioactive heterocycles based on the works of his group are retrospectively reviewed. The content is believed to attracts much interest of synthetic chemists as well as heterocyclic chemists and researchers in life science fields.

I hope that our readers find this series to be a useful guide to modern heterocyclic chemistry. As always, I encourage both suggestions for improvements and ideas for review topics.

Nagoya, March 2006 Shoji Eguchi

Contents

Contents of Volume 2

Heterocyclic Antitumor Antibiotics

Volume Editor: Mosews Lee
ISBN: 3-540-30982-9

Top Heterocycl Chem (2006) 6: 1–37
DOI 10.1007/7081_035

Published online: 21 April 2006

Directed Synthesis of Biologically Interesting Heterocycles with Squaric Acid (3,4-Dihydroxy-3-cyclobutene-1,2-dione) Based Technology

Masatomi Ohno[1] (✉) · Shoji Eguchi[2]

[1]Department of Advanced Science and Technology, Toyota Technological Institute, 2-12-1 Hisakata, Tempaku, 468-8511 Nagoya, Japan
ohno@toyota-ti.ac.jp

[2]Department of Molecular Design & Engineering, Graduate School of Engineering, Nagoya University, Furo-cho, Chikusa, 464-8603 Nagoya, Japan

Abstract A variety of methods for organic transformation starting from squaric acid have been developed in this decade. These are based on conversion of pseudoaromatic 3,4-dihydroxy-3-cyclobutene-1,2-dione into the more reactive 4-hydroxy-2-cyclobutenone by introduction of the required (or desired) functional groups followed by key ring transformation, the rearrangement being stimulated thermally or induced by a reactive intermediate. These strategies can construct a variety of bioactive heterocycles when functional groups contain heteroatoms or heterocycles. Interestingly, squaric acid is rendered as an acid part, for example, of an amino acid, and this bioisostere concept is extended to various heterocycle-containing squaramides (3,4-diamino-3-cyclobutene-1,2-dione derivatives) as bioactive conjugate compounds. This review article covers biologically interesting heterocyclic compounds accessible with the squaric acid based technology.

Keywords Bioisostere · Cyclobutenone · Electrocyclic reaction · Reactive intermediate · Squaric acid

1
Chemistry of Squaric Acid with 3-Cyclobutene-1,2-Dione Skeleton

Squaric acid (**1**) is categorized as an oxocarbon having a four-membered ring [1] (Fig. 1). Despite being a small molecule, it possesses unique 2π-pseudoaromaticity [2–5], which brings high acidity ($pK_{a1} = 0.52$, $pK_{a2} = 3.48$) as an organic acid, and polyfunctionality, including two hydroxyl and two carbonyl groups conjugated across a double bond. Peculiar hydrogen-bonded network and chelated structures in some acid derivatives have been occasionally discussed [6–12]. The unique structure is utilized in electronic devices, for example, as a donor acceptor triad called "squaraine" (**2**) [1, 13–15]. The dicationic nature of the cyclobutene ring necessary for aromatic character is combined with the donating nature of aromatic and heteroaromatic rings to produce SHG properties, for example [16]. Dimer **3** is a new candidate designed for extension of conjugation plane [17, 18].

On the other hand, the unique structure of **1** has also been applied in organic synthesis as an attractive C_4-synthon. The relief of ring strain can serve as a significant driving force in its ring-transformation reaction and this is in fact accomplished by two processes. The first is conversion of the stable aromatic cyclobutenedione system to the more reactive hydroxycyclobutenone system; where required or desired substituents can be introduced into the ring system. The second is ring expansion from a four-membered ring to five ~ seven-membered rings in either concerted or stepwise manner. This methodology has been exploited in the synthesis of various bioactive carbo- and heterocycles [19–23]. Another feature of using **1** to develop bioactive compounds is based on variation of substituents on squaric acid esters and amides, where the cyclobutenedione ring is still retained. In fact, semisquaric acid (**4**), which is known as moniliformin, is a primitive derivative with biological activity (mycotoxin) [24]. According to the concept, for **1** to play

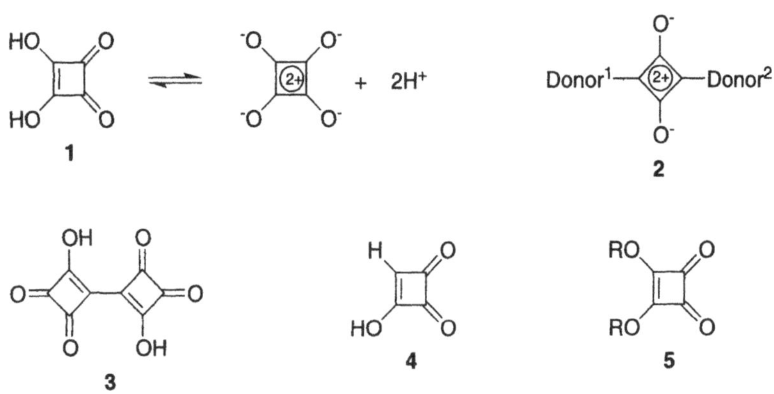

Fig. 1 Squaric acid and its derivatives [1–18]

a role of bioisostere (e.g., semisquaric acid is considered to have similarity to pyruvic acid in structure), research has pursued the possibility of **1** as a replaceable moiety for an acid part of natural products such as amino acids and nucleic acids, and for a certain part of pharmacoactive compounds (see later). As a synthetic tool, the use of **1** has recently been demonstrated in asymmetric reduction adopting the cyclobutenedione as a ligand with a chiral center [25, 26].

2
Derivatization of Squaric Acid to 4-Hydroxy-2-Cyclobutenone Skeleton

Squaric acid itself is almost useless for this aim because of its intrinsic aromatic stability and difficult solubility in organic solvents. Instead, its esters **5** are the most convenient compounds from which derivatization reactions start. While acid **1** and its esters are now commercially available, (cf. **1** is now produced on a commercial basis by Kyowa Hakkou Kogyo Co. Japan [27]). The esterification method for **1** is improved [28] and preparation of dimethyl squarate **5** (R = CH_3) is recorded in *Organic Synthesis* [29].

There are several approaches for derivatization of squaric acid (**1**). The traditional major route relies on the nucleophilic reaction of the eligible esters **5** with organolithium and organomagnesium reagents; their addition to **5** is known to be sufficiently selective to give 1,2-addition products **6** (4-substituted 4-hydroxy-2-cyclobutenones) from the former and 1,4-addition products **7** (3-substituted cyclobutene-1,2-diones) from the latter [30]. When acid-catalyzed rearrangement of **6** to form **7** with one combined substituent (R^1) at C-3 is followed by the repeated nucleophilic addition to combine another substituent (R^2), this series of reactions gives rise to 2,4-doubly substituted 4-hydroxy-2-cyclobutenone **8** [31, 32]. Trisubstituted 4-hydroxy-2-cyclobutenone **11** can be prepared via acetal intermediates **9** and **10** [33] (Scheme 1).

Organotin and copper species are also used for the derivatization of **5** by cross-coupling reactions (typically shown as **5** → **12** → **13**) [34–38]. The α-carbanion generated at the position adjacent to cyclobutenedione ring is a different derivatization route via nucleophilic addition (**14** → **15** → **16**) [39] (Scheme 2).

The derivatization method is compensated by the electrophilic addition reaction using organosilicon reagents [40–43]. Thus, the squaric acid family of derivatives, e.g., dichloride **17**, methyl ester chloride **18**, amide chloride **19**, and diester **5** are the partners of the reactions with allylsilanes, silyl enol ethers, and silyl ketene acetals (Scheme 3). In this case, 1,2- and 1,4-addition to **20** and **21**, respectively, are regulated by the substitution pattern of unsaturated organosilanes, kind of Lewis acid catalysts, and the reactivity of acid derivatives. The less congested is the reaction site, the more preferable is 1,2-

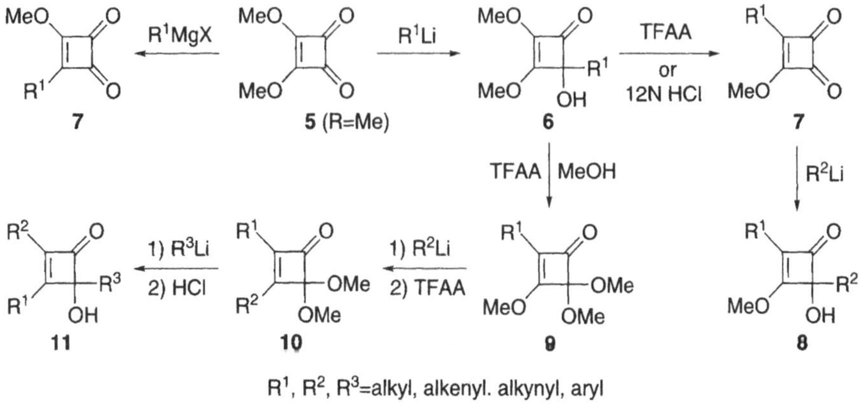

Scheme 1 Derivatization of squaric acid: traditional nucleophilic conditions [30–33]

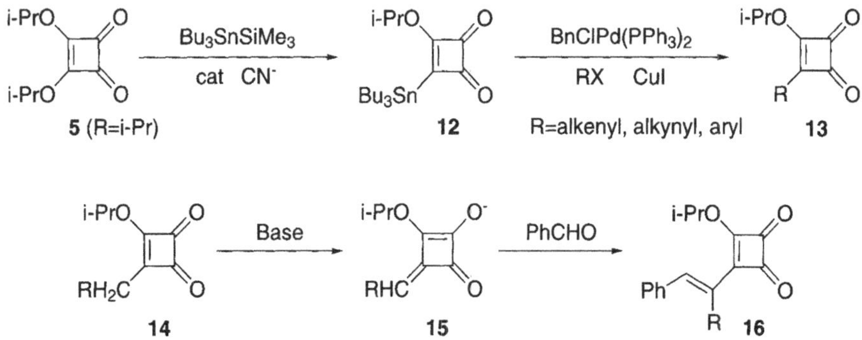

Scheme 2 Derivatization of squaric acid: organometallic routes via coupling reactions [34–39]

addition in allylsilanes and silyl enol ethers. $TiCl_4$ catalyzes 1,2-addition and $ZnCl_2$ 1,4-addition in silyl ketene acetals regardless of the substitution pattern. Only silyl enol ethers and silyl ketene acetals are reactive with diester **5** via 1,4-addition. In addition to the above carbonyl group activation, the acetal **10** is also a useful candidate for generating the electrophilic center under these conditions. Thus, besides typical organosilanes, azide functions can be introduced with BF_3-catalysis (vide infra).

Similar electrophilic Friedel–Crafts-like reactions allow the most reactive dichloride **17** to furnish 1,4-diarylcyclobutenedione derivatives [44, 45]; for example, 1,4-thieno[3,2-*b*]pyrrole-substituted cyclobutenedione **23** was prepared by this method and an oxygen-inserted conjugation system **24** was attained as a photochromic devise [46] (Scheme 4).

Apart from these methods based on the squaric acid family, direct formation of cyclobutenedione rings by [4 + 2] and [2 + 2] cycloaddition reactions is a plausible approach to variably substituted 4-hydroxy-2-cyclobutenone systems [47–54].

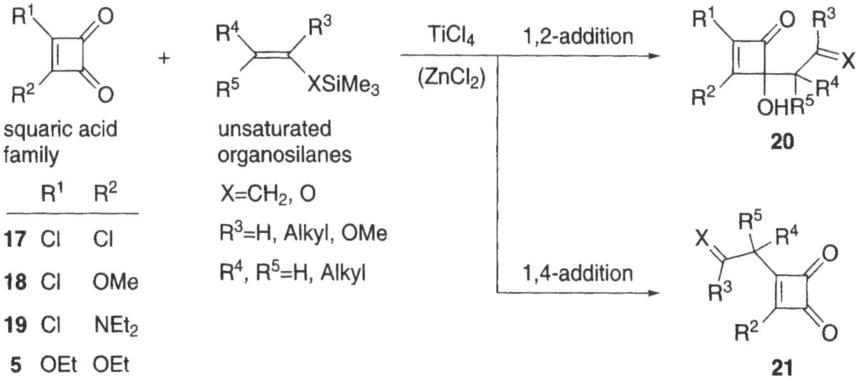

Scheme 3 Derivatization of squaric acid: electrophilic conditions using unsaturated organosilanes [40–43]

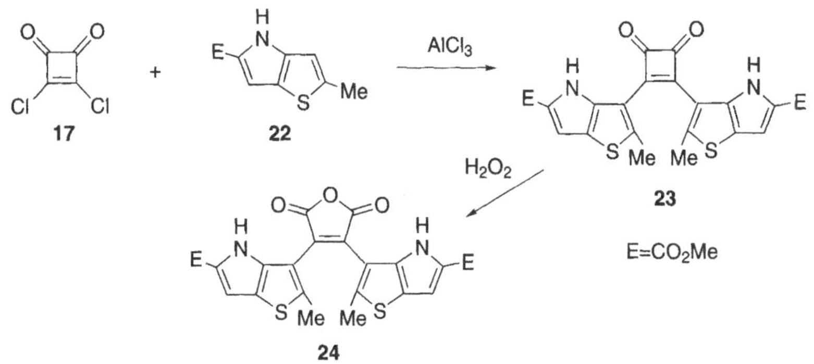

Scheme 4 Derivatization of squaric acid by Friedel–Crafts-like reaction: an example [46]

3 Ring Transformation of the Derivatized Cyclobutenone

3.1 Varied Reactivity in Ring Opening and Ring Closure

The intrinsic reactivity of small rings is ascribable to ring strain relief in nature, and in squaric acid chemistry it is accomplished by conversion of rather stable cyclobutene-1,2-dione to the more reactive 4-hydroxy-2-cyclobutenone [55, 56] as described in the previous section. At the same time, this conversion step fulfills the regiospecific introduction of substituents required for the targeted heterocyclic structure. Thereby, the set-up four-membered ring is now subjected to directed synthesis through variable ring transformation reactions.

These involve tandem ring opening and ring closure steps, which are concerted or non-concerted. The typical concerted process is 4π-electrocyclic ring opening of cyclobutene to 1,3-butadiene. This was discussed in terms of torquoselectivity by Houk [57–64]. According to his theory, π-donor substituents (R = O^-, OH, NH_2) prefer outward rotation while π-acceptor substituents (R = BMe_2, CHO) should rotate inwardly on the thermal process (Fig. 2). Recent discovery has extended this concept; a silyl substituent accelerates and promotes inward rotation despite the resulting steric congestion, and a stannyl substituent does similarly [65, 66].

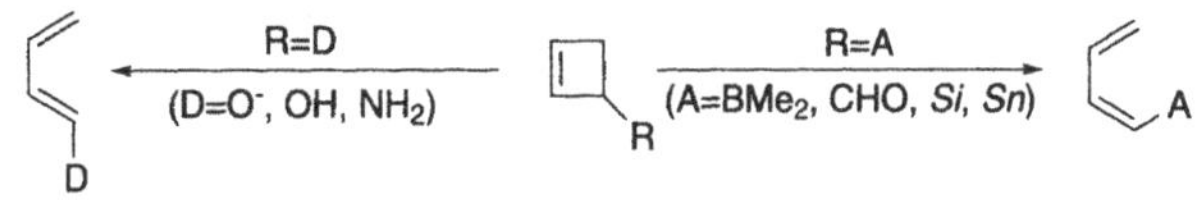

Fig. 2 Torquoselectivity in 4π electrocyclic ring opening (thermal conditions) [57–64]

4-Hydroxy-2-cyclobutenone adheres to the above prediction [55–64]. In this case, it is important that the inwardly-directed substituent (i.e., OH is an outward-directing group) is capable of participating within the molecule. Moreover, a highly reactive vinylogous ketene function occurs instead of butadiene formation to assist efficient ring-closure through intramolecular interaction. When an unsaturated bond is located at the 4-position, the consecutive process is thermally allowed 6π-electrocyclization (Fig. 3).

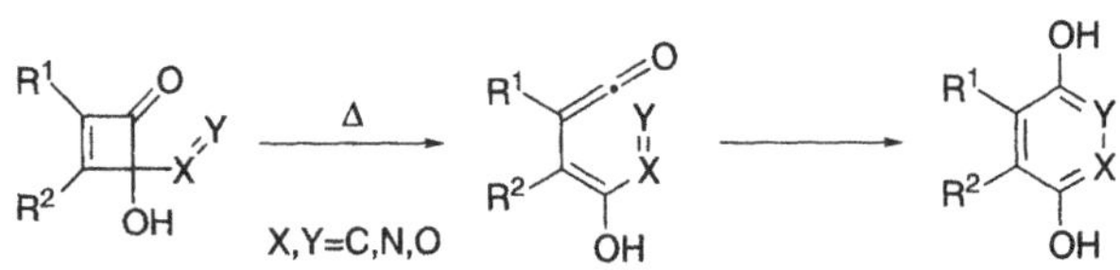

Fig. 3 Sequence of 4π–6π electrocyclic ring opening and ring closure [19–23]

This strategy is very powerful and fruitful for the directed synthesis of both carbo- and heterocycles, and successful examples have cumulatively been reported until now [19–23]. The major contribution has come from the Moore and Liebeskind groups. Among many efforts devoted in this area, the recent typical example [polysubstituted naphthoquinone, Echinochrome A (**30**)] constitutes a characteristic feature for the method of directed synthesis including **26** → **27** and **28** → **29** as key steps [67]. Ferrocenyl quinone and 5-*O*-methylembelin were also synthesized according to this methodology [68, 69] (Scheme 5).

In the case of monosubstituted cyclobutenone **31**, the adduct with lithiovinylsuofone **32** was reported to undergo an extraordinarily facile tandem 4π–6π electrocyclic process (**33** → **34**) at – 78 °C to give cyclohexenone **36** [70]. The photochemical process may oblige the opposite direction on a hydroxyl group to be oriented inwardly; actually cyanohydrin **37** was reported to give butenolide **39** as a result of an intramolecular addition reaction of (*Z*)-hydroxyvinylketene **38** [71] (Scheme 6).

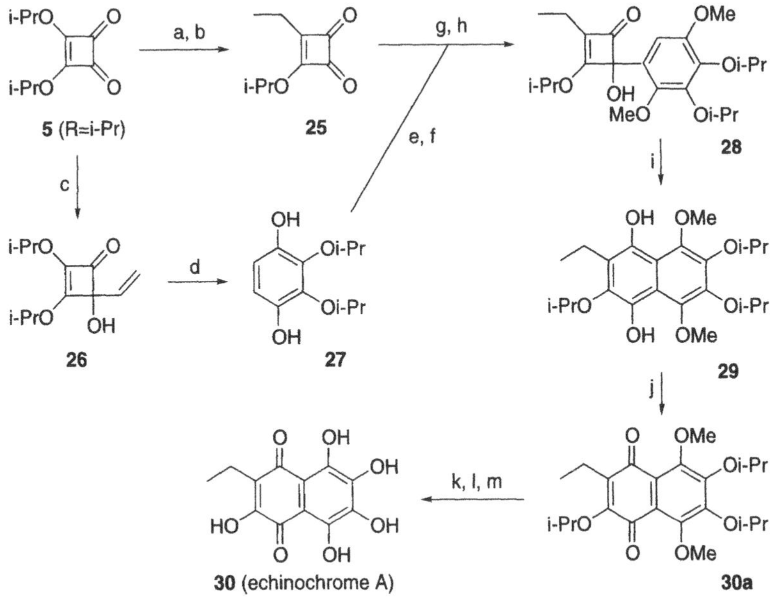

Reaction conditions: (a) EtMgBr, -78 °C; (b) cat H^+, 25 °C; (c) CH_2=CHMgBr, -78 °C; (d) 160 °C/15 min; (e) MeI/KOH; (f) *s*-BuLi; (g) THF, -78 °C; (h) H^+, -78 °C~25 °C; (i) 160 °C/15 min; (j) CAN; (k) air; (l) BBr_3; (m) H^+, -78 °C

Scheme 5 Synthesis of echinochrome A: a typical example for 4π–6π electrocyclic ring opening and ring closure [67]

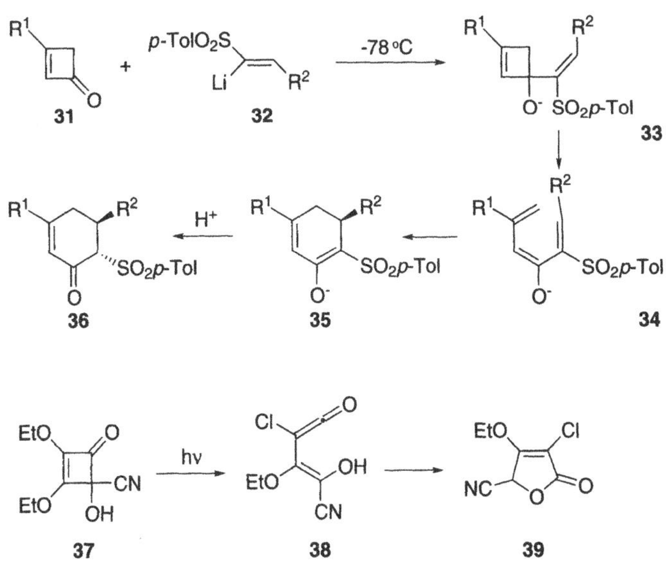

Scheme 6 Tandem 4π–6π electrocyclic concerted process at low temperature [70] and under photochemical conditions [71]

If the substituent is allylic and hence homoconjugative, intramolecular [2 + 2] cycloaddition is progressive to form bicyclo[3.2.0]heptenone as shown in Fig. 4 [19–23]. This is the second pattern of the ring transformation based on squaric acid.

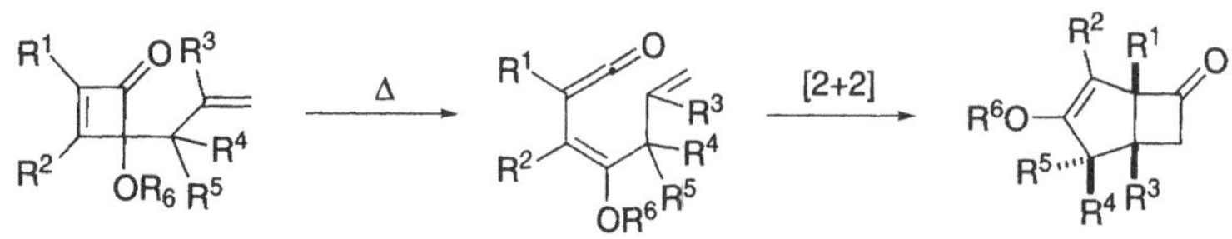

Fig. 4 Intramolecular [2 + 2] cycloaddition to bicyclo[3.2.0]heptenone [21]

The third pattern was developed by Paquette, who studied the possibility of the concerned reaction extensively and established the cascade rearrangements route [72]. The scenario of the cascade rearrangement is:

1. 1,2-Addition of a pair of alkenyl anions (either the same or different) to a squarate ester in an *anti* and/or *syn* manner
2. Charge-driven 4π conrotatory opening to coiled 1,3,5,7-octatetraene followed by 8π recyclization for *anti*-adduct, and straightforward oxy-Cope rearrangement for *syn*-adduct
3. Transannular Aldol condensation for the final ring closure to bicycles

The prototype as shown in Fig. 5 was extended to more sophisticated molecular design such as a polyquinane skeleton.

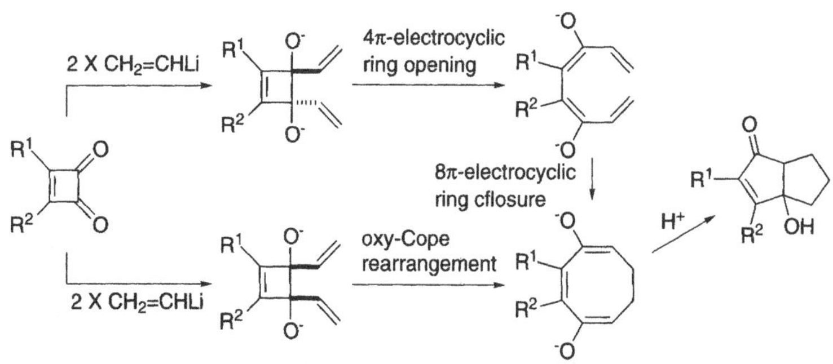

Fig. 5 Cascade rearrangements following twofold addition of alkenyl anions to squarate esters [72]

On the other hand, ring strain relief is triggered by reactive intermediates, and as a matter of fact, this is the alternative option. The reactive intermediates are carbocation, carbon radical and carbene, and their hetero-analogs. Once generated at the position adjacent to C-4, they mediate sequential ring expansion to a five-membered ring [23]. The similar story may be depicted by metal catalyses [73, 74].

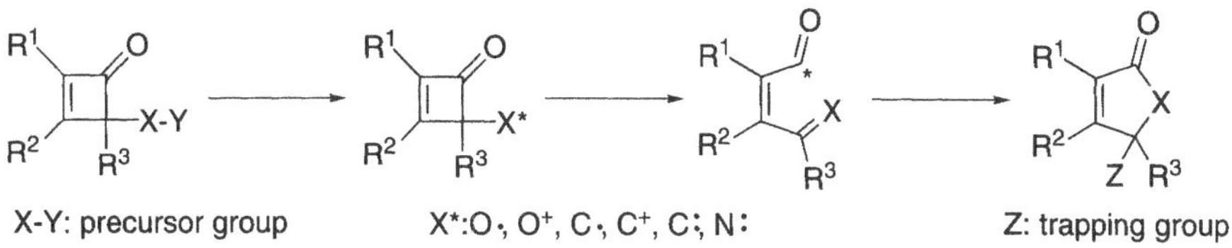

Fig. 6 Ring transformation induced by reactive intermediates [23]

The above ring transformation strategies have also been investigated by beginning with preparation of cyclobutenedione and cyclobutenone skeletons rather than employment of squaric acid [75–77].

3.2 Thermal Concerted Process

Thermal ring expansion of polysubstituted 4-hydroxy-2-cyclobutenones, which can be prepared from squaric acid ester (see the previous section), has been extensively studied and its synthetic value has now been confirmed. The early works have been reviewed several times for the cases of cyclobutenones that have unsaturated bonds at the 4-position, such as (cyclo)alkenyl, alkynyl, and aromatic groups [19–23].

Especially, directed synthesis of heterocycles is feasible by placing heteroatoms such as nitrogen and oxygen at the appropriate position. When hetero double bonds are located at the 4-position, tandem 4π–6π electrocyclic reactions (Fig. 3) can afford six-membered heterocycles. This is the case for a C = O bond to give α-pyrone. Thus, treatment of cyclobutenedione [e.g., **5** (R = i-Pr)] at – 78 °C with *O*-TBDMS-cyanohydrin/LiHMDS followed by low-temperature quench and workup directly gave α-pyrone **41**, which is often found in bioactive compounds [37, 78]. A particularly interesting aspect is the ability of the intermediate 4-acylcyclobutenone **40** to rearrange to **41** at or below room temperature as most ring expansions of 4-aryl or 4-vinylcyclobutenones require heating at higher than 100 °C. This is attributed to greater polarization of the C = O bond (Scheme 7).

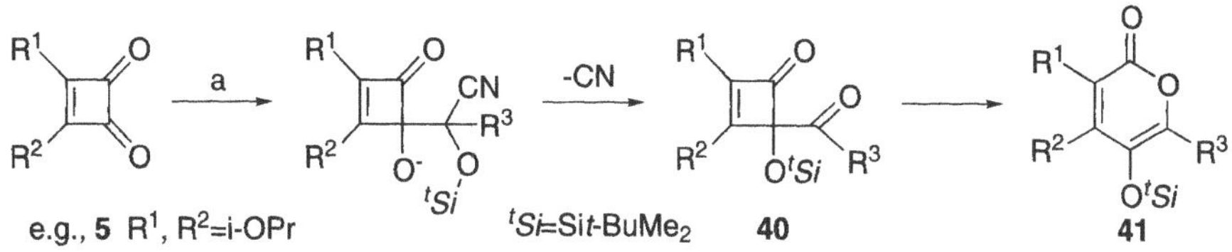

Reaction Conditions: (a) LiC(OSi*t*-BuMe$_2$)R^{3}CN, -78 °C

Scheme 7 Synthesis of α-pyrone: tandem 4π–6π electrocyclic concerted process with a C = O function [78]

The C = N version was realized by using azaheteroaryl substituents at the 4-position [79]. The required cyclobutenones **43** were prepared by the addition of the corresponding 2-lithioheteroaromatics (or Pd-catalyzed cross-coupling with 2-stannylheteroaromatics). The usual ring opening followed by intramolecular cyclization of the C = N bond of azaheteroaromatics onto the vinylketene end occurred faithfully to give quinolizin-4-ones **45a**, imidazo[1,2-a]pyridin-5-ones **45b**, and thiazolo[3,2-a]pyridin-5-ones **45c** (Scheme 8).

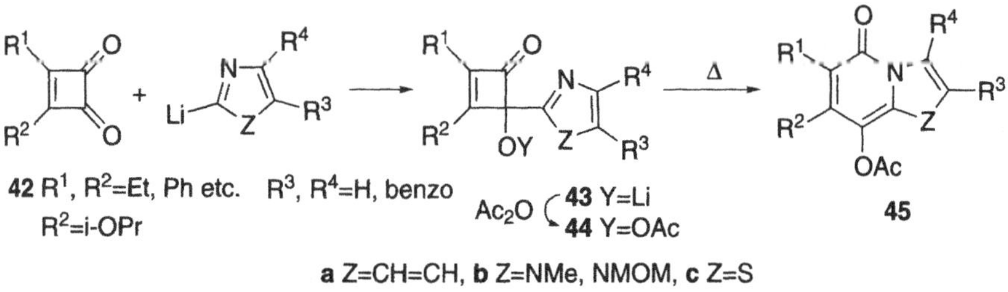

Scheme 8 Synthesis of fused pyridones: tandem 4π–6π electrocyclic concerted process with C = N functions [79]

Furan and thiophene have also been utilized in this type of transformation as building blocks. In the same manner, prerequisite structures prepared by cross-coupling as well as traditional carbanion addition were converted expectedly into benzo- and dibenzofurans and their thiophene analogs (i.e., **47** → **48**) [80]. Likewise, sesquiterpene furanoquinone **51** was synthesized [81], and the total synthesis of the indolizidine alkaloid, septicine **54**, was performed with the key step **52** → **53** for the pyrrole case [82] (Scheme 9).

Dihydropyridine is another building block. Construction of the pyridoacridone ring system, which is found in marine alkaloids and often exhibits an array of biological activities (e.g., amphimedine), was accessible from condensation of **5** (R = i-Pr) with 1-BOC-2-lithio-1,4-dihydropyridine **55** (note: the N atom has no nucleophilicity toward the ketene group that is formed transiently upon thermolysis). Neat thermolysis of the 1,2-adduct **56** produced an oxazolone-fused dihydroquinoline **57** as a result of the expected tandem 4π–6π electrocyclizations. The subsequent removal of the protecting pyrrole group and oxidative aromatization, with loss of the oxazolone ring, afforded the aimed-at heteropolycycle **58** [83] (Scheme 10).

The xanthone core is present in a large family of natural products with broad biological activities. Highly substituted xanthone systems with linear and angular fusion were designed along the cyclobutenedione route [84, 85]. First, the requisite benzopyrone-fused cyclobutenedione structure (such as **61**) was constructed by addition of dithiane anion **59b** of salicylaldehyde **59a** to dimethyl squarate **5**, followed by acid-catalyzed cyclization with elimination of methanol. The next step of adding unsaturated organolithium (**61** → **62**)

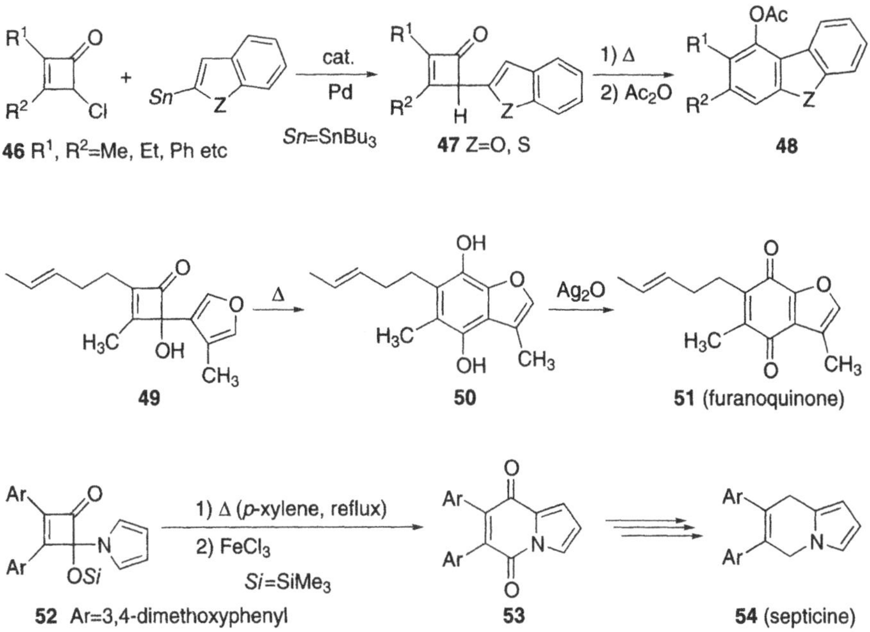

Scheme 9 Tandem 4π–6π electrocyclic concerted process with furan, thiophene, and pyrrole rings [80–82]

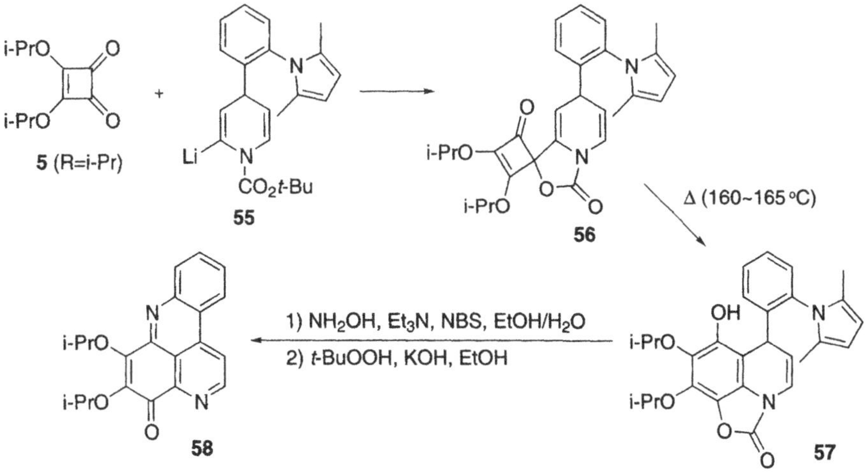

Scheme 10 Tandem 4π–6π electrocyclic concerted process with a dihydropyridine ring [83]

occurred selectively at the carbonyl group opposite the bulky dithiane moiety. The key ring opening step (**62** → **63**) proceeded even at room temperature (practically reflux was applied in THF) to give the target molecule after deprotection. Another method for obtaining angularly fused xanthones was done by successive treatment of 3-anisoylcyclobutene-1,2-dione with heteroaryl-lithium and methyl triflate and prolonged heating of the adduct (mesitylene, reflux). In a related work using 3-benzoylcyclobutene-1,2-dione, the adduct **64** having a *p*-dimethylaminophenyl group at the 4-position underwent unusual rearrangement to a furan derivative **66** due to participation of an allenylketene iminium ion intermediate **65** [86] (Scheme 11).

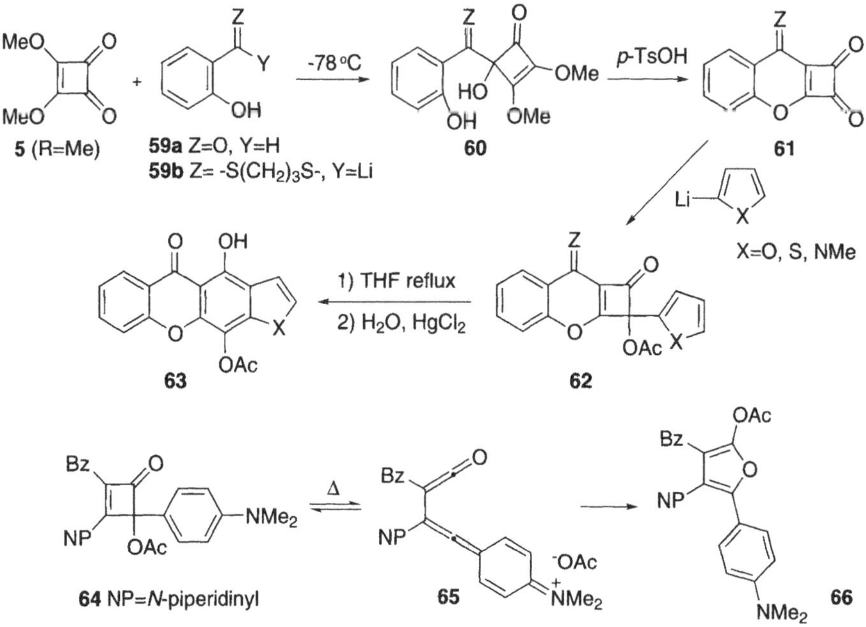

Scheme 11 Synthesis of xanthone core by tandem 4π–6π electrocyclic concerted process [84, 85] and an unusual rearrangement to form furan [86]

Alternative synthesis of six-membered oxygen-heterocycles was demonstrated in examples of chroman and pyranoquinone, which constitute a large class of biologically active natural products. Here, allenyl and alkynyl groups were utilized as the substituents at the 4-position. An approach to chroman depends on combination of the 4π–6π electrocyclic reactions and an intramolecular hetero Diels–Alder reaction [87]. The prerequisite structure was constructed by introduction of a designed alkyne **67** to **5** (R = Me) followed by F^--promoted isomerization to the allenyl function at C-4 (**68** → **69**). The key thermal reactions (50 °C, 36 h) involving both electrocyclic ring opening of **69** and consecutive intramolecular [4 + 2] cycloaddition of *o*-quinone methide **70** afforded the hexahydrocannabinol analog **71** (Scheme 12).

In the case of an alkynyl substituent, the targeted pyranoquinone **76** was obtained by thermolysis (toluene reflux 1.5 h) of the adduct **72** prepared

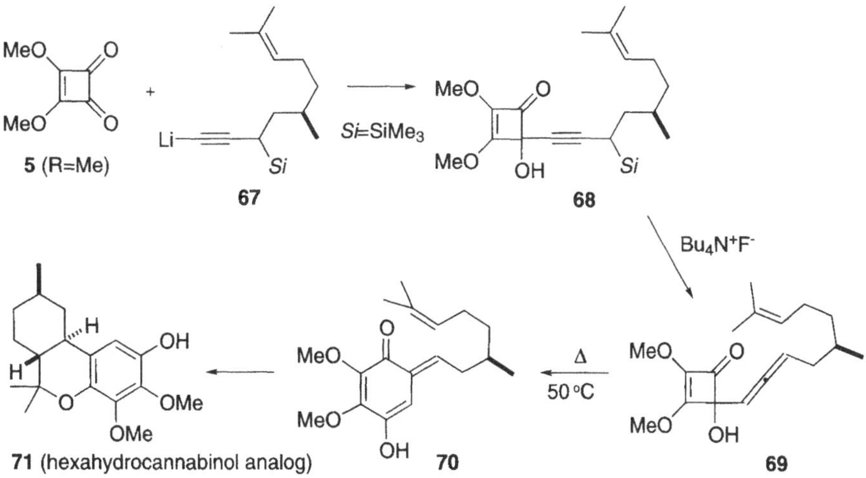

Scheme 12 Synthesis of hexahydrocannabinol analog by combination of 4π–6π electrocyclic reactions and hetero Diels–Alder reaction [87]

from lithiated alkynyl glycopyranoside and **5** (R = Me), involving a more complicated route (Scheme 13). The rearrangements featuring the alkynyl substituent were envisaged to proceed via the mechanism in which ketene **73** and diradical intermediates **74** and **75** participate. The electrocyclization was succeeded by 6-exo radical cyclization and H-abstraction to lead to the quinone ring [88]. Interestingly, it was pointed out that such diradical intermediates formed during thermolysis of 4-alkynylcyclobutenones contributed

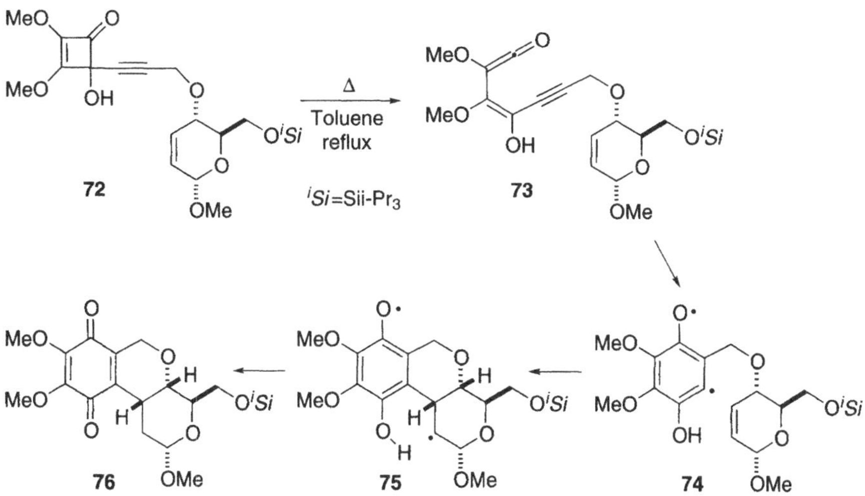

Scheme 13 Synthesis of pyranoquinone by 4π–6π electrocyclic reactions followed by cyclization of the diradical intermediate [88, 89]

DNA cleavage (mimic of esperamicin) [89]. The related diradical mechanism was operated in the photoannulation reaction of 2-aryl-3-isopropoxy-1,4-naphthoquinone available from **5** (R = i-Pr) and was fruitful in synthesis of dimethylnaphthogeranine E [90].

Furaquinocins are a class of antibiotics composed of naphthoquinone fused with an angular five-membered oxygen ring bearing the isoprenoid side chain. After the enantio- and diastereoselective construction of dihydrobenzofuran and introduction of an unsaturated side chain via the Horner–Wadsworth–Emmons reaction, assembly of the naphthoquinone was achieved by the squaric acid based technology (addition of designed organolithium **78** to **77a**/hydrolysis/heating of **79a** at 110 °C/air oxidation/desilylation) to give furaquinocin E (**80a**). Different substitution patterns for biological testing were performed by changing those of squaric acid; the regioisomer (**80b**) of natural product E was accomplished by reversing the chemoselectivity from imine (i.e., **77a**) to pristine carbonyl group (i.e., **77b**) and structural isomers (**80c**, **80d**) by placing the same substituents [91] (Scheme 14).

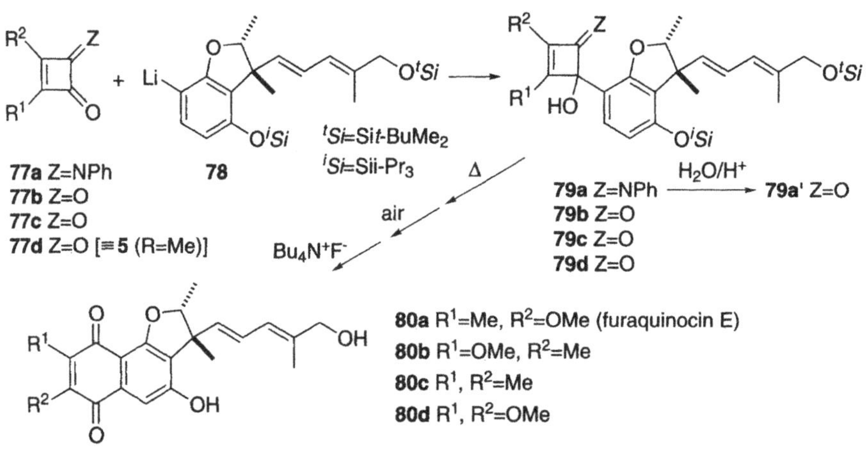

Scheme 14 Synthesis of furaquinocins: assembly of furanonaphthoquinone by 4π–6π electrocyclic reactions [91]

In analogy with the foregoing construction of six-membered oxygen-heterocycles, several nitrogen versions were also developed. Synthesis of benzophenanthridine and isoquinoline resembles that of **76** as shown in Scheme 13, and alkynyl substituents bearing a nitrogen atom incorporated in the side chain were applied in these reactions with a similar mechanism. For example, chelilutine analog **82** (related to antitumor NK109) was produced by the use of an *N*-propargylnaphtylamine block, and isoquinolinetrione **84** by the use of an *N*-propargylacrylamide block [92, 93] (Scheme 15).

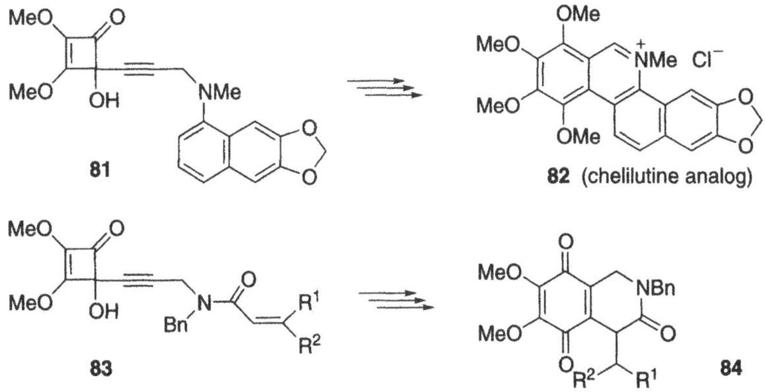

Scheme 15 Synthesis of six-membered nitrogen heterocycles in analogy with the oxygen version [92, 93]

Synthetic strategy for tetrahydroquinolinequinone **87** is similar to that for **63** (Scheme 11). Fusion with a piperidine ring at C-3/C-4 by introduction of hydroxypropyl and *N*-benzylamino groups at these sites, and subsequent cyclization by the Mitsunobu reaction (**5** → **85**) was followed by the usual sequence (**85** → **86** → **87**) [94] (Scheme 16).

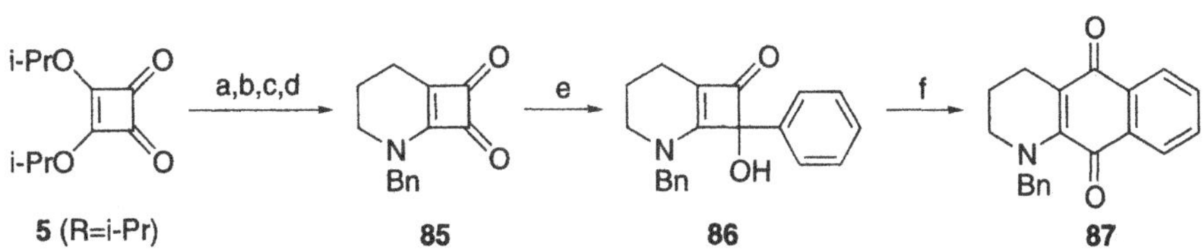

Reaction conditions: (a) $THPO(CH_2)_3MgBr$; (b) $PhCH_2NH_2$; (c) H_3O^+; (d) DEAD/PPh_3; (e) PhLi; (f) heat in air

Scheme 16 Synthesis of tetrahydroquinolinequinone in analogy with the xanthone core [94]

Finally noted in this type of ring transformation is construction of porphyrin–quinone architectures. This was performed by introduction of a porphyrin residue at C-3 by the coupling reaction of **12** with some bromoporphyrins (cf. Scheme 2) and the prescribed conversion to quinone derivatives, fascinating as potential anticancer agents [95, 96].

The electrocyclic reaction in which unsaturated substituents participate at C-2 of cyclobutenedione provides a somewhat different cyclization mode. The thermal rearrangements of 2-dienylcyclobutenones **88** and 3-(*o*-vinylphenyl)-cyclobutenediones **92** underwent well-precedented 4π–6π electrocyclic reactions, but within the diene moiety to phenolic intermediates **90** and **94**. These were allowed to react intramolecularly to give benzofurans **91** and naphthofuranones **95**, respectively [97, 98]. This work represents a new aspect of squaric acid chemistry (Scheme 17).

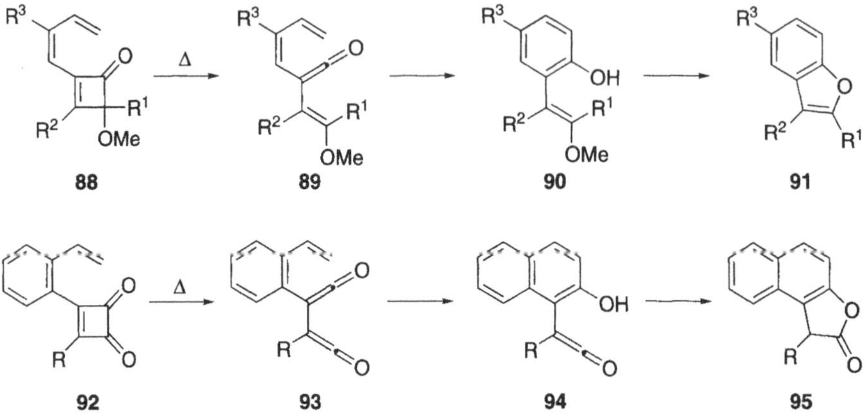

Scheme 17 Thermal rearrangements of 2-dienylcyclobutenones and 3-(*o*-vinylphenyl)-cyclobutenediones [97, 98]

When the substituents of *o*-vinylphenyl and isobutenyl groups were placed at C-2 and C-3, respectively, such as **96**, thermolysis (70 °C) preferred tandem 8π–6π electrocyclic reactions between these substituents to give a tetracyclic cyclobutenone **98**. This underwent the precedented ring transformation/oxidation and, ultimately, photofragmentation with expulsion of isobutylene to give angular furoquinone **99**, for example, by the use of 2-lithiofuran. In the case of an alkynyl group at C-3, thermolysis gave an alkenylidenefuran [99–101] (Scheme 18).

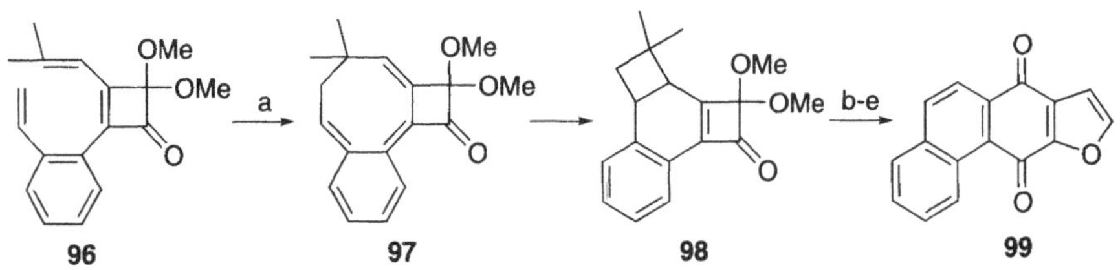

Reaction conditions: (a) 70 °C; (b) 2-lithiofuran; (c) benzene reflux; (d) Ag_2O; (e) hν (visible light)

Scheme 18 Tandem 8π–6π electrocyclic reactions of 2-(vinylphenyl)-3-isobutenylcyclobutenone [99–101]

The cascade rearrangements following double addition of alkenyl anions to the squarate ester was initiated by Paquette from the clue of Asensio's finding that twofold addition of organolithium (MeLi, PhLi, etc.) leads to the facile electrocyclic ring opening to 1,4-diketones [102]. This expedient method for construction of complex polycycles is achieved by a simple one-pot process amenable to regioselective operation, stereochemical control, self-immolative chirality transfer, 1,5-asymmetric induction, and chemical modulation [72]. In this regard, synthesis of oxa- and aza-heterotricycles was exemplified in the case where 2-lithiated dihydrofuran and dihydropyrrole were used as the alkenyl anions. The stereochemical course was completely dissected on the basis of cascades as shown in Scheme 19 (see also Fig. 5) [103, 104]. Total synthesis of hypnophilin (triquinane epoxide) has proved the method to be a valuable synthetic tool [105, 106].

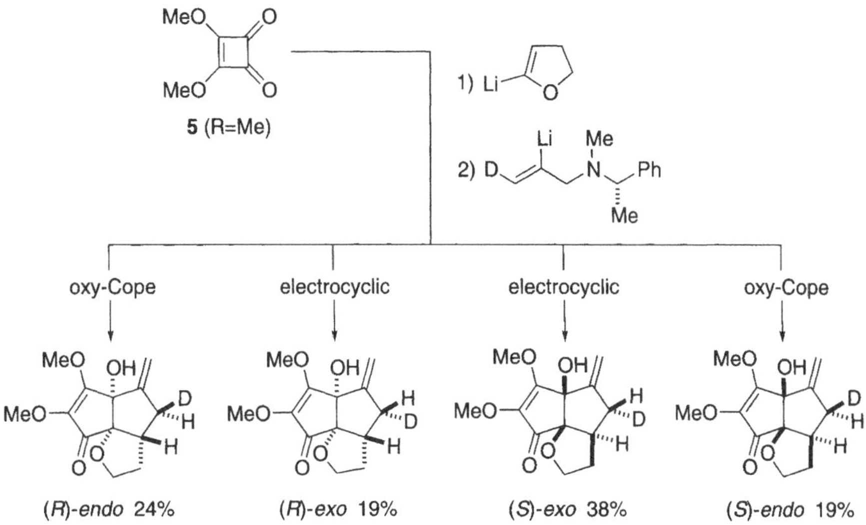

100

Scheme 19 Dissected cascade rearrangements with 2-lithiodihydrofuran as an alkenyl anion [103, 104]

A different approach to triquinane was made by Moore [107–110]. His synthetic route includes the intramolecular [2 + 2] cycloaddition of the vinylketene intermediate as shown in Fig. 4. If the bicyclo[3.2.0]heptenone from this reaction is designed to have an alkenyl-substituent at the bridgehead (i.e., **101** → **102** → **103**), the next oxy-Cope rearrangement is satisfied by adding another alkenyl group (i.e., **104**) to give the triquinane **106** (Scheme 20).

As a matter of fact, the above preparative reaction to obtain the framework of bicyclo[3.2.0]heptenone is already in hand. Indeed, the ring closure step after electrocyclic ring opening of 4-hydroxy-2-cyclobutenone is not limited to fully conjugated systems; synthetic variants are realizable with other proximally placed ketenophiles. When an allyl group was located at C-4, the ketene underwent an intramolecular [2 + 2] cycloaddition reaction with this double bond to give the bicyclo[3.2.0]heptenone derivatives [111, 112].

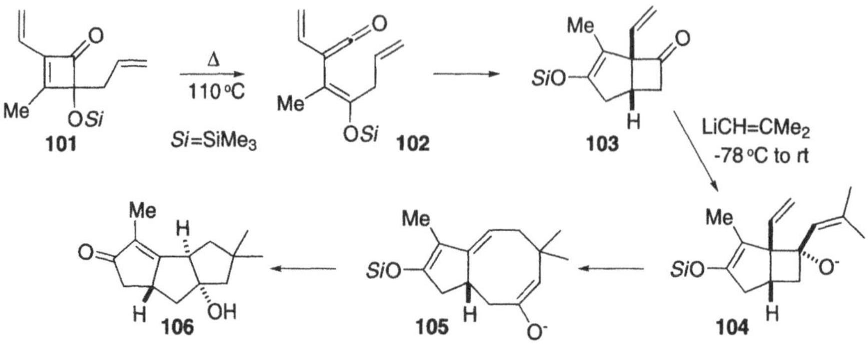

Scheme 20 Different route to triquinane via oxy-Cope rearrangement of bicyclo[3.2.0] heptenone [107–110]

While an allylic portion has hitherto been introduced under usual nucleophilic conditions, allylsilanes **108** are the reagent of choice as an alternative under electrophilic conditions. The electrophilic center was generated from cyclobutenedione monoacetal **107** with BF_3 catalyst, being allowed to react smoothly to give regiospecifically allylated product **109**. This was obtained as a protected form and utilized directly for the following thermal ring opening to give the expected [2 + 2] cycloadducts **111** in a high yield. A triquinane framework **112** was also accessible by this route from one-carbon homologation of the adduct with diazoacetate, when cyclopentylmethyltrimethylsilane (**108**, $R^1 - R^2 = CH_2CH_2CH_2$) was employed as a starting reagent. A tricyclic oxygen-heterocycle **114** was constructed by the same sequential reactions using 6-hydroxy-2-hexenyltrimethylsiane (**108**, $R^1 = H$, $R^2 = CH_2CH_2CH_2OH$) [42] (Scheme 21). Interestingly, reactivity of cyclobutenones having both phenyl and allyl groups at C-4 (obtained by BF_3-catalyzed reaction of 4-phenyl-4-hydroxycyclobutenone with allylsilane) was judged to be competitive between the thermal [2 + 2] cycloaddition and 6π-electrocyclic ring closure under equilibrated conditions (cyclobutenone $\rightleftharpoons$ vinylketene), although inward rotation is preferred for the allyl substituent on the basis of torquoselectivity arguments [43].

The 2-chloro-4-hydroxy-2-cyclobutenone with an acylmethyl substituent at C-4 (**115**) was available from the electrophilic reaction of ester chloride **18** with silyl enol ether and silyl ketene acetal (Scheme 3). This was also found to be thermolabile to give a rearranged product, γ-acylmethylenetetronate **118** [113]. In this case, the cyclization occurred by choosing the hydroxyl function as a proximal ketenophile from an equilibrated mixture. Although the [2 + 2] cycloaddition mode might be a possible route to a β-lactone according to the favored outward rotation of a hydroxyl group (**115** → **116a**), the equilibrium could be shifted by lactonization and dehydrochlorination to thermodynamically stable (*Z*)-tetronate (**116b** → **117** → **118**) (Scheme 22). Photochemistry of the same compound resulted in the formation of chlorine-

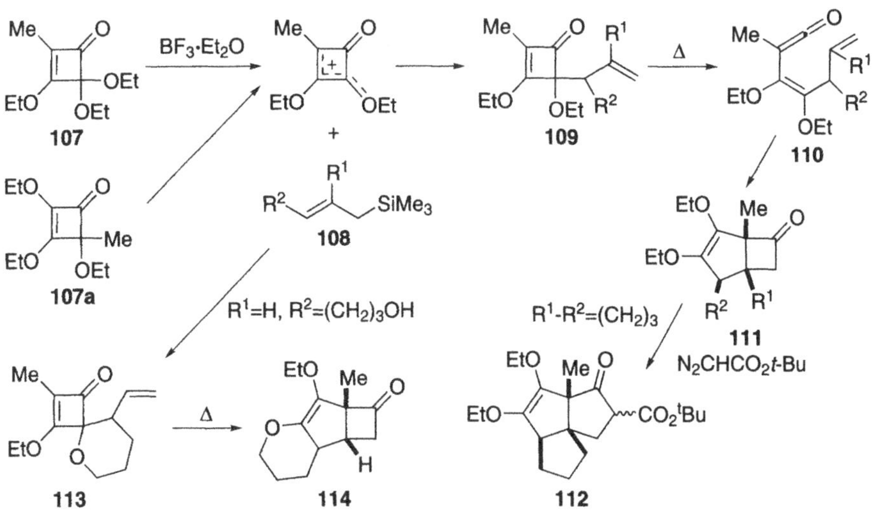

Scheme 21 Allylation under electrophilic conditions and thermal rearrangement to bicyclo[3.2.0]heptenone [42]

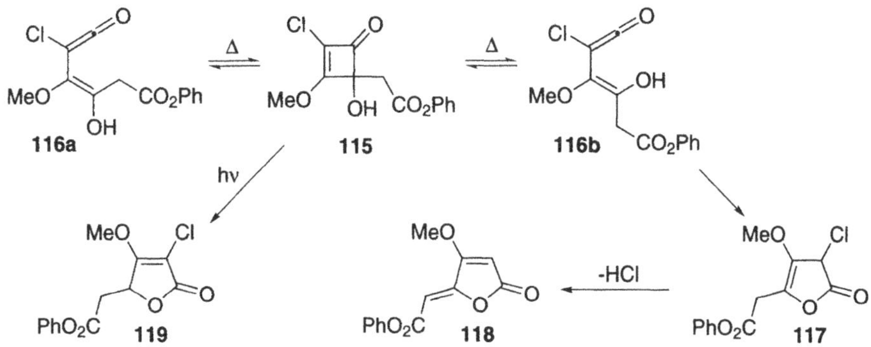

Scheme 22 Thermal and photochemical conversion of 2-chloro-4-hydroxy-2-cyclobutenone [113]

retained 2(5*H*)-furanone **119** from the sequence of favored inward rotation of a hydroxyl group, lactonization, and 1,3-hydrogen shift.

This protocol was applied to the total synthesis of antibacterial and antitumor (*E*)-basidalin. The precursor **120** was made by $TiCl_4$-catalyzed addition of dichloride **17** to silyl enol ether of 3,3-dimethyl-4-penten-2-one (used as a protected form of aldehyde function). Then, the more reactive chlorine atom at C-3 was replaced with an amino group, and key ring expansion was successfully performed by heating (reflux, xylene/pyridine) to give tetronamide **122**. This is formed stereospecifically in *E*-form due to intramolecular hydrogen-bonding, mimicking the biogenetic route of naturally occurring 5-ylidene-2(5*H*)-furanone (in contrast to thermodynamically controlled *Z*-form as observed above). Final deprotection afforded (*E*)-basidalin (**123**) [113] (Scheme 23).

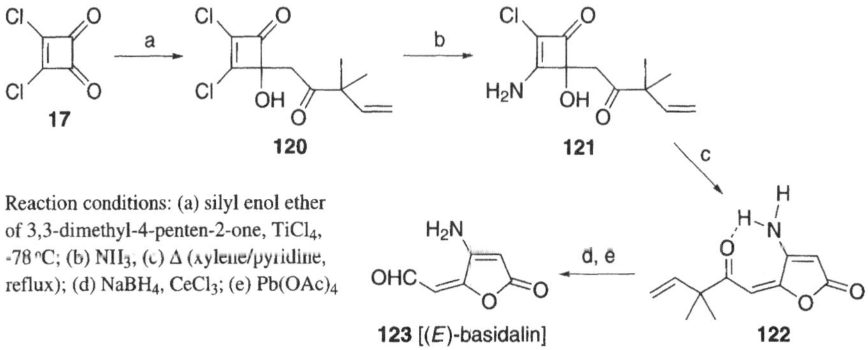

Scheme 23 Synthesis of (*E*)-basidalin mimic to the biogenetic route [113]

For the preparation of 4-amino-2(5*H*)-furanone as above, the disfavored outward rotation of hydroxyl group was compensated for by adding an acid such as trifluoroacetic acid to assist cyclization [114]. Notably, 4-aryl-4-hydroxy-2-cyclobutenone bearing *o*-carboranyl substituent at C-2 also gave rise to the corresponding 2(5*H*)-furanone (**124** → **125**) rather than the usual quinone, indicating that direction of rotation is affected even by substituent at C-2 [115]. Anyhow, the product has potential utility for boron neutron therapy. Analogously, in a cyclobutenone system other than squaric acid, intramolecular addition of hydroxyl group to in situ formed ketene was also reported to give a lactone ring (**126** → **127**) [116] (Scheme 24).

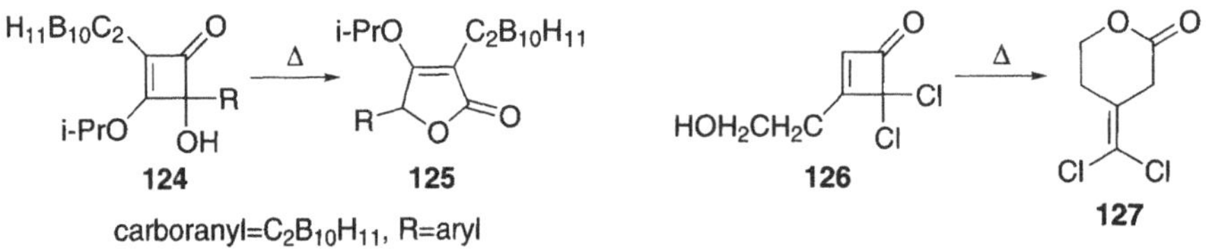

Scheme 24 Analogous thermal rearrangements of cyclobutenones to form lactone rings [115, 116]

Dithiane and oxirane substituents at C-4 are documented to be other ketenophiles, yet the thermolytic products were not composed of the expected medium rings but of contracted rings because of further reactions under the reaction conditions [117].

3.3
Reactive Intermediate Induced Process

Small ring compounds, in general, have considerable strain within the molecule and the ring strain relief is capable of driving the ring opening reactions to lead to the formation of ring expansion products. These trends are described in the foregoing thermal electrocyclic process. However, such chemical behavior is not limited to the concerted manner. As is well known, the reactive intermediate generated at the position adjacent to a strained ring induces ring opening to other transient intermediates, which are fated to fall into the ring expansion products. This scheme is represented by four- to five-membered expansion in squaric acid chemistry, and candidates of reactive intermediates range from radical and cation to carbene and nitrene (Fig. 6).

For heterocycle synthesis, first investigated was the oxygen radical-mediated reaction. The cycloalkoxy radical is a fascinating intermediate suitable for 4-hydroxy-2-cyclobutenone, since it can be readily generated from a parent alcohol, and the formed oxy radical is so reactive that C–C bonds adjacent to the radical center are efficiently cleaved to produce a carbonyl and an unsaturated acyl radical (β-scission). Recyclization via addition of the radical to the radicophilic carbonyl end constitutes an effective approach to ring transformation. The action of lead tetraacetate on the alcohol is a preferable method for the aimed reaction. Thus, simple treatment of 4-hydroxy-2-cyclobutenone **128** with this reagent at room temperature gave acetoxy-substituted 2-(5*H*)-furanone **134** as a ring expansion product; 5-alkylidene-2-(5*H*)-furanone **135** was accomplished when the 4-substituent has an α-hydrogen to eliminate [118]. The outcome is explained by the sequence of β-scission of the initial 4-oxo-2-cyclobutenoxy radical **130**, 5-*endo*-*trig*-cyclization of the resulting acyl radical **131**, and final reductive elimination of lead(II) acetate from **133** (Scheme 25). The fact that the same

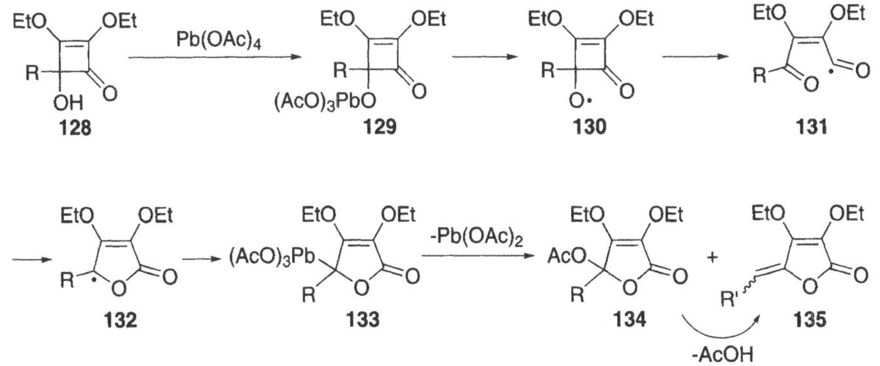

Scheme 25 Ring expansion to 2-(5*H*)-furanone induced by oxy-radical intermediate [118]

reaction took place with HgO/I_2 indicated the distinct participation of a free radical mechanism.

The versatility of the present furanone synthetic method was demonstrated in the stereoselective synthesis of the *Z*-isomer of multicolanate (**139**) [118] (Scheme 26). The prerequisite compound **137** was prepared by successive treatment of appropriate organomagnesium and organolithium reagents, and it was transformed smoothly with lead tetraacetate to the target molecule (the incomplete acetate product **138** was subjected to elimination reaction with DBU).

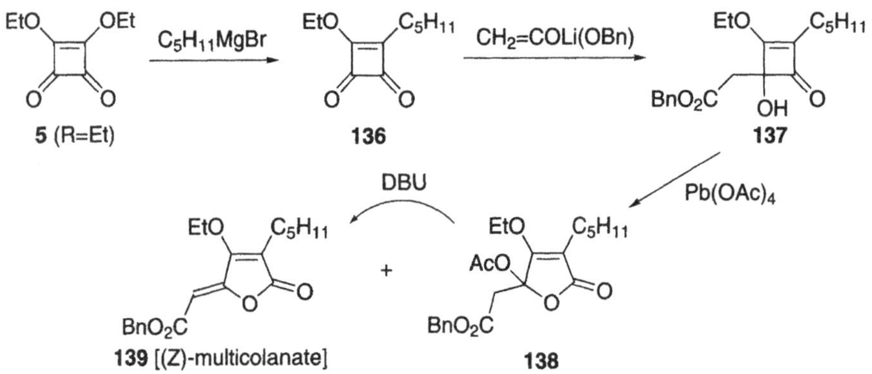

Scheme 26 Synthesis of (*Z*)-multicolanate by simple treatment with $Pb(OAc)_4$ [118]

In connection with this search, the chemical behavior of the carbon-centered radical was also examined [118]. The similar hydroxycyclobutenone **140** bearing Barton's ester at C-4 was photolyzed (W-lamp) to again give a 5-*endo*-cyclized product, 4-cyclopentene-1,3-dione **145**, prior to enol-keto tautomerization (Scheme 27).

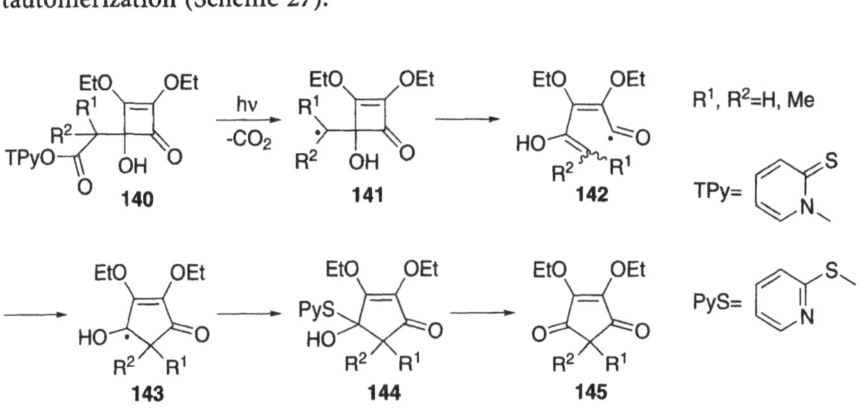

Scheme 27 Ring expansion to 4-cyclopentene-1,3-dione induced by carbon-centered radical [118]

Although this ring does not include a hetero atom, biological activity is known in some derivatives. Human chymase inhibitor methyllinderone (**148**) was obtained from 4-alkynyl-4-hydroxycyclobutenone **146** by Shionogi's group [119], and antimicrobial gloiosiphone A (dimethyl derivative, **151**) was elaborated by Paquette with new methodology involving BF_3-catalyzed ring expansion of hydroxycyclobutenone **149** having an acetal-functionalized cyclopentenyl group at C-4 [120, 121] (Scheme 28). New routes to iodo- and silylalkylidenecyclopentene-1,3-diones have also been developed by us [122, 123].

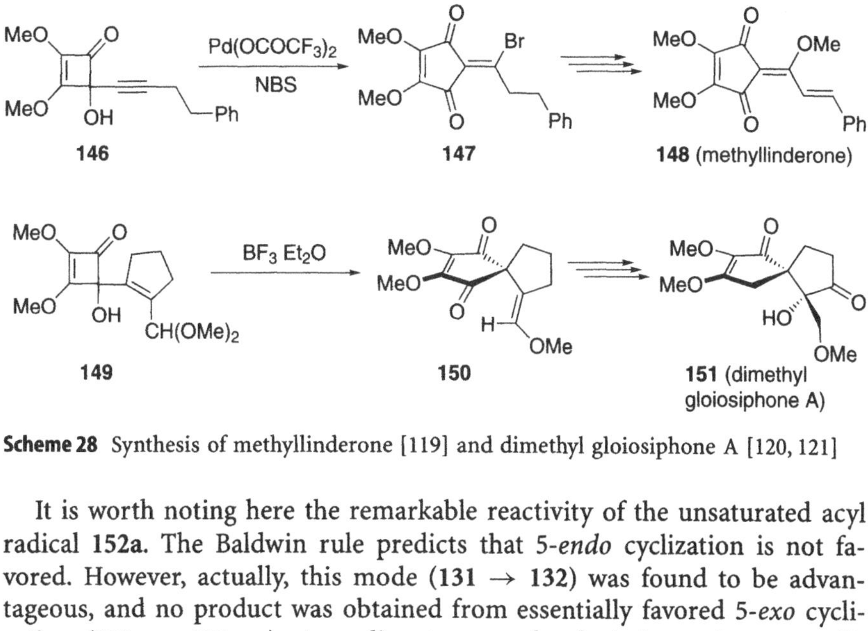

Scheme 28 Synthesis of methyllinderone [119] and dimethyl gloiosiphone A [120, 121]

It is worth noting here the remarkable reactivity of the unsaturated acyl radical **152a**. The Baldwin rule predicts that 5-*endo* cyclization is not favored. However, actually, this mode (**131** → **132**) was found to be advantageous, and no product was obtained from essentially favored 5-*exo* cyclization (**131** → **132exo**). According to several calculations, the net cyclization is best explained by non-radical ring closure from ketene-substituted α-carbonyl radical **152b** to the cyclized radical **153** (i.e., nucleophilic attack of OH on C = C = O with a dipolar π-radical-stabilized transition structure **154TS**) [124]. Independently, the similar chemical behavior of the ketenyl radical was reported by Pattenden [125, 126] (Scheme 29).

Additional reactions involving the electron-deficient oxygen center were carried out by employing a hypervalent iodine reagent, because the facile displacement on iodine with nucleophiles (e.g., NH_2 and OH) and the super-leaving ability of newly formed iodine intermediates endows the electron-deficient center of these heteroatoms. Whereas this type of rearrangement has already been found for nitrogen, the case for oxygen was provided for the first time by the reaction of 4-hydroxy-2-cyclobutenone to 2-(5*H*)-furanone [127]. $PhI(OAc)_2$ is the reagent of choice, and a better result was attained in refluxing methanol to give the 5-methoxy-2-(5*H*)-furanone **158**; Scheme 30 illus-

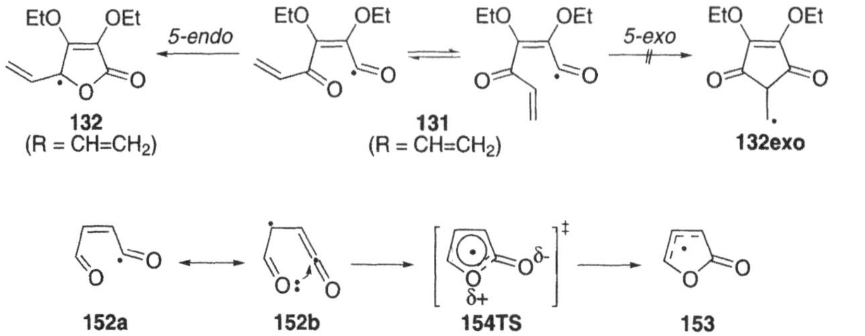

Scheme 29 Preferential 5-*endo* cyclization and estimated unusual cyclization mode (nucleophilic attack and π-radical stabilization) [124]

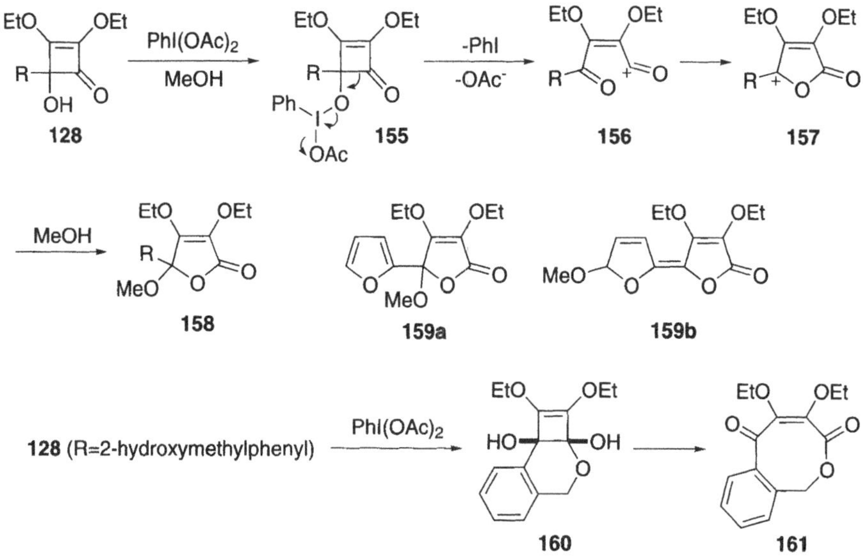

Scheme 30 Ring expansion to 2-(5*H*)-furanone induced by electron-deficient oxygen center [127]

trates a plausible mechanism. First, nucleophilic displacement on a hypervalent iodine with the hydroxyl group of the substrate **128** generates another hypervalent iodine intermediate **155**, which generates an electron-deficient oxygen center susceptible to eliminative ring opening to an acyl cation intermediate **156**. Second, recyclization of this acyl cation with carbonyl oxygen is a facile process for giving a furanone cation **157**, which is trapped with the solvent nucleophile to give the final product **158**. With R = furyl group in **128**, the product was a mixture of the usual furanone **159a** and furylidenefuranone **159b** (2 : 1). It should be noted here that the carbocation version of this type of ring transformation has already been substantiated in the reaction of **149** → **150** (Scheme 28, by Paquette). In the 4-(2-hydroxymethylphenyl)-substituted case, eight-membered lactone **161** was obtained via a different mechanism involving glycol cleavage of hemiacetal **160**.

Carbene (carbenoid)-mediated ring transformation was exemplified in the reaction of diazo-functionalized hydroxycyclobutenone **162** [129]. While furanone **165** and cyclopentenedione **166** were found to be products from the reaction of **162** depending on the decomposition conditions (acid-catalysis: TfOH, BF_3 etc., metal-catalysis: $Rh_2(OAc)_4$, photolysis, thermolysis), two facts are worthy of note. In the $Rh_2(OAc)_4$-catalyzed reaction, a metallacyclic intermediate **169** was suggested to cause selective formation of **166**; a similar one was proposed in the reaction of cyclobutenedione with ferrocenyl chromium carbene complex, from which potentially antitumor-active ferrocenylidenefuranone was obtained despite much lower yield [128]. In the thermal reaction, a new type of ring transformation was observed; relatively stable diazoacetate derivative (**162**: R = O*t*-Bu) underwent 8π electrocyclic ring closure (1,7-dipolar cycloaddition) and prototropy to give the 1,2-diazepinedione **172** after usual 4π-electrocyclic ring opening to diazovinylketene intermediate **170** (Scheme 31). Seven-membered carbon ring formation based on squaric acid technology was precedented with the use of a cyclopropyl substituent [130].

Along with the carbene case, nitrene-mediated ring transformation took advantage of azido-functionalization (Ohno et al. unpublished data). Intro-

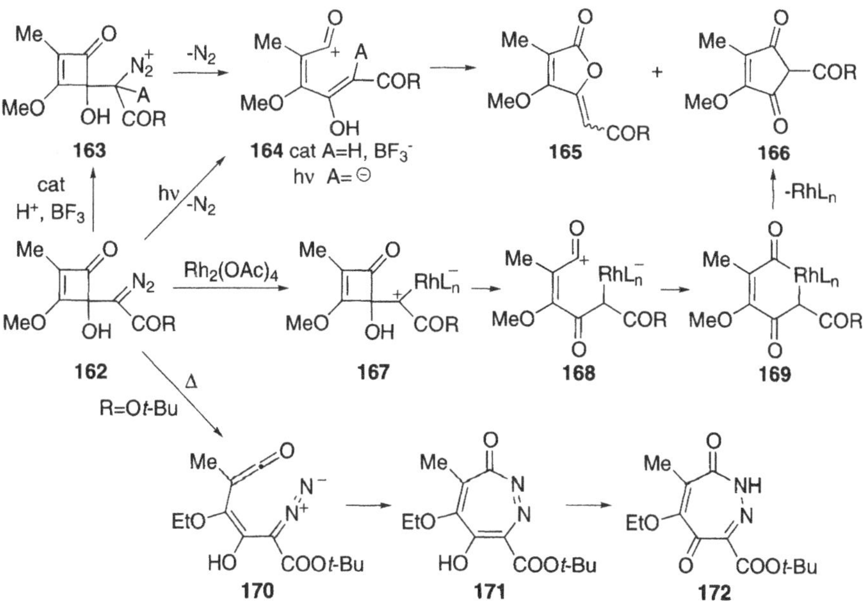

Scheme 31 Ring expansion of diazo-functionalized cyclobutenones: 1,7-dipolar cycloaddition to diazepinedione [129]

duction of an azido group was accomplished by an electrophilic substitution reaction using the acetal **173** and trimethylsilyl azide catalyzed with $BF_3 \cdot Et_2O$. Typically, the 2-phenyl substituted case was examined for thermal decomposition. Thus, heating azide **174** for 30 min in refluxing xylene gave a maleimide derivative **180** after treatment of the primary product with water. This maleimide is likely to be formed from 2-aza-2,4-cyclopentadienone **177**, to which there are two possible routes, either via extrusion of nitrogen followed by nitrene-induced ring expansion (**175** → **176**) or via consecutive 4π–8π electrocyclic rearrangements followed by extrusion of nitrogen (**178** → **179**). More importantly, the above experiment indicates that polysubstituted 2-aza-2,4-cyclopentadienone **177** can survive even at higher temperatures. In fact, without addition of water, it could be isolated as a yellow crystal after concentration of the solution. Whereas the parent 2-aza-2,4-cyclopentadienone is known to be anti-aromatic (life time: ca. 2 s at 30 °C) [131, 132], the observed extreme stability of **177** is attributed to dou-

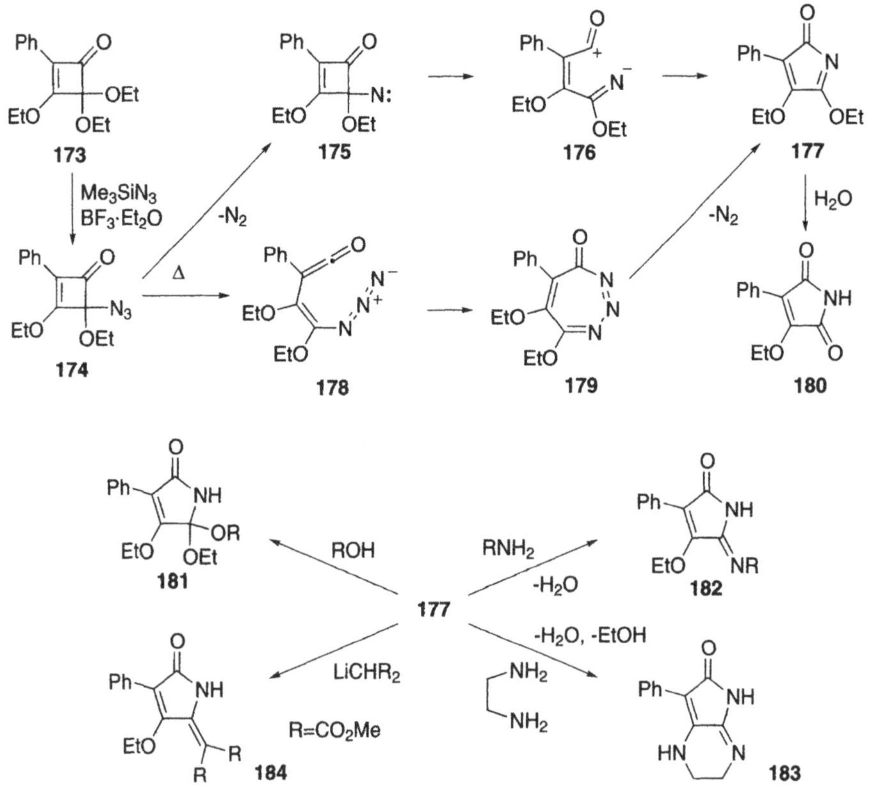

Scheme 32 Ring expansion of azido-functionalized cyclobutenones: formation of stable 2-aza-2,4-cyclopentadienone and its reaction with some nucleophiles (Ohno et al. unpublished data)

ble resonance between the carbonyl group and both ethoxy groups. Despite such thermodynamic stability, **177** is reactive enough to give nitrogen heterocycles with some nucleophiles: maleimide acetals **181** (with alcohol), cyclic amidines **182** (with amine), a bicyclic nitrogen-heterocycle **183** (for example, with ethylenediamine), and a tetramic acid analog **184** (for example, with lithium malonate) (Scheme 32).

4
Squaric Acid Bioisostere

In the previous section, the utility of squaric acid is described from the viewpoint of developing biologically active substances, especially focusing on synthetic strategies for ring transformation to various heterocycles. The structure of cyclobutenedione itself is also intriguing and occasionally exploited for studies on biologically interesting molecular design. One reason is the peculiar acidic functionality of squaric acid. Therefore, "bioisostere" with this ring investigates how the acid derivatives show their activity relative to naturally occurring products or pharmacologically interesting products. These compounds are mainly attained by functional interconversion of acid ester to acid amide (i.e., squarate to squaramide). This section deals with chemistry of biologically active cyclobutenediones bearing heterocycles at the C-2 and/or C-3 position.

Amino squaric acids **186** and **188**, the relatives of natural amino acids, represent an intuitive example of bioisostere, although the heterocycle is absent in these molecules. These were obtained by radical reaction of 2-stannylcyclobutenedione [133] and carbanion addition to squarate ester [134, 135] (Scheme 33).

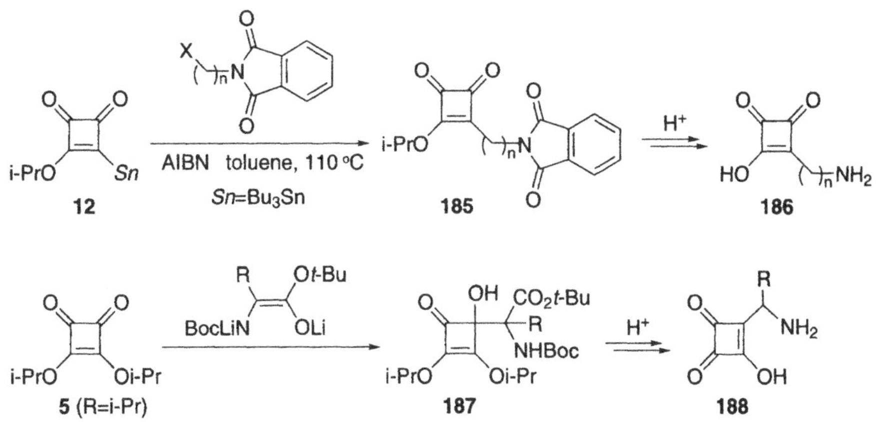

Scheme 33 Squaric acid bioisostere: synthesis of amino squaric acid [133–135]

A new type of modified oligonucleotides containing a squaryl group was also developed as a novel mimic of the phosphate group [136, 137]. A modified thymidine dimer derivative $T_{3'}sq_{5'}T$ (**189**) having a squaramide linkage was synthesized by two-step substitution with **5** (R = Et) and incorporated into oligodeoxynucleotides. The DNA duplex [5′-d(CGCATsqTAGCC)-3′/5′-d(GGCTAATGCG)-3′] was found to be distorted at the modified dimer site but preserved the base pairing ability. Continued work with $U_{2'}sq_{5'}T$ (**190**) indicated unique base-pairing ability for oligonucleotides with G at the site opposite to the T of $U_{2'}sq_{5'}T$ (Fig. 7). These results suggest that there is the possibility to create a new class of antisense/antigene molecules and nucleotide analogs based on squaric acid–nucleoside (base) conjugates.

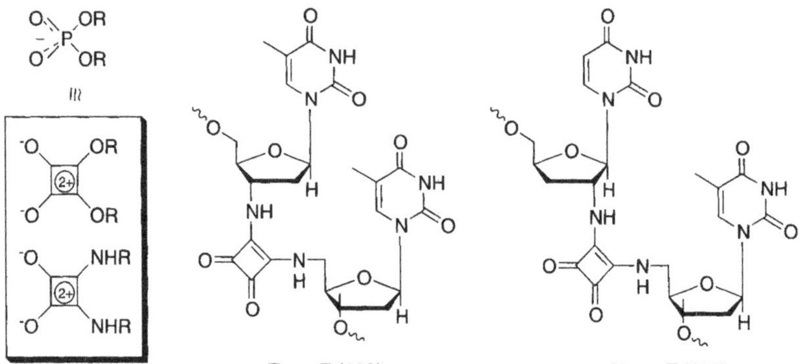

Fig. 7 Structural similarity of squaric acid ester and amide and the structures of oligonucleotide analogs containing a squaramide linkage [136, 137]

Prior to these findings, the key coupling reaction using squarate esters [**5** (R = Et) → **191** → **192**] had been put into practical use by the pioneering work of Tietze. This featured use of a slight excess of amines, excellent yield, stepwise introduction of two amine components, and easy analysis by UV spectroscopy [138]. This methodology has been applied to the synthesis of (neo)glycoconjugates, in which the cyclobutene-1,2-dione skeleton is used as a linker between saccharide and protein. In a recent work hydrophobic didecyl squarate **5** ($R = C_{10}H_{21}$) was reported to be a reagent to handle with ease for purification, as shown in conjugation to BSA (**193** → **194**), for example [139] (Fig. 8).

Among recent developments in glycoconjugates applied for drug immunotherapy and so on [140], a variety of oligosaccharide–squaramide–protein (e.g., bovine, human, and chicken serum albumins) conjugates have been prepared with the Tietze's squarate ester methodology and tested for their bioactivity [141–157]. The glycoconjugates with special functions were also documented: for example, 1,4,7,10,13-pentaazacyclopentadecane

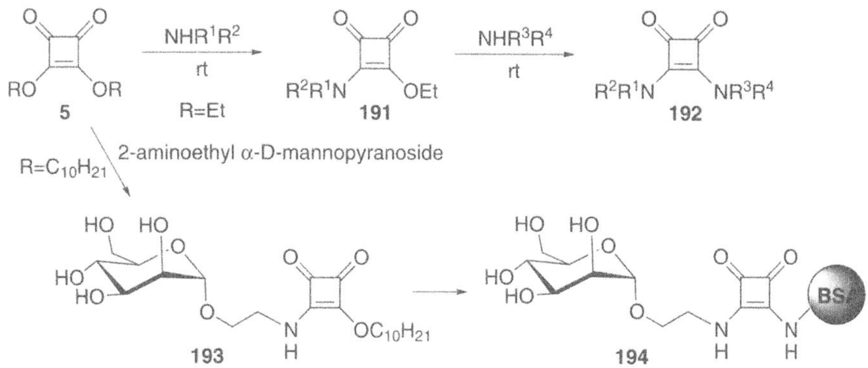

Fig. 8 Tietze's method for conjugation between two amines with squarate ester and an example for squaric decyl ester glycoside conjugation to BSA (bovine serum albumin) [138, 139]

core **195** with five squaramide-linking glycoconjugates as a high affinity pentavalent receptor-binding inhibitor for cholera toxin due to the 1 : 1 association [158, 159]; Gd^{III} complex of inulin-1,4,7,10-tetraazacyclododecan-1,4,7-triacetic acid conjugate **196** linked with squaramide as a contrast agent in magnetic resonance imaging [160]; anthracycline glycoside **197** with a squaramide linker as a modified antibiotic [161] (Fig. 9); cholesterol glycoside with a squaramide linker as an additive to cationic liposome formulation [162]; and photophore (Nakanishi diazirine)–ligand (moenomycin derivative)–biotin with a squaramide linker as a photoaffinity label [163, 164]. It is interesting that such squaramide linking was also used for networking of chitosan to form a hydrogel [165].

Along this line, pharmacologically interesting cyclobutenedione derivatives are also synthesized and screened in relation to the lead compounds. Diazabicyclic amino acid phosphonate **199** was identified as *N*-methyl-D-aspartate antagonist for the treatment of neurological disorders such as stroke and head trauma; this was found out from bioactive model 4-(3-phosphonopropyl)-2-piperazinecarboxylic acid (**198**) [166]. Bioisosteric replacement of the *N*-cyanoguanidine moiety of pinacidil (**200**) with diaminocyclobutenedione template afforded the prototype **201** of a novel series of adenosine 5′-triphosphate-sensitive potassium channel openers with unique selectivity for bladder smooth muscle in vivo [167, 168]. In a series of very late antigen-4 integrin antagonists, incorporation of the cyclobutenedione ring as an amino acid isostere (i.e., **203**) was designed in relation to existing active thioproline CT5219 (**202**) [169, 170]. Pibutidine hydrochloride (or IT-066) (**205**), developed from the lead lafudidine (**204**) was launched as a potent H_2-receptor antagonist for the treatment of peptic ulcers [171–173]. Indole-based derivative BMS-181885 (**206**) was recorded as a potential antimigraine agent, elicited by binding to 5-HT receptor [174] (Fig. 10).

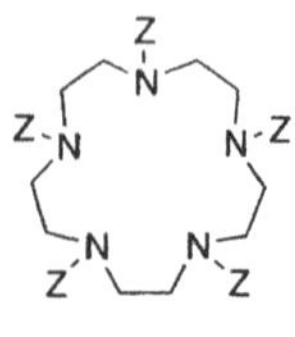

195

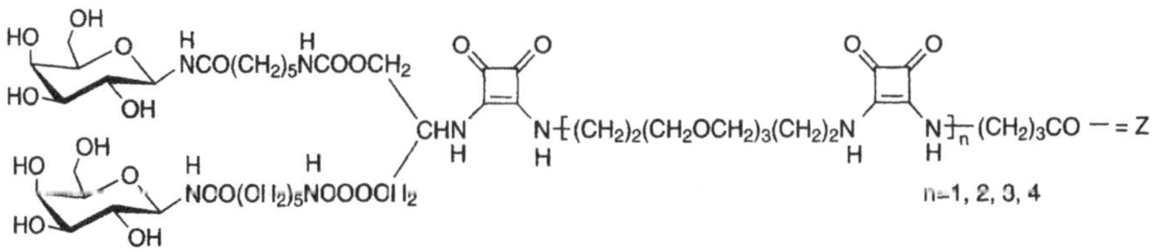

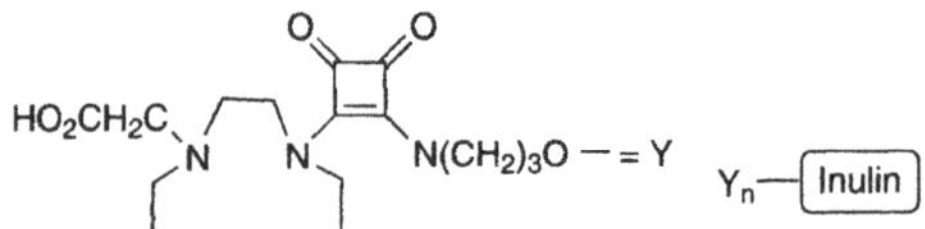

196

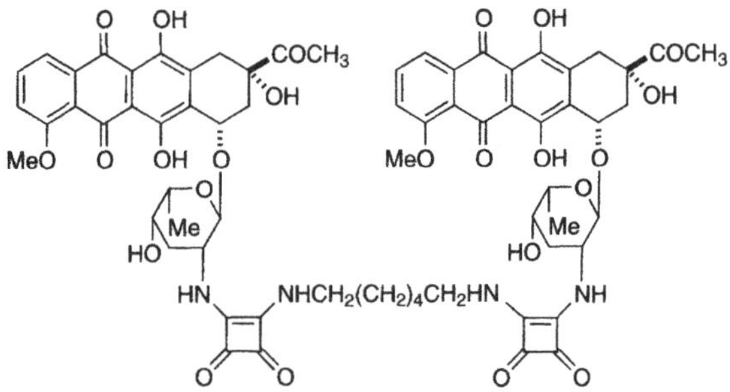

197

Fig. 9 Some squaramide-linked glycoconjugates with special functions [158–161]

In the squaraine chemistry, biologically interesting heterocyclic squarylium dyes are also being explored (Fig. 11). Potential usefulness in photodynamic therapy was envisioned by effective singlet oxygen generation of benzothia(selena)zole-derived squarylium dye **207** and halogenated squarylium dye [175, 176]. Some chemosensors for some metal ions were based on the red-fluorophore of azacrown-appended squarylium dyes **208**. Related podand-based (H-aggregation) squarylium dyes **209** were demonstrated to detect alkaline and earth-alkaline metal ions (Na^+, K^+, Mg^{2+}, Ca^{2+}) selectively [177, 178]. Carbohydrates and proteins were monitored by labeling with properly functionalized symmetric and asymmetric dyes (e.g., **210**) [179–182].

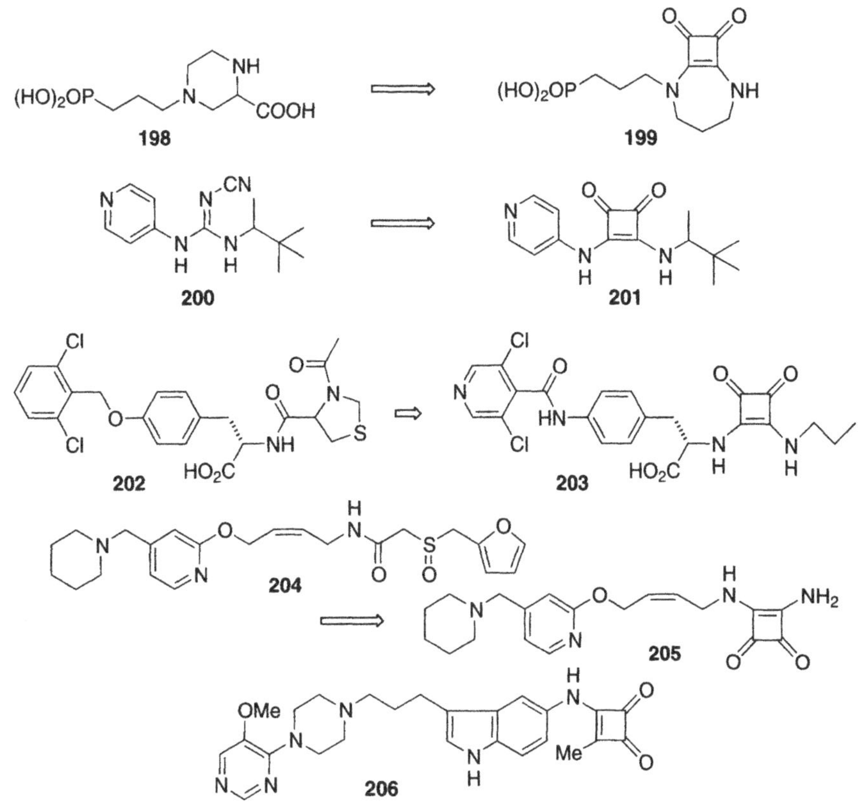

Fig. 10 Some heterocyclic derivatives of squaramide designed in relation to the lead compounds [166–174]

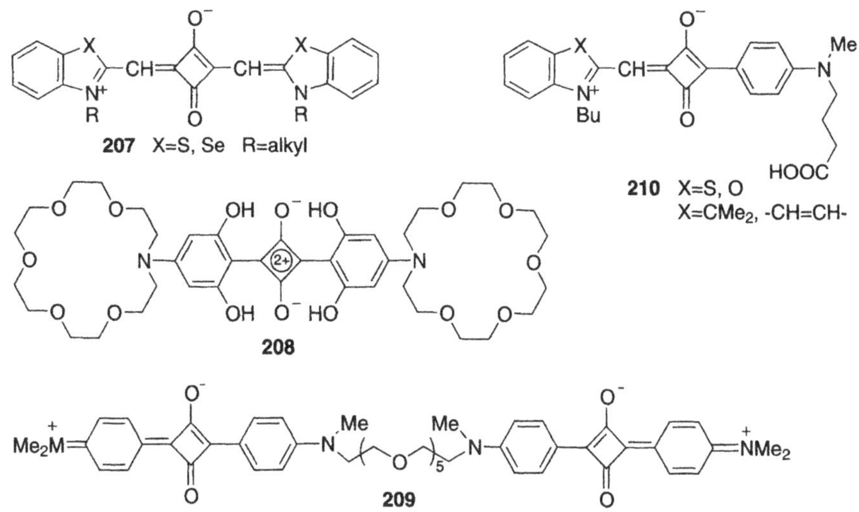

Fig. 11 Biologically interesting squaraines conjugated with heterocycles [175–182]

5 Concluding Remarks

Squaric acid is in principle a pseudoaromatic and therefore stable four-membered compound designated as 3,4-dihydroxy-3-cyclobutene-1,2-dione. Characteristic square planarity brings it high acidity, a peculiar network, and a donor–acceptor triad. Nevertheless, its intrinsically strained and highly oxygenated character makes it extraordinarily useful in synthesis. Once the cyclobutenedione skeleton is converted into the hydroxycyclobutenone skeleton, it can undergo further ring transformation either by thermal concerted rearrangements (chiefly depending on electrocyclic ring opening and ring closure) or by reactions induced by a reactive intermediate (including cation, radical, and divalent species). The directed synthesis of various heterocycles is accomplished by virtue of incorporation of heteroatom(s) on one or more of four possible sites of the four-membered ring and execution of the above ring transformation. Fortunately, the methods developed for introduction of substituents range from nucleophilic to electrophilic conditions and from ionic to radical and organometallic coupling reactions. It is evident from the cumulative successes shown in this review article that squaric acid plays the role of a useful synthetic C-4 building block for construction of biologically interesting oxygen, nitrogen, and sulfur heterocycles, if combined appropriately with corresponding heteroatoms. Bioactive heterocycles can also be provided by derivatization of the cyclobutenedione ring as retained. This utilization comes from the bioisostere concept and from its versatility as a linker of bioconjugates. In these cases, a squaramide form is often employed in the acid interconversion. Conclusively, the small molecule of squaric acid is able to produce a big effect on the synthesis of biologically interesting heterocycles.

References

1. West R (1980) Oxocarbons. Academic, New York
2. Cerioni G, Janoschek R, Rappoport Z, Tidwell TT (1996) J Org Chem 61:6212
3. Quiñonero D, Frontera A, Ballester P, Deyà PM (2000) Tetrahedron Lett 41:2001
4. Quiñonero D, Grau C, Frontera A, Ballester P, Costa A, Deyà PM (2002) Chem Eur J 8:433
5. Zhou L, Zhang Y, Wu L, Li J (2000) J Mol Struct (Theochem) 497:137
6. Davis AP, Draper SM, Dunner G, Ashton P (1999) Chem Commun 2265
7. Terao H, Sugawara T, Kita Y, Sato N, Kaho E, Takeda S (2001) J Am Chem Soc 123:10468
8. May E, Destro R, Gatti C (2001) J Am Chem Soc 123:12248
9. Lim NC, Morton MD, Jenkins HA, Brückner C (2003) J Org Chem 68:9233
10. Rotger MC, Pinã MN, Frontera A, Martorell G, Ballester P, Deyà PM, Costa A (2004) J Org Chem 69:2302

11. Tomàs S, Prohens R, Vega M, Rotger MR, Deyà PM, Ballester P, Costa A (1996) J Org Chem 61:9394
12. Prohens R, Rotger MR, Pinà MN, Deyà PM, Morey J, Ballester P, Costa A (2001) Tetrahedron Lett 42:4933
13. Law KY (1993) Chem Rev 93:449
14. Ajayaghosh A (2003) Chem Soc Rev 32:181
15. Hyodo Y, Nakazumi H, Yagi S, Nakai K (2001) J Chem Soc Perkin Trans 1 2823
16. Ashweil GJ, Jefferies G, Hamilton DG, Lynch DE, Roberts MPS, Bahra GS, Brown CR (1995) Nature 375:385
17. Liebeskind LS, Yu MS, Yu RH, Wang L, Hagen KS (1993) J Am Chem Soc 115:9048
18. Wang J, Jiang X, Chen M, Hu Y, Hu H (2001) J Organometal Chem 629:213
19. Moore HW, Decker OHW (1986) Chem Rev 86:821
20. Liebeskind LS (1989) Tetrahedron 45:3053
21. Moore HW, Yerxa BR (1992) Chemtracts Org Chem 5:273
22. Bottari PQ, Battiste MA (1998) Cyclobutenediones as versatile four-carbon synthons in organic synthesis. In: Hudlicky T (ed) Organic synthesis: theory and applications, vol 4. JAI, Greenwich, p 79
23. Ohno M, Yamamoto Y, Eguchi S (1998) Synlett 1167
24. Pirrung MC, Nauhaus SK (1996) J Org Chem 61:2592
25. Zhou H, Lü S, Xie R, Chan ASC, Yang T (2001) Tetrahedron Lett 42:1107
26. Zhang J, Zhou H, Lü S, Luo M, Xie R, Choi MCK, Zhou Z, Chan ASC, Yang T (2001) Tetrahedron Asymm 12:1907
27. Shimizu I (1991) J Syn Org Chem Jpn [Yuki Gosei Kagaku Kyokaishi] 53:330
28. Liu H, Tomooka CS, Moore HW (1997) Syn Commun 27:2177
29. Liu H, Tomooka CS, Xu SL, Yerxa BR, ZSullivan RW, Moore HW (1998) Org Syn 76:189
30. Kraus JL (1985) Tetrahedron Lett 26:1867
31. Reed MW, Pollart DJ, Perri ST, Foland LD, Moore HW (1988) J Org Chem 53:2477
32. Liebeskind LS, Fengl RW, Wirtz R, Shawe TT (1988) J Org Chem 53:6382
33. Gayo LW, Winters MP, Moore HW (1992) J Org Chem 57:6896
34. Liebeskind LS, Fengl RW (1990) J Org Chem 55:5359
35. Krysan DJ, Gurski A, Liebeskind LS (1992) J Am Chem Soc 114:1412
36. Edwards JP, Krysan DJ, Liebeskind LS (1993) J Am Chem Soc 115:9868
37. Liebeskind LS, Wang JY (1993) Tetrahedron 49:5461
38. Sidduri A, Budries N, Laine RM, Knochel P (1992) Tetrahedron Lett 33:7515
39. Shinada T, Ooyama Y, Hayashi K, Ohfune Y (2002) Tetrahedron Lett 43:6755
40. Ohno M, Yamamoto Y, Shirasaki Y, Eguchi S (1993) J Chem Soc Perkin Trans 1:263
41. Yamamoto Y, Nunokawa K, Okamoto K, Ohno M, Eguchi S (1995) Synthesis, p 571
42. Yamamoto Y, Ohno M, Eguchi S (1996) Bull Chem Soc Jpn 69:1353
43. Tuedemann R, Turnbull P, Moore HW (1999) J Org Chem 64:4030
44. Schmidt AH, Thiel SH, Gaschler O (1996) J Chem Soc Perkin Trans 1:495
45. Schmidt AH, Kircher G, Maus S, Bach H (1996) J Org Chem 61:2085
46. Krayushkin MM, Yarovenko VN, Semenov SL, Shirinyan VZ, Martynkin AY, Uzhinov BM (2002) Rus J Org Chem 38:1331
47. Schmidt AH, Lechler KO, Pretz T, Franz I (1996) J Chem Soc Perkin Trans 1:497
48. Schmidt AH, Kircher G, Zylla J, Veit S (1999) J Chem Soc Perkin Trans 1:409
49. Schmidt AH, Kircher G, Willems M (2000) J Org Chem 65:2379
50. Periasamy M, Radhakrishnan U, Brunet JJ, Chauvin R, Zaizi AWE (1996) Chem Commun, p 1499

51. Periasamy M, Rameshkumar C, Radhakrishnan U, Brunet JJ (1998) J Org Chem 63:4930
52. Rameshkumar C, Periasamy M (2000) Organometallics 19:2400
53. Periasamy M, Mukkanti A, Raj DS (2004) Organometallics 23:619
54. Hamura T, Kakinuma M, Tsuji S, Matsumoto T, Suzuki K (2002) Chem Lett p 748
55. Nguyen MH, Ha T-K, Moore RA, O'Ferrall RAM (1990) J Org Chem 55:3251
56. McAllister MA, Tidwell TT (1994) J Am Chem Soc 116:7233
57. Niwayama S, Kallel EA, Sheu C, Houk KN (1996) J Org Chem 61:2517
58. Niwayama S, Kallel EA, Spellmeyer C, Sheu C, Houk KN (1996) J Org Chem 61:2813
59. Niwayama S (1996) J Org Chem 61:640
60. Dolbier WR Jr, Koroniak H, Houk KN, Sheu C (1996) Acc Chem Res 29:471
61. Leigh WJ (1995) J Am Chem Soc 117:1688
62. Ding W-J, Fang D-C (1999) J Mol Struct (Theochem) 468:119
63. Regenhardt W, Schaumann E, Moore HW (2001) Synthesis p 1076
64. Matyuya Y, Sasaki K, Nemoto H (2003) ARKIVOC (8)79
65. Murakami M, Miyamoto Y, Ito Y (2001) J Am Chem Soc 123:6441
66. Murakami M, Hasegawa M, Igawa H (2004) J Org Chem 69:587
67. Peña-Cabrera E, Liebeskind LS (2002) J Org Chem 67:1689
68. Zora M, Yucel B, Acikalin S (2003) Tetrahedron Lett 44:2237
69. Miles DH, Payne M (2001) Tetrahedron 57:5769
70. Magomedov NA, Riggiero PL, Tang Y (2004) J Am Chem Soc 126:1624
71. Yamamoto Y, Nunokawa K, Ohno M, Eguchi S (1993) Synlett, p 781
72. Paquette LA (1998) Eur J Org Chem, p 1709
73. Liebeskind LS, Bombrun A (1994) J Org Chem 59:1149
74. Zora M, Herndon JW (1994) J Org Chem 59:699
75. Hamura T, Tsuji S, Matsumoto T, Suzuki K (2002) Chem Lett, p 750
76. Hamura T, Morita M, Matsumoto T, Suzuki K (2003) Tetrahedron Lett 44:167
77. Inoue M, Sato T, Hirama M (2003) J Am Chem Soc 125:10772
78. Mingo P, Zhang S, Liebeskind LS (1999) J Org Chem 64:2145
79. Birchler AG, Liu F, Liebeskind LS (1994) J Org Chem 59:7737
80. Liebeskind LS, Wang J (1993) J Org Chem 58:3550
81. Liu H, Gayo LM, Sullivan RW, Choi AYH, Moore HW (1994) J Org Chem 59:3284
82. Yerxa BR, Yang K, Moore HW (1994) Tetrahedron 50:6173
83. Zhang D, Llorente I, Liebeskind LS (1997) J Org Chem 62:4330
84. Sun L, Liebeskind LS (1996) J Am Chem Soc 118:12473
85. Sun L, Liebeskind LS (1997) Tetrahedron Lett 38:3663
86. Sun L, Liebeskind LS (1995) J Org Chem 60:8194
87. Taing M, Moore HW (1996) J Org Chem 61:329
88. Xiong Y, Xia H, Moore HW (1995) J Org Chem 60:6460
89. Sullivan RW, Coghlan VM, Munk SA, Reed MW, Moore HW (1994) J Org Chem 59:2276
90. Onofrey TJ, Gomez D, Winters M, Moore HW (1997) J Org Chem 62:5658
91. Trost BM, Thiel OR, Tsui H-C (2003) J Am Chem Soc 125:13155
92. Hergueta A, Moore HW (1999) J Org Chem 64:5979
93. Wipf P, Hopkins CR (1999) J Org Chem 64:6881
94. Liebeskind LS, Zhang J (1991) J Org Chem 56:6379
95. Shi X, Amin SR, Liebeskind LS (2000) J Org Chem 65:1650
96. Shi X, Liebeskind LS (2000) J Org Chem 65:1665
97. Heileman MJ, Moore HW (1998) Tetrahedron Lett 39:3643
98. Turnbull P, Heileman MJ, Moore HW (1996) J Org Chem 61:2584

99. Heileman MJ, Tiedemann R, Moore HW (1998) J Am Chem Soc 120:3801
100. Tiedemann R, Heileman MJ, Moore HW (1999) J Org Chem 64:2170
101. Hergueta AR, Moore HW (2002) J Org Chem 67:1388
102. Varea T, Grancha A, Asensio G (1995) Tetrahedron 51:12373
103. Paquette LA, Tae J (1998) J Org Chem 63:2022
104. Tae J, Paquette LA (2000) Can J Chem 78:689
105. Geng F, Liu J, Paquette LA (2002) Org Lett 4:71
106. Geng F, Paquette LA (2002) J Am Chem Soc 124:9199
107. MacDougall JM, Santora VJ, Verma SK, Turnbull P, Hernandez CR, Moore HW (1998) J Org Chem 63:6905
108. MacDougall JM, Moore HW (1999) J Org Chem 64:7445
109. Verma SK, Nguyen QH, MacDougall JM, Fleischer EB, Moore HW (2000) J Org Chem 65:3379
110. Verma SK, Fleischer EB, Moore HW (2000) J Org Chem 65:8564
111. Xu SL, Moore HW (1989) J Org Chem 54:6018
112. Xu SL, Xia H, Moore HW (1991) J Org Chem 56:6094
113. Yamamoto Y, Ohno M, Eguchi S (1994) Tetrahedron 50:7783
114. Wang J, Jiang X, Chen M, Ge Z, Hu Y, Hu H (2001) J Chem Soc Perkin Trans 1:66
115. Goudgaon NM, Shi J, Schinazi RF (1998) Tetrahedron Lett 39:1869
116. Dillon JL, Gao Q, Dillon EA, Adams N (1997) Tetrahedron Lett 38:2231
117. Lee KH, Moore HW (1995) J Org Chem 60:735
118. Yamamoto Y, Ohno M, Eguchi S (1995) J Am Chem Soc 117:9653
119. Aoyama Y, Konoike T, Kanda A, Naya N, Nakajima M (2001) Bioorg Med Chem Lett 11:1695
120. Paquette LA, Sturino CF, Doussot P (1996) J Am Chem Soc 118:1996
121. Sturino CF, Doussot P, Paquette LA (1997) Tetrahedron 53:8913
122. Yamamoto Y, Ohno M, Eguchi S (1995) Tetrahedron Lett 36:5539
123. Yamamoto Y, Noda M, Ohno M, Eguchi S (1997) J Org Chem 62:1292
124. Yamamoto Y, Ohno M, Eguchi S (1996) J Org Chem 61:9264
125. Hayes CJ, Pattenden G (1996) Tetrahedron Lett 37:271
126. De Boeck B, Herbert N, Pattenden G (1998) Tetrahedron Lett 39:6971
127. Ohno M, Oguri I, Eguchi S (1999) J Org Chem 64:8995
128. Ohno M, Noda M, Yamamoto Y, Eguchi S (1999) J Org Chem 64:707
129. Zora M, Yucel B, Peynircioglu B (2002) J Organomet Chem 656:11
130. Huffman MA, Liebeskind LS (1993) J Am Chem Soc 115:4895
131. Gaviña F, Costero AM, Andreu MR, Carda M, Luis SV (1988) J Am Chem Soc 110:4017
132. Yokotsuji DLS, Dailey WP, Kende AS, Birzan L, Liu K (1995) J Phy Chem 99:15870
133. Campbell EF, Park AK, Kinney WA, Fengl RW, Liebeskind LS (1995) J Org Chem 60:1470
134. Shinada T, Hayashi K-I, Hayashi T, Yoshida Y, Horikawa M, Shinamoto K, Shigeri Y, Yumoto N, Ohfune Y (1999) Org Lett 1:1663
135. Ishida T, Shinada T, Ohfune Y (2005) Tetrahedron Lett 46:311
136. Sato K, Seio K, Sekine M (2002) J Am Chem Soc 124:12715
137. Sato K, Tawarada R, Seio K, Sekine M (2004) Eur J Org Chem, p 2142
138. Tietze LF, Arlt M, Beller M, Glüsenkamp K-H, Jähde E, Rajewsky MF (1991) Chem Ber 124:1215
139. Bergh A, Magnusson B-G, Ohlsson J, Wellmar U, Nilsson UJ (2001) Glycoconj J 18:615
140. Davis BG (1999) J Chem Soc Perkin Trans 1:3215

141. Nitz M, Bundle DR (2001) J Org Chem 66:8411
142. Kitov PI, Bundle DR (2001) J Chem Soc Perkin Trans 1:838
143. Kitov PI, Shimizu H, Homans SW, Bundle DR (2003) J Am Chem Soc 125:3284
144. Kitov PI, Bundle DR (2003) J Am Chem Soc 125:16271
145. Vermeer HJ, Halkes KM, Albert van Kuik J, Kamerling JP, Vliegenthart JFG (2000) J Chem Soc Perkin Trans 1:2249
146. Vermeer HJ, Kamerling JP, Vliegenthart JFG (2000) Tetrahedron Asym 11:539
147. Lefeber DJ, Kamerling JP, Vliegenthart JFG (2001) Chem Eur J 7:4411
148. Mawas F, Niggemann J, Jones C, Corbel MJ, Kamerling JP, Vliegenthart JFG (2002) Infection and Immunity 70:5107
149. Zhang J, Yergey A, Kowalak J, Kováč P (1998) Carbohydr Res 313:15
150. Chernyak A, Karavanov A, Ogawa Y, Kováč P (2001) Carbohydr Res 330:479
151. Ma X, Saksena R, Chernyak A, Kováč P (2003) Org Biomol Chem 1:775
152. Pozsgay V, Dubois EP, Pannell L (1997) J Org Chem 62:2832
153. Blixt O, Norberg T (1999) Carbohydr Res 319:80
154. Chernyak A, Oscarson S, Turek D (2000) Carbohydr Res 329:309
155. Bergh A, Bhattacharyya S, Nilsson UJ (2002) Carbohydr Res 337:947
156. Auzanneau F-I, Pinto BM (1996) Bioorg Med Chem 4:2003
157. Wang JY, Chang AHC, Guttormsen H-K, Rosas AL, Kasper DL (2003) Vaccine 21:1112
158. Merritt EA, Zhang Z, Pickens JC, Ahn M, Hol WGJ, Fan E (2002) J Am Chem Soc 124:8818
159. Zhang Z, Merritt EA, Ahn M, Roach C, Hou Z, Verlinde CLMJ, Hol WGJ, Fan E (2002) J Am Chem Soc 124:812991
160. Corsi DM, Elst LV, Muller RN, Bekkum H, Peters JA (2001) Chem Eur J 7:64
161. Tevyashova A, Sztaricskai F, Batta G, Herczegh P, Jeney A (2004) Bioorg Med Chem Lett 14:4783
162. Düffels A, Green LG, Ley SV, Miller AD (2000) Chem Eur J 6:1416
163. Daghish M, Henning L, Findeisen M, Giesa S, Schumer F, Henning H, Beck-Sickinger AG, Welzel P (2002) Angew Chem Int Ed 41:2293
164. Buchynskyy A, Kempin U, Vogel S, Henning L, Findeisen M, Müller D, Giesa S, Knoll H, Welzel P (2002) Eur J Org Chem p 1149
165. Crescenzi V, Dentini M, de Nooy AEJ (2000) Polym Repr 41:718
166. Kinney WA, Abou-Gharbia M, Garisson DT, Scmid J, Kowal DM, Bramlett DR, Miller TL, Tasse RP, Zaleska MM, Moyer JA (1998) J Med Chem 41:236
167. Butera JA, Antane MM, Antane SA, Argentieri TM, Freeden C, Graceffa RF, Hirth BH, Jenkins D, Lennox JR, Matelan E, Norton NW, Quagliato D, Sheldon JH, Spinelli W, Warga D, Wojdan A, Woods M (2000) J Med Chem 43:1187
168. Gilbert AM, Antane MM, Argentieri TM, Butera JA, Francisco GD, Freeden C, Gundersen EG, Graceffa RF, Herbst D, Hirth BH, Lennox JR, McFarlane G, Norton NW, Quagliato D, Sheldon JH, Warga D, Wojdan A, Woods M (2000) J Med Chem 43:1203
169. Porter JR, Archibald SC, Childs K, Critchley D, Head JC, Linsley JM, Parton TAH, Robinson MK, Shock A, Taylor RJ, Warrellow GJ, Alexander RP, Langham B (2002) Bioorg Med Chem Lett 12:1051
170. Brand S, de Candole BC, Brown JA (2003) Org Lett 5:2343
171. Proudfoot JR (2002) Bioorg Med Chem Lett 12:1647
172. Kato K, Jingu S, Ogawa N, Higuchi S (2000) J Chromatogr B 740:187
173. Naito Y, Yoshikawa T, Matsuyama K, Yagi N, Arai M, Nakamura Y, Kaneko T, Yoshida N, Kondo M (1995) Eur J Pharm 294:47
174. Srinivas NR, Shyu WC, Soong CW, Greene D (1998) J Pharm Sci 87:1170

175. Arun KT, Epe B, Ramaiah D (2002) J Phys Chem B 106:11622
176. Santos PF, Reis LV, Almeida P, Oliveira AS, Ferreira LFV (2003) J Photochem Photobiol A: Chemistry 160:159
177. Oguz U, Akkaya EU (1997) Tetrahedron Lett 38:4509
178. Ajayaghosh A, Arunkumar E, Daub J (2002) Angew Chem Int Ed 41:1766
179. Kukrer B, Akkaya EU (1999) Tetrahedron Lett 40:9125
180. Oswald B, Lehmann F, Simon L, Terpetschnig E, Wolfbeis OS (2000) Anal Biochem 280:272
181. Oswald B, Patsenker L, Duschl J, Szmacinski H, Wolfbeis OS, Terpetschnig E (2000) Bioconjugate Chem 10:925
182. Nakazumi H, Colyer CL, Kaihara K, Yagi S, Hyodo Y (2003) Chem Lett 32:804

Top Heterocycl Chem (2006) 6: 39–76
DOI 10.1007/7081_023

Published online: 24 March 2006

Manganese(III)-Based Peroxidation of Alkenes to Heterocycles

Hiroshi Nishino

Department of Chemistry, Faculty of Science, Kumamoto University, Kurokami 2-39-1, 860-8555 Kumamoto, Japan
nishino@sci.kumamoto-u.ac.jp

Abstract Synthesis of endoperoxides and hydroperoxides using tris(2,4-pentanedionato) manganese(III) or manganese(III) acetate is described. The reaction of various alkenes with tris(2,4-pentanedionato)manganese(III) at ambient temperature in air gives 4-acetyl-1,2-dioxan-3-ols. The manganese(III) acetate-catalyzed aerobic oxidation of various alkenes in the presence of various 1,3-dicarbonyl compounds also gives the corresponding 1,2-dioxan-3-ols. The basic skeleton of naturally occurring heterocycle-fused endoperoxides can be produced by the manganese(III)-catalyzed aerobic oxidation. On the other hand, a similar reaction using barbituric acids, pyrazolidinediones, and hydroxyquinolinones gives bis(hydroperoxyethyl) derivatives. In addition, the direct hydroperoxidation of the cyclic amides occurs in the absence of alkenes. The spectroscopic features, stereochemistry, mechanism for the formation of 1,2-dioxan-3-ols, and synthetic application are also mentioned.

Keywords 1,2-Dioxan-3-ols · Aerobic oxidation · Endoperoxidation · Hydroperoxidation · Manganese(III) acetate

1 Naturally Occurring 1,2-Dioxane Derivatives

Unbelievably, the 1,2-dioxane scaffold is sometimes found in nature [1,2]. Chemists have imagined that the 1,2-dioxane derivatives would be unstable in air since the oxygen–oxygen bond dissociation energy is estimated to be only 34 kcal/mol [3]. However, artemisinin isolated from *Artemishia annua* [4–11] and yingzhaosu isolated from *Artabotrys unciatus* [12–16] are the most famous 1,2-dioxane analogs and have very strong antimalarial activity like quinine, chloroquine, and mefloquine [17] (Fig. 1). The 1,2-dioxane derivatives, found in *Eucalyptus grandis* and related species in Australia, inhibit root formation in cuttings [18–20] and also possess an auxin-like activity (promontory) as well as an abscissic acid-type activity (inhibitory) [21,22] (Fig. 2). Marine sponges contain the antitumor cyclic peroxy ketals and a series of related ketals, such as chondrillin and plakorin, isolated from *Plakortis lita* [23–33]. *Plakortis* sp also produces diverse 1,2-dioxolanes, such as plakinic acids [34–36] (Fig. 2). Many further 1,2-dioxane derivatives may be found in the natural environment in the near future.

2 History of the Discovery of the 1,2-Dioxane Synthesis Using Manganese(III) Complexes

Manganese compounds are used as oxidant and catalyst in the various electron-transfer reactions in vivo involving the Mn(II), Mn(III), or Mn(IV) oxidation states. In particular, manganese complexes play an important role as oxygen carrier, similar to iron and copper porphyrins such as hemoglobin and hemocyanin. Manganese compounds have a peculiar nature for the reaction with molecular oxygen.

Manganese(III) acetate dihydrate, $Mn(OAc)_3 \cdot 2H_2O$, was first used in an organic synthesis by Bush and Finkbeiner [37] and Heiba and Dessau [38],

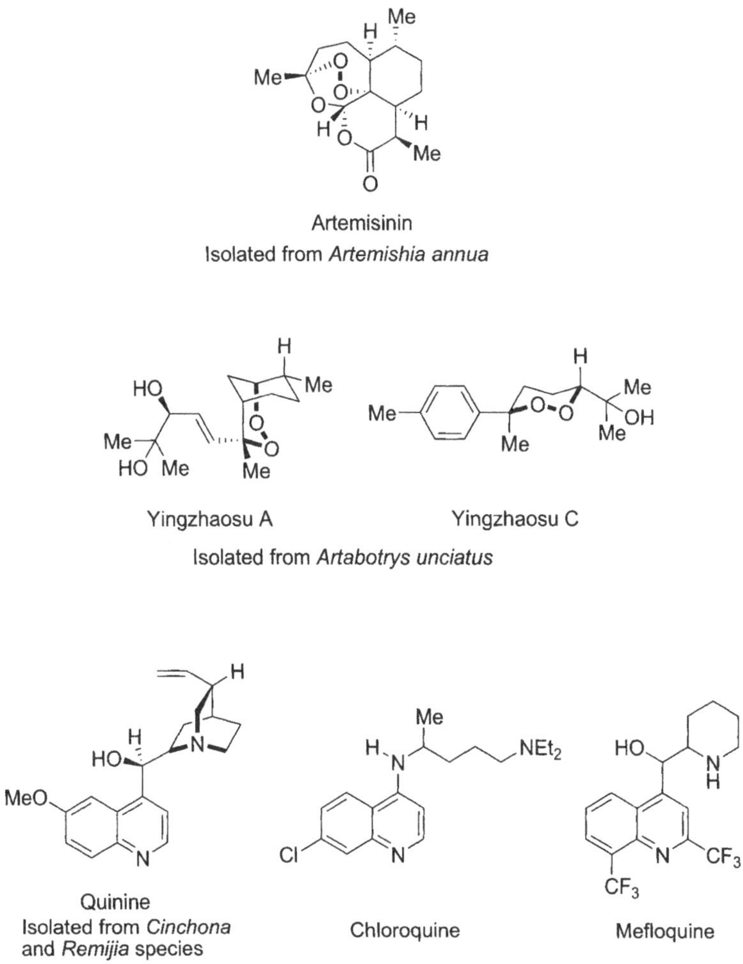

Fig. 1 Natural and artificial antimalarial agents of cyclic peroxides and quinoline alkaloids

producing 4-butanolides (Scheme 1). Thereafter, many chemists applied manganese(III) acetate to organic syntheses and developed synthetic techniques [39–46]. However, there is a significant disadvantage, manganese(III) acetate dihydrate is totally soluble only in hot acetic acid. In fact, the manganese(III) acetate dihydrate is prepared and recrystallized from glacial acetic acid. In most reactions, acetic acid is used as the solvent and the best results are obtained, although some exceptions have been reported [47, 48]. Furthermore, the manganese(III) oxidation complicates the reaction since the direct inner- or outer-sphere one-electron-transfer oxidation of the substrate (for example, producing 1,2-glycol monoacetates and/or ketones [49–59]) and the free-radical reaction of enolizable ligands (producing butanolides) competitively take place at the same time [60–72] (Scheme 1). Therefore, the use

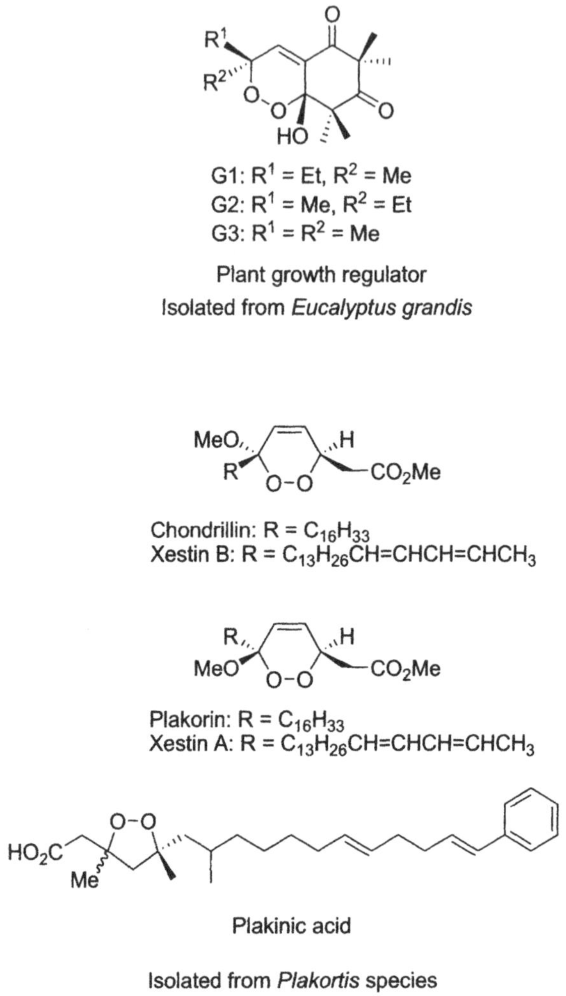

Fig. 2 Naturally occurring cyclic peroxides

of manganese(III) acetate might not be convenient from the viewpoint of organic synthesis, except for the 4-butanolide synthesis.

On the other hand, tris(2,4-pentandionato)manganese(III) (manganese(III) acetylacetonate), $Mn(acac)_3$, is soluble in both polar and nonpolar solvents, although its oxidation ability is weaker than that of manganese(III) acetate [73]. Tris(2,4-pentandionato)manganese(III) is slightly soluble even in water, but disproportionates into manganese(II) and (IV), similarly to manganese(III) acetate. The complex is stable in organic solvents, however, the ligand-exchange reaction takes place since the complex is labile in a protic solvent such as acetic acid. It is expected that the corresponding ligand radicals may be formed when the reaction accompanies a one electron-transfer

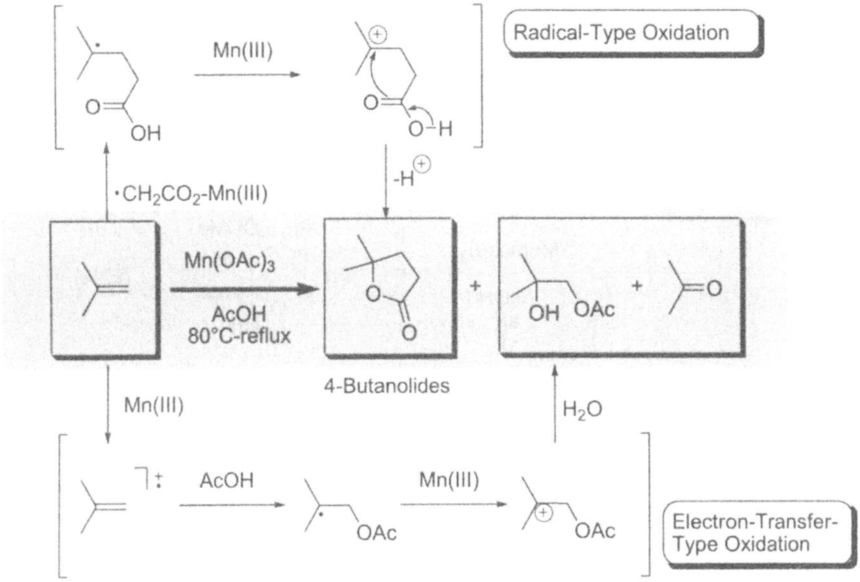

Scheme 1 Typical oxidation of alkenes with manganese(III) acetate in acetic acid

oxidation. Therefore, it is possible to prepare a new carbon–carbon bond using the ligand radicals [74].

On the basis of the character mentioned above, the oxidation of alkenes with tris(2,4-pentandionato)manganese(III) was carried out in acetic acid at room temperature, even though the reaction is generally carried out at elevated temperatures in order to increase the oxidizing ability of the manganese oxidant. Incidentally, the mild reaction in air led to the discovery of the 1,2-dioxane synthesis [73]. A liquid 4,5-dihydrofuran was obtained in good yield by the reaction of 1,1-diphenylethene at an elevated temperature (normally reflux temperature), whereas the same reaction at room temperature in air gave a colorless solid product in high yield (Scheme 2). The NMR spectra of both products were quite different. Spectroscopic data and elemental analysis indicated that the colorless solid has a hydroxyl functionality and one more oxygen than the corresponding 2-hydroxytetrahydrofuran. At that time, it was postulated that the structure of the colorless solid product must be either a hydroperoxide or endoperoxide (Fig. 3). However, in the 1980s, high-resolution FT-NMR spectroscopy and FAB mass spectrometry were not commonly used in the organic laboratory, even an expensive X-ray single crystal diffraction analysis was not generally employed to characterize organic compounds. Furthermore, although a qualitative test using potassium iodide-starch paper for hydroperoxides, oxidation using lead(IV) acetate, and reduction with triphenylphosphine were examined, the results were not clear enough to de-

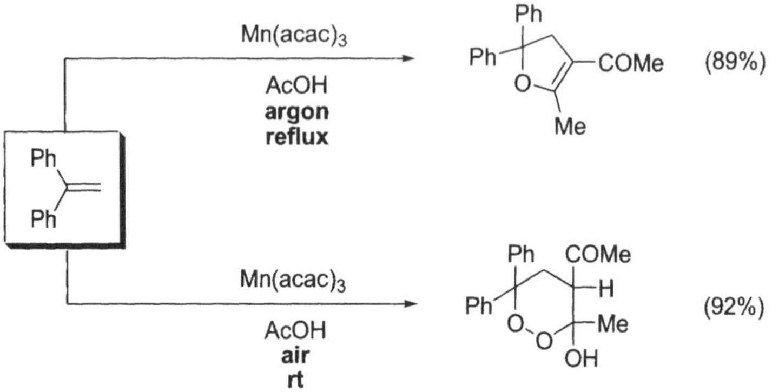

Scheme 2 Oxidation of 1,1-diphenylethene with tris(2,4-pentanedionato)manganese(III)

termine the peroxide structure. Accordingly, the structure was temporary deduced to be hydroperoxytetrahydrofuran since it was known that most endoperoxides formed by the addition of alkenes with singlet oxygen were usually unstable (Fig. 3) [73].

In September 1989, a single crystal of the colorless solid product recrystallized from ethanol was subjected to X-ray crystallography at the University of Houston, USA, and the exact structure was determined to be 4-acetyl-3-methyl-6,6-diphenyl-1,2-dioxan-3-ol (Fig. 4). Thus, the products obtained by the reaction of the alkenes with tris(2,4-pentanedionato)manganese(III) at room temperature in air were concluded to be functionalized endoperoxides [75].

Since manganese(III) complexes are labile in coordination solvents, it was then conceived that manganese(III) acetate, which has an oxygen-centered triangular structure, must cause a ligand-exchange reaction with 1,3-dicarbonyl compounds to produce new manganese(III)-1,3-dicarbonyl enolate complexes in situ, and a similar reaction, such as the 1,2-dioxane synthesis using tris(2,4-pentanedionato)manganese(III), might take place. In fact, manganese(III) acetate is insoluble in glacial acetic acid at room temperature, however, it gradually dissolved in a solution of acetic acid containing 1,3-dicarbonyl compounds. This strongly suggested the formation of a new manganese(III) complex. In this way, it was inevitably found that

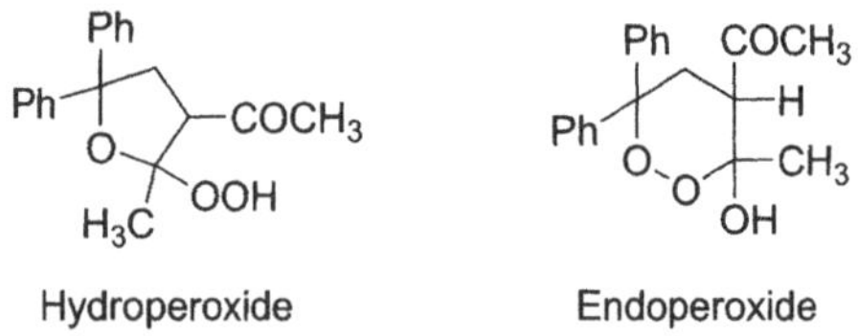

Fig. 3 Deduced hydroperoxide and endoperoxide structures

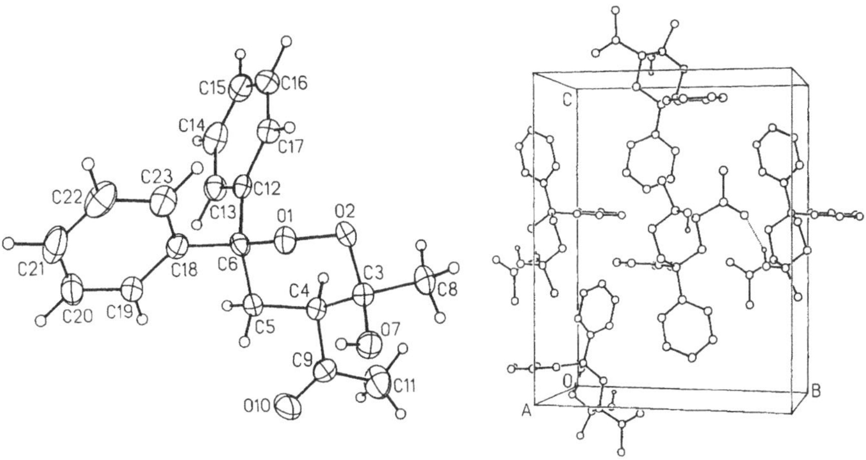

Fig. 4 ORTEP drawing and stereoscopic view of 4-acetyl-3-methyl-6,6-diphenyl-1,2-dioxan-3-ol

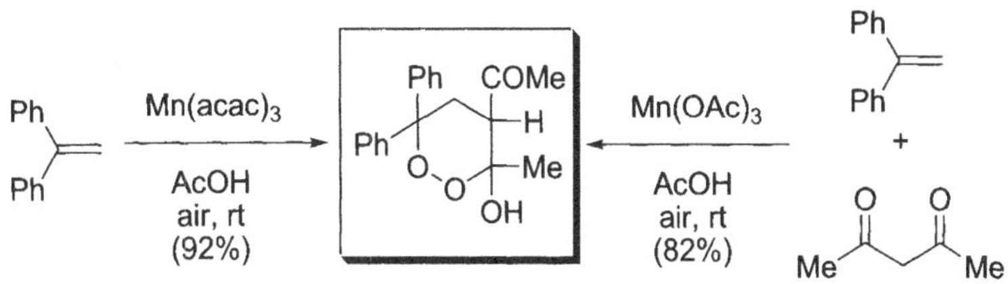

Scheme 3 Synthesis of 1,2-dioxan-3-ol using $Mn(acac)_3$ or $Mn(OAc)_3$

the 1,2-dioxane derivatives were also formed by the reaction of alkenes with 1,3-dicarbonyl compounds in the presence of manganese(III) acetate under mild conditions (Scheme 3). It was a significant discovery that octahedral manganese(III)-1,3-dicarbonyl enolate complexes also serve as molecular oxygen carriers analogous to the square planar manganese(III)-porphyrin and -salen complexes.

Thereafter, the discovery led to the synthesis of many functionalized 1,2-dioxane derivatives and dioxane-fused heterocyclic analogs.

3
Formation of 1,2-Dioxanes Using Tris(2,4-pentanedionato)manganese(III)

Tris(2,4-pentanedionato)manganese(III) was easily prepared by the direct oxidation of 2,4-pentanedione with potassium permanganate [76]. The synthesis of 1,2-dioxane was normally carried out as follows: To a solution of 1,1-diphenylethene in glacial acetic acid was added freshly prepared tris(2,4-

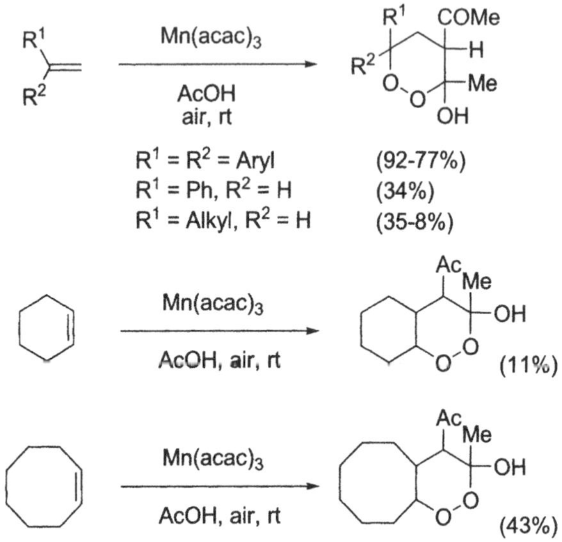

Scheme 4 Reaction of alkenes with tris(2,4-pentanedionato)manganese(III) in acetic acid at room temperature in air

pentanedionato)manganese(III) at a molar ratio of 1 : 1, and the solution stirred at ambient temperature in a round-bottomed flask equipped with a calcium chloride drying tube until the brown color of the Mn(III) species turned a transparent pale yellow. The solvent was removed in vacuo and the residue triturated with water followed by extraction. After separation by silica gel TLC or column chromatography, crystalline 1,2-dioxan-3-ol was obtained in excellent yield (Scheme 4) [77]. Other 1,1-disubstituted ethenes also gave similar results. However, the yields from the reaction of aliphatic terminal alkenes and cyclic alkenes were poor. In addition, under an oxygen atmosphere, formally produced ligand radicals were preferentially oxidized before reacting with the alkenes, and the formation of the 1,2-dioxan-3-oles was consequently prevented. It was found that the best results were obtained by the reaction in air at room temperature (23 °C). The formation of 1,2-dioxane is characteristic for manganese(III). Other transition metal complexes, such as Cr(III), Fe(III), Co(III), Ni(II), and Cu(II), do not produce the reaction [77].

For the synthesis using tris(2,4-pentanedionato)manganese(III), the 1,2-dioxan-3-ol scaffold was limited to bear only an acetyl and methyl functionality derived from 2,4-pentanedione without alkenes. Therefore, in order to apply the synthesis of various endoperoxide analogs, the behavior of manganese(III) acetate in acetic acid at ambient temperature in the presence of 1,3-dicarbonyl compounds needed attention.

4 Combination of Manganese(III) Acetate with 1,3-Dicarbonyl Compounds

4.1 Reaction Using 1,3-Diketones

Instead of tris(2,4-pentanedionato)manganese(III), manganese(III) acetate dihydrate was used in the reaction of alkenes with 1,3-dicarbonyl compounds [75, 77, 78]. The adopted reaction conditions were similar to those for the reaction using tris(2,4-pentanedionato)manganese(III). A mixture of 1,1-diphenylethene, 2,4-pentanedione, and manganese(III) acetate was stirred in acetic acid at room temperature in air until the brown color of Mn(III) turned colorless, giving the same endoperoxide in good yield. This method threw light on the synthesis of the functionalized 1,2-dioxan-3-ols by allowing the choice of combination of various alkenes and 1,3-diketones (Scheme 5).

1-Hydroxy-2,3-dioxabicyclo[4.4.0]decan-7-one, the analog of the root formation inhibitor (G factor) [18–20], was produced by the reaction with 1,3-cyclohexanedione. Other cyclic diketones also afforded the corresponding bicyclic endoperoxides (Scheme 6) [79].

Interestingly, for the reaction of the 1,3-cyclopentanedione, a double endoperoxidation took place and octahydro-3,4,7,8-tetraoxabenz[*c*]indene-4a,6a-diols were obtained (Scheme 7) [79]. The most characteristic feature in the spectral data was the absence of any carbonyl signals in the ^{13}C NMR spectrum as well as in the IR spectrum. The double endoperoxidation could not stop the corresponding monocyclization stage.

Since treatment of 4-acetyl-3-methyl-6,6-diphenyl-1,2-dioxan-3-ol and 1-hydroxy-4,4-diphenyl-2,3-dioxabicyclo[4.4.0]decan-7-one with a mixture of 1,1-diphenylethene and manganese(III) acetate did not afford any spirodiper-

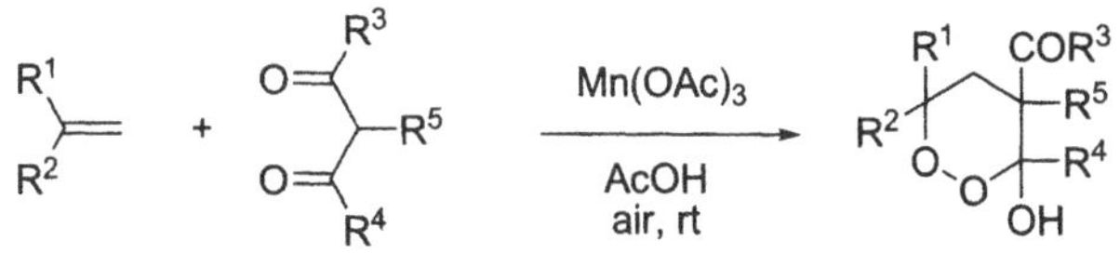

$R^1 = R^2$ = Aryl, $R^3 = R^4$ = Me, R^5 = H (95-44%)
R^1 = Aryl, $R^2 = R^3 = R^4$ = Me, R^5 = H (73-40%)
$R^1 = R^2$ = Ph, $R^3 = R^4$ = Me, R^5 = Alkyl (88-71%)
$R^1 = R^2$ = Ph, R^3 = Me, R^4,R^5 = -$(CH_2)_4$- (67%)
$R^1 = R^2 = R^3$ = Ph, R^4 = Me, R^5 = H (43%)
$R^1 = R^2 = R^3 = R^4$ = Me, R^5 = H (20%)

Scheme 5 Reaction of various alkenes with 1,3-diketones in the presence of manganese(III) acetate in acetic acid at room temperature in air

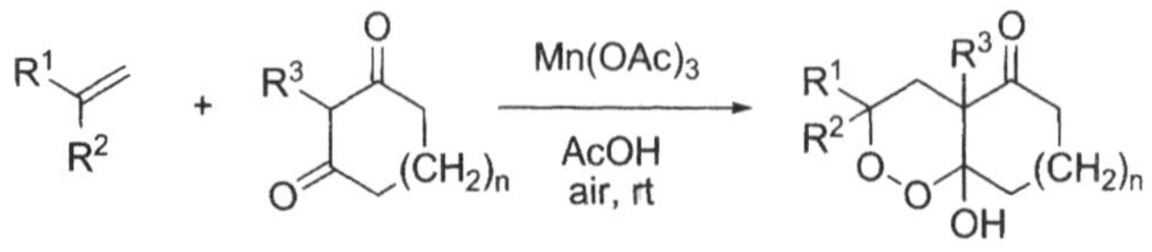

$R^1 = R^2 =$ Aryl, $R^3 =$ H, n = 1 (79-45%)
$R^1 = R^2 =$ Aryl, $R^3 =$ Me, n = 0,1 (93-54%)

Scheme 6 Reaction of alkenes with cyclic 1,3-diketones

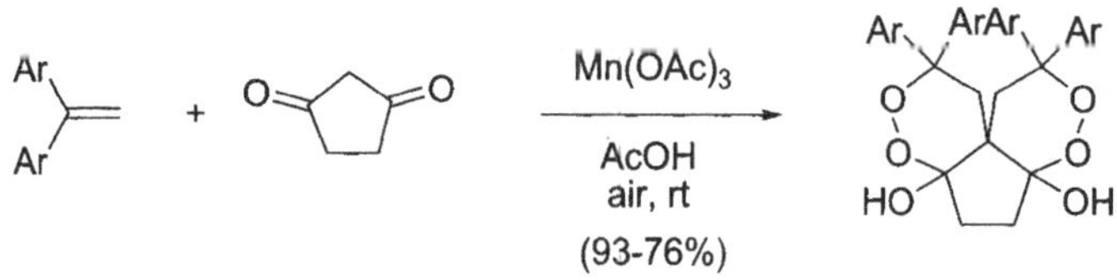

Scheme 7 Reaction of alkenes with 1,3-cyclopentanedione

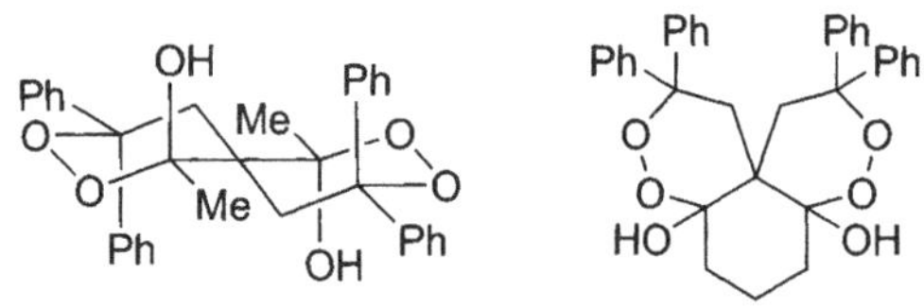

Spirodiperoxide Cyclic Diperoxide

Fig. 5 Postulated double endoperoxidation products

oxides or cyclic diperoxides (Fig. 5), the double endoperoxidation is characteristic of the 1,3-cyclopentanedione.

4.2 Reaction Using *β*-Ketoesters

The reaction using *β*-ketoesters resulted in 1,2-dioxan-3-ols in a comparable or better yield than that using the 1,3-diketones (Scheme 8) [80–82]. The cyclic peroxidation occurred at the more electrophilic ketocarbonyl carbon than the ester carbonyl carbon.

Cyclic *β*-ketoesters and aryl-substituted cycloalkenes also afforded the corresponding bicyclic endoperoxides and spirodioxanes (Scheme 9).

When the reaction was carried out using more than one equivalent of manganese(III) acetate, the yield of the cyclic peroxides was sometimes medium or poor, and a considerable amount of by-products was formed. The major by-products were dihydrofurans, benzophenones, and glycol monoacetates.

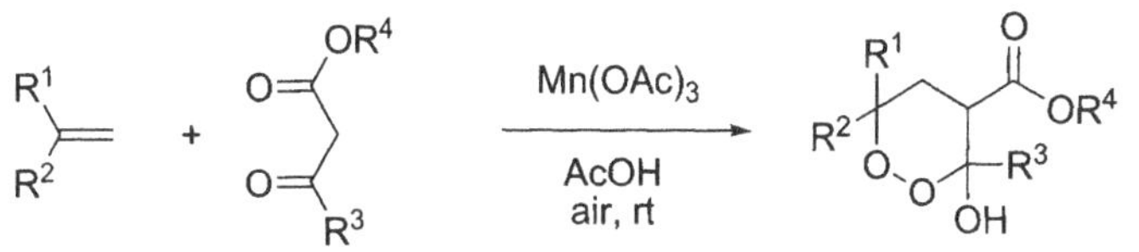

$R^1 = R^2 =$ Aryl, $R^3 =$ Me, $R^4 =$ Alkyl (95-65%)
$R^1 = R^2 = R^3 =$ Aryl, $R^4 =$ Et (74-29%)
$R^1 =$ Ph, $R^2 =$ H, $R^3 =$ Me, $R^4 =$ Et (61%)

Scheme 8 Reaction of various alkenes with *β*-ketoesters in the presence of manganese(III) acetate in acetic acid at room temperature in air

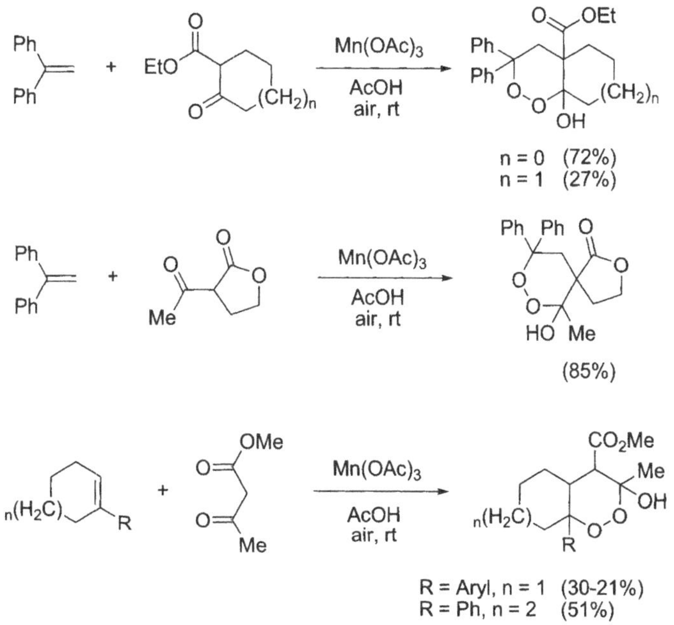

Scheme 9 Synthesis of bicyclo- and spiro-dioxanes from *β*-ketoesters

The dihydrofurans derived from the overoxidation of the intermediate carbon radicals and the others occurred from the direct oxidation of the alkenes during the electron-transfer mechanism (Scheme 1). In order to depress the side reactions, it was found that a catalytic amount of manganese(III) acetate was enough to produce the cyclic peroxidation reaction. The reaction proceeded only in the presence of manganese(II) acetate in air, although it took a longer reaction time [79–82]. Since the active species in the aerobic oxidation is manganese(III), the presence of various cooxidants, such as cobalt(III) acetate, potassium permanganate, lead(IV) acetate, copper(II) acetate, chromium(VI) trioxide, thallium(III) acetate, ammonium cerium(IV)

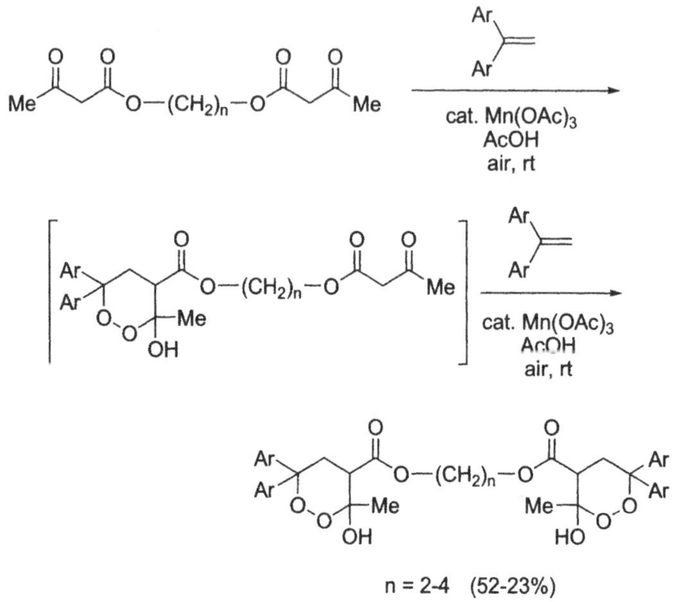

Scheme 10 Synthesis of methylene-tethered bisdioxanes from oligomethylene di(3-oxobutanoate)s

nitrate, and iron(III) perchlorate, accelerated the oxidation of Mn(II) to Mn(III). The combination of manganese(II) acetate and a catalytic amount of manganese(III) acetate was the most effective for the cyclic peroxidation. Consequently, a 95% yield of the 1,2-dioxan-3-ol was achieved along with 3% 4,5-dihydrofuran in the reaction of 1,1-diphenylethene with ethyl 3-oxobutanoate using one equivalent of manganese(II) acetate containing a 0.1 equivalent of manganese(III) acetate.

When the reaction was applied to oligomethylene di(3-oxobutanoate)s, oligomethylene-tethered terminal bis(1,2-dioxan-3-ol)s were obtained in moderate yields [82](Scheme 10).

When a similar reaction of 1,1-diphenylethene with ethylene di(3-oxobutanoate) was stopped for 2.5 h, an intermediate monodioxane was obtained in 60% yield. The reaction that continued for 20 h resulted in the corresponding bis(1,2-dioxan-3-ol) in 52% yield together with the monodioxane (2%). Furthermore, the isolated monodioxane was allowed to react with the alkene under the same reaction conditions to produce the bis(1,2-dioxan-3-ol). Accordingly, it was suggested that the formation of the bis(1,2-dioxan-3-ol)s apparently occurred during the stepwise mechanism.

4.3
Reaction Using Acetoacetamides

The manganese(III)-based oxidative endoperoxidation was applied to the *β*-ketoamide system. The reaction using various acetoacetamides was carried out in the presence of 1,1-disubstituted ethenes, producing crystalline carbamoyl-substituted 1,2-dioxanes in quantitative yields (Scheme 11) [83]. Interestingly, when the ^{1}H NMR spectrum of the carbamoyl-substituted dioxane was measured in DMSO-d_6, the signal intensity of the equatorial methyl group at the C-3 position that appeared at the higher field gradually decreased, while that of the axial methyl group shown at lower field increased The mixture reached equilibrium after 12 h at room temperature. This phenomenon obviously shows that the carbamoyl-substituted 1,2-dioxanes exist as an equilibrium mixture of two epimers at the C-3 position (Scheme 12). Normally, the hydroxyl group of the 1,2-dioxan-3-ol ring system is in an *axial*-like orientation due to the anomeric effect (Fig. 4) [75, 77]. However, in the carbamoyl-substituted dioxane derivatives, hydrogen bonds can be formed between the hydroxyl group at the C-3 position and the *equatorial* carbamoyl group at the adjacent position, and therefore, the equilibrium as shown in Scheme 12 lies very far to the right. In particular, using the *N*,*N*-disubstituted acetoacetamides or *ortho*-substituted *N*-arylacetoacetamides resulted in the predominant production of the carbamonyl-substituted 1,2-dioxanes having an *equatorial* hydroxyl group.

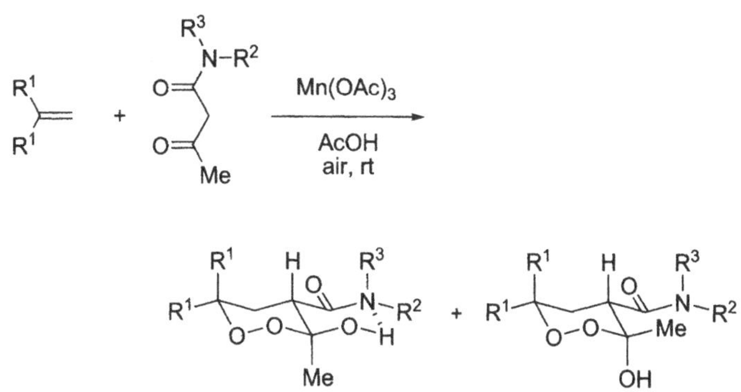

R^1 = Aryl, R^2 = 4-Sub-C_6H_4, R^3 = H	(96-82%)	*ca.* 60:40
R^1 = Aryl, R^2 = 2-Sub-C_6H_4, R^3 = H	(100-83%)	*ca.* 90:10
R^1 = Aryl, R^2 = Alkyl, R^3 = H	(96-88%)	*ca.* 55:45
R^1 = Aryl, R^2 = R^3 = Alkyl	(89-59%)	*ca.* 95:5
R^1 = Alkyl, R^2 = Aryl, R^3 = H	(68-62%)	*ca.* 55:45
R^1 = Ph, R^2, R^3 = -CH_2CH_2-O-CH_2CH_2-	(88%)	100:0

Scheme 11 Reaction using acetoacetamides

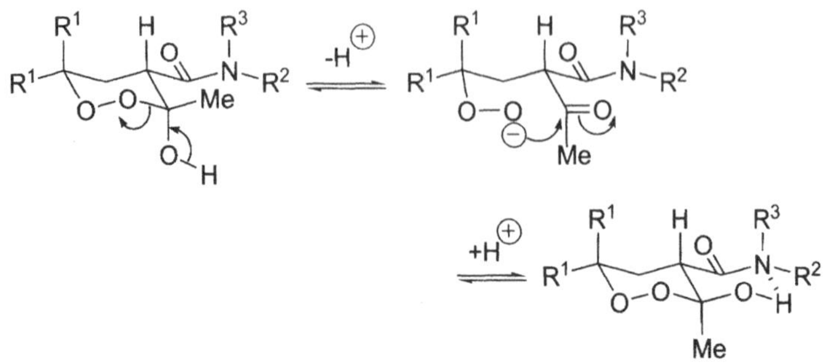

Scheme 12 Equilibrium of carbamoyl-substituted 1,2-dioxane stereoisomers

4.4 Reaction Using Sulfinyl-, Sulfonyl-, and Phosphinoyl-2-propanones

It was found that the 1,3-dicarbonyl compounds were effective for use in the synthesis of 1,2-dioxan-3-ols. In order to pioneer its synthetic utility, the endoperoxidation of alkenes using sulfinyl-, sulfonyl-, and phosphinoyl-substituted 2-propanones, which are the congener of the 1,3-dicarbonyl

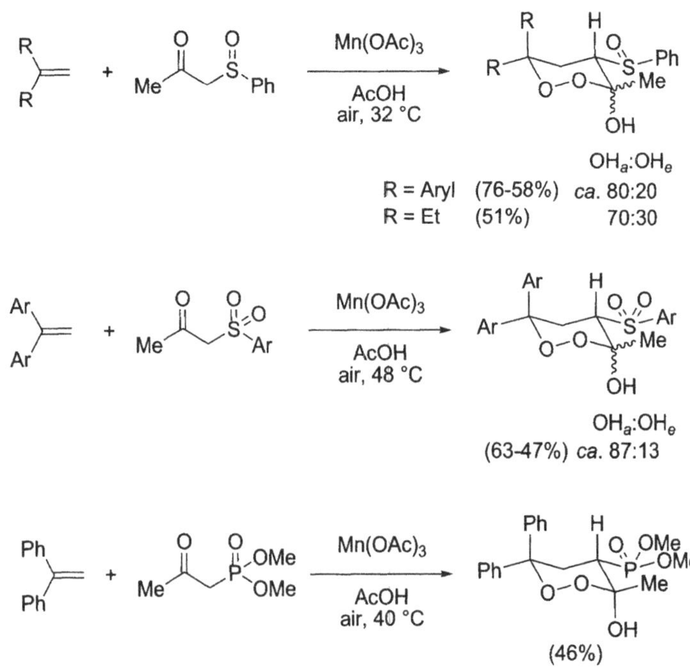

Scheme 13 Synthesis of 1,2-dioxan-3-ols using sulfinyl-, sulfonyl-, and phosphinoyl-2-propanones

compounds, were investigated. The reaction at room temperature (23 °C) practically did not take place, but needed a higher reaction temperature (32–48 °C) for the reaction to progress. The reaction of phenylsulfinyl-2-propanone with alkenes was carried out at 32 °C in the presence of manganese(III) acetate to give the corresponding 4-(phenylsulfinyl)-1,2-dioxan-3-ols together with a small amount of benzophenones and glycol monoacetates (Scheme 13) [84]. A similar reaction at 48 °C using phenylsulfonyl-2-propanone, (4-methylphenyl)sulfonyl-2-propanone, and dimethoxyphosphinoyl-2-propanone also afforded the corresponding dioxanes, along with a small amount of by-products.

The sulfinyl- and sulfonyl-substituted dioxanes were obtained as a mixture of two stereoisomers. The C-3 hydroxyl group of the major stereoisomer was axially oriented, and the sulfinyl, sulfonyl, and phosphinoyl groups at the C-4 position were placed equatorially, based on the ^{1}H and ^{13}C NMR spectra [84].

4.5 Reaction Using Acylacetonitrile Building Blocks

Since it is known that α-protons of the acylacetonitriles are more acidic than those of the methyl ketones, it was logical to use the acylacetonitrile building blocks during the first stage of the aldol-like condensation for the synthesis of heterocyclic compounds such as the 4*H*-pyran, 2-pyridone, and furan derivatives [85–95]. Therefore, it was expected that acylacetonitriles would be oxidized by manganese(III) acetate in a similar manner to the oxidation of α-cyanoacetic acid [65, 96–99] and 1,3-dicarbonyl compounds [100] to give acylcyanomethyl radicals, $^{\bullet}$CH(COR)CN, which would attack the alkenic double bonds to produce heterocyclic compounds in one step [73, 75, 77, 80, 101–

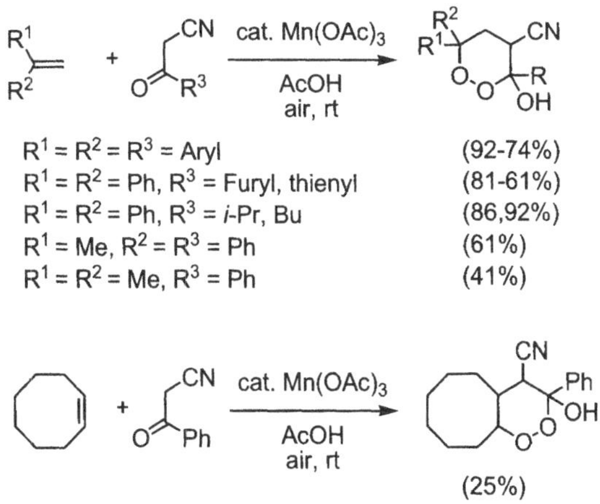

Scheme 14 Synthesis of cyanodioxanes using acylacetonitrile building blocks

104]. Benzoylacetonitrile was oxidized with manganese(III) acetate in the presence of 1,1-diphenylethene at room temperature in air and gave 4-cyano-3,6,6-triphenyl-1,2-dioxan-3-ol in 86% yield (Scheme 14) [105]. This was a very simple and convenient reaction for forming substituted 1,2-dioxanes without any by-products and with only simple temperature monitoring in air as a stipulation. Accordingly, the reaction was applied to many alkenes and acylacetonitriles, and the corresponding cyanodioxanes were obtained in satisfactory yields, except for two cases using styrene and 2-(5-methylfuryl)-substituted acylacetonitrile (R^3 = 2-(5-methylfuryl)) (Scheme 14) [106].

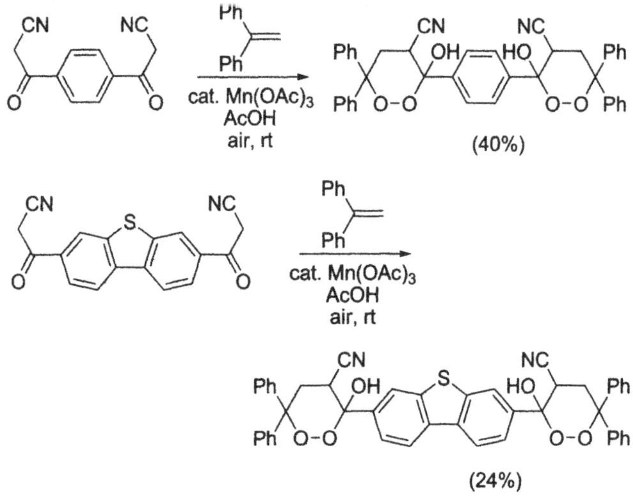

Scheme 15 Synthesis of bis(dioxanyl)arenes using acylacetonitrile building blocks

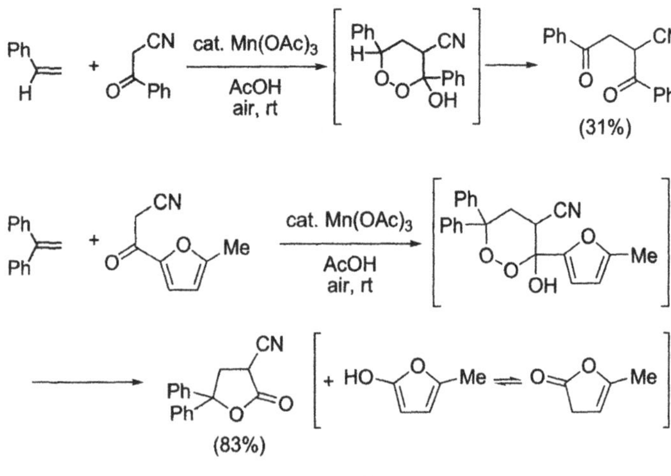

Scheme 16 Two exceptions using acylacetonitriles

Using aromatic compounds containing two acylacetonitrile functionalities led to the production of bis(dioxanyl)arenes at room temperature in moderate yields as shown in Scheme 15.

For the reaction of styrene with benzoylacetonitrile, 2-cyano-1,4-diphenylbutane-1,4-dione was isolated in 31% yield, while the reaction of 1,1-diphenylethene with 2-(2-cyano-1-oxoethyl)-5-methylfuran gave 2-cyano-4,4-diphenylbutanolide in 83% yield and no 1,2-dioxanes were detected. Both exceptions are explained by the fact that the corresponding dioxane, once formed during the reaction, would be decomposed in acidic media (Scheme 16) [78, 80, 106].

4.6
Reaction Using Cyclic Amides: 2,3-Pyrrolidinediones, 2,4-Pyrrolidinediones, 4-Piperidine-3-carboxylates, and 2,4-Piperidinediones

Nitrogen heterocycles widely occur in nature and as structural subunits in many families of alkaloids, and possess wide-ranging biological and pharmaceutical activities. The 2,3-pyrrolidinediones, 2,4-pyrrolidinediones (tetramic acids), 4-piperidine-3-carboxylates, and 2,4-piperidinediones are a particularly important subgroup, and are well known for their potent antibiotic, antiviral, antifungal and cytotoxic activities [107–116]. The pyrrolidinedione derivatives were also synthesized as endothelin receptor antagonists [117, 118]. The 1,2-dioxane scaffold containing nitrogen heterocycles is particularly important in nature as a metabolite and biologically active substance [1, 2, 119–121]. Some plants produce 1,2-dioxane derivatives in order to protect themselves and their territory (Fig. 2) [4–11, 13, 18–21, 122, 123]. Artemisinin (quighaosu) is a unique antimalarial drug consisting of the 1,2,4-trioxane skeleton (Fig. 1) [4–16], and it was reported that the synthesized *N*-substituted azaartemisinins were more active than the naturally occurring artemisinin [5]. Since the 2,3-pyrrolidinediones, 2,4-pyrrolidinediones, 4-piperidine-3-carboxylates, and 2,4-piperidinediones exist as the *enol* form, they could be effective candidates for forming carbon radicals during the manganese(III) acetate oxidation [99, 124]. Therefore, the Mn(III)-catalyzed aerobic oxidation of these heterocycles would be expected to produce azadioxabicyclic compounds. The synthesis of the heterocycle-fused endoperoxide framework has attracted considerable attention from the viewpoint of the development of new antibiotics, in spite of its instability.

Alkenes and 2,3-pyrrolidinediones were treated with manganese(III) acetate to give the desired 1-hydroxy-8-aza-2,3-dioxabicyclo[4.3.0]nonan-9-ones, which were obtained in good yields (Scheme 17) [125, 126].

A similar reaction of various alkenes with 2,4-pyrrolidinediones was carried out in acetic acid at room temperature under a dry air stream in the presence of manganese(III) acetate to give the 1-hydroxy-8-aza-2,3-dioxabicyclo[4.3.0]nonan-7-ones in good to quantitative yields [127,

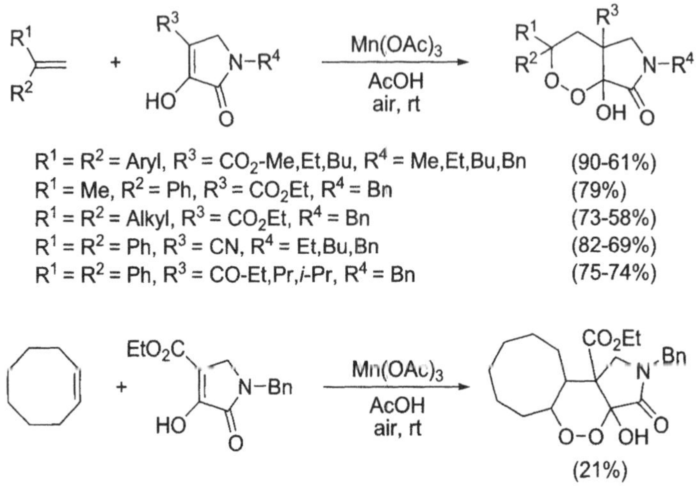

Scheme 17 Synthesis of 1-hydroxy-8-aza-2,3-dioxabicyclo[4.3.0]nonan-9-ones

128]. The reaction normally finished within 12 h depending on the 2,4-pyrrolidinediones used (Scheme 18).

In order to prepare structurally interesting azabicyclic peroxides, the synthesis of dumbbell-type compounds, which connected two azabicyclic peroxides in the methylene or oxamethylene chain, was attempted since such bioactive peroxides have been isolated from some marine sponges (Fig. 2) [1, 2, 24, 25]. Dialkyl *N,N'*-oligomethylenebis(2,3-pyrrolidinedione-4-carboxylate)s were allowed to react with 1,1-diarylethenes in the presence of manganese(III) acetate in a dry air stream. Although a long reaction

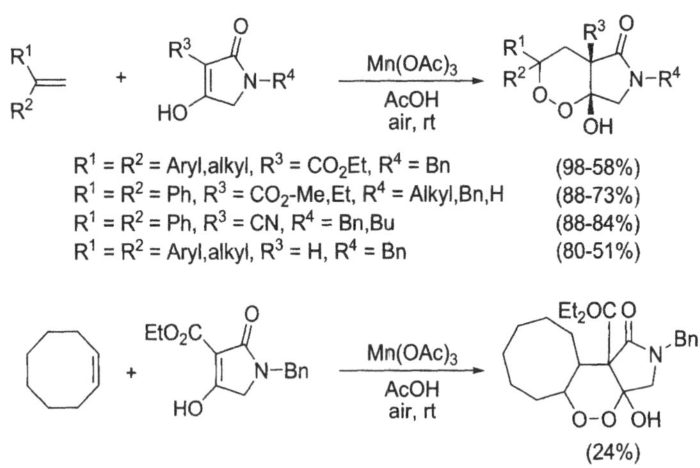

Scheme 18 Synthesis of 1-hydroxy-8-aza-2,3-dioxabicyclo[4.3.0]nonan-7-ones

time was needed to complete the aerobic oxidation (24–54 h), since the substrates were not soluble in acetic acid, the corresponding dialkyl *N,N'*-oligomethylenebis(1-hydroxy-4,4-diaryl-8-aza-2,3-dioxabicyclo[4.3.0]nonan-9-one-6-carboxylate)s were produced in good yields. It was observed that the longer the methylene chain length of the *N,N'*-oligomethylenebis(2,3-pyrrolidinedione)s, the higher the yield of the corresponding bis(8-aza-2,3-dioxabicyclo[4.3.0]nonan-9-one)s (Scheme 19) [129]. A similar reaction of the dialkyl *N,N'*-oligooxamethylenebis(2,3-pyrrolidinedione-4-carboxylate)s also gave the corresponding azabicyclic peroxides connected by an oligooxamethylene chain in acceptable yields (Scheme 20).

As good results were obtained using the five-membered nitrogen-heterocycles, the reaction was applied to the six-membered nitrogen-heterocycles, the 4-piperidine-3-carboxylates, and 2,4-piperidinediones. Since free 4-piperidine-3-carboxylates and 2,4-piperidinediones were too sensitive for the manganese(III) oxidant to survive under the reaction conditions [130–132], the *N*-protected 4-piperidine-3-carboxylates and 2,4-piperidinediones were used. When the 4-piperidone-3-carboxylates were allowed to react with the 1,1-disubstituted alkenes in the presence of a catalytic amount of manganese(III) acetate in air at room temperature, the 1-hydroxy-8-aza-2,3-dioxabicyclo[4.4.0]decane-6-carboxylates were produced in good to moderate yields (Scheme 21) [133].

A similar reaction of alkenes with 2,4-piperidinediones was carried out using a catalytic amount of manganese(III) acetate until the 2,4-

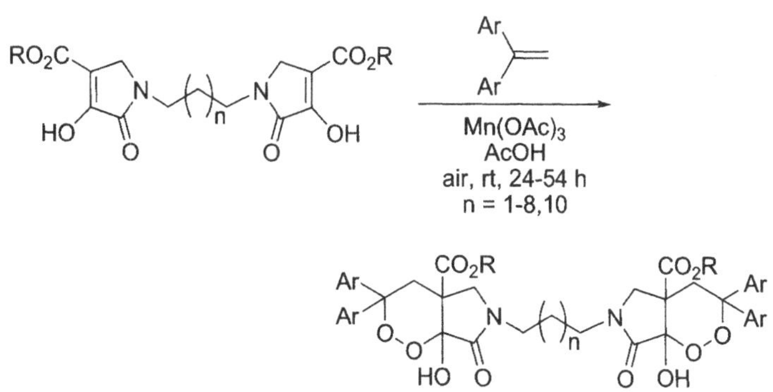

n = 1, R = Et, Ar = Ph	(46%)
n = 2, R = Et, Ar = Ph	(61%)
n = 3, R = Et, Ar = Ph	(73%)
n = 4, R = Et,Me, Ar = Aryl	(93-64%)
n = 5, R = Et, Ar = Ph	(82%)
n = 6, R = Et,Me,Bu, Ar = Ph	(83-54%)
n = 7, R = Et, Ar = Ph	(83%)
n = 8, R = Et, Ar = Ph	(87%)
n = 10, R = Et, Ar = Ph	(91%)

Scheme 19 Manganese(III)-catalyzed reaction of *N,N'*-oligomethylenebis(2,3-pyrrolidinedione-4-carboxylate)s with alkenes

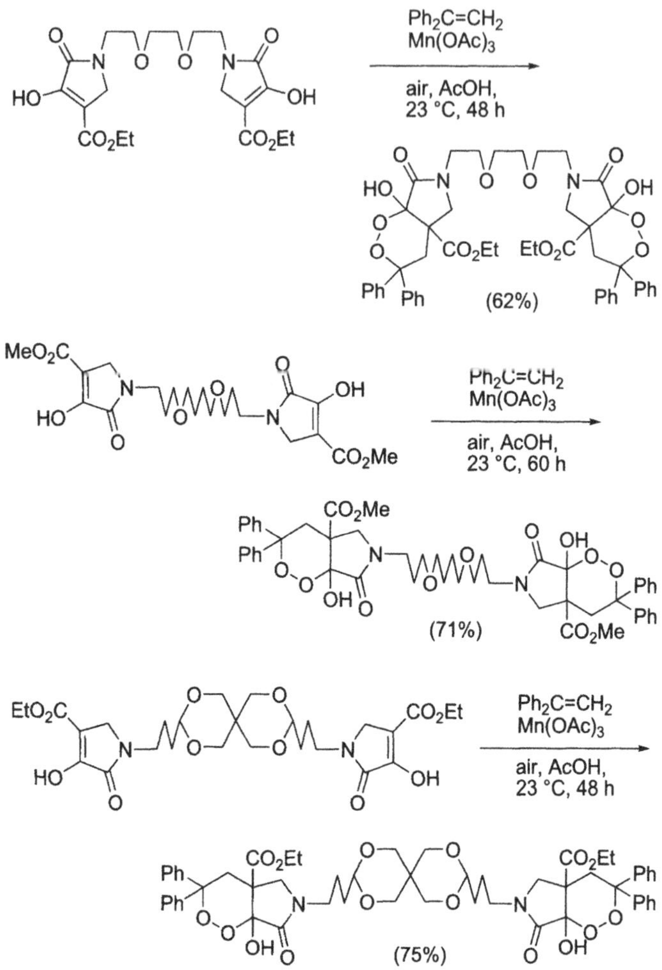

Scheme 20 Manganese(III)-catalyzed reaction of bis(2,3-pyrrolidinedione-4-carboxylate)s with 1,1-diphenylethene

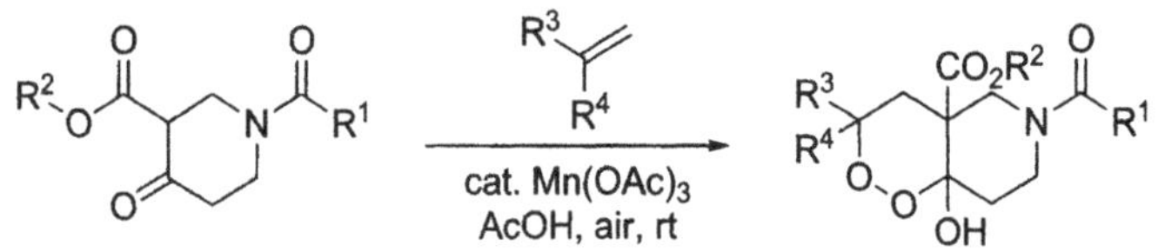

R^1 = Ph,Me,Et,NEt_2, R^2 = Et,Bu, R^3 = R^4 = Aryl (84-44%)
R^1 = Ph, R^2 = R^3 = R^4 = Et (22%)
R^1 = R^3 = Ph, R^2 = Et, R^4 = Me (33%)
R^1 = R^3 = Ph, R^2 = Et, R^4 = H (35%)

Scheme 21 Manganese(III)-catalyzed aerobic oxidation of 4-piperidone-3-carboxylates in the presence of alkenes

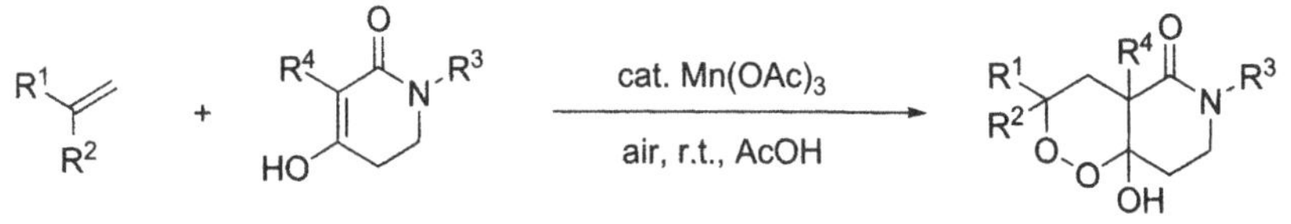

$R^1 = R^2$ = Aryl, R^3 = Bn, $R^4 = CO_2Et$
R^1 = Aryl, R^2 = Me, R^3 = Bn, $R^4 = CO_2Et$
$R^1 = R^2$ = Alkyl, R^3 = Bn, $R^4 = CO_2Et$
R^1 = Ph, R^2 = H, R^3 = Bn, $R^4 = CO_2Et$
R^1 = Heptyl, R^2 = H, R^3 = Bn, $R^4 = CO_2Et$
$R^1 = R^2$ = Ph, R^3 = Alkyl,Ph, $R^4 = CO_2$-Et,*i*-Pr
$R^1 = R^2$ = Aryl, R^3 = Bn,Me,Et,Pr,*i*-Pr,Ph, R^4 = H

(88-71%)
(95-82%)
(77-75%)
(75%)
(17%)
(93-50%)
(99-60%)

Scheme 22 Manganese(III)-catalyzed aerobic oxidation of a mixture of alkenes and 2,4-piperidinediones

piperidinediones were completely consumed. After chromatographic separation, the corresponding 3-dioxabicyclo[4.4.0]decan-7-ones were isolated in excellent to good yields, except for the reaction using 1-nonene (Scheme 22) [134, 135].

5
Hydroperoxidation Using Manganese(III) Acetate

Since the nitrogen-heterocycle-fused dioxabicyclic compounds were obtained by the reaction of 2,3-pyrrolidinediones, 2,4-pyrrolidinediones, 4-piperidine-3-carboxylates, and 2,4-piperidinediones, the manganese(III)-catalyzed endoperoxidation was applied to other pharmaceutically important cyclic amide derivatives, such as the barbituric acids, pyrazolidine-3,5-diones, and 4-hydroxy-1*H*-quinolin-2-ones. However, the corresponding azadioxabicyclic compounds were not formed, but the hyroperoxidation unexpectedly occurred.

5.1
Reaction Using Barbituric Acids

The barbituric acid derivatives, which have been known as hypnotic sedatives [136], were chosen as the first of the heterocyclic 1,3-dicarbonyl compounds for study because they consist of a cyclic diamide structure. The reaction of barbituric acid with 1,1-diphenylethene was conducted in acetic acid at room temperature in air using a catalytic amount of manganese(III) acetate. Surprisingly, solid products were easily obtained as a diastereomixture, meso and racemic. Although the diastereomers could not be

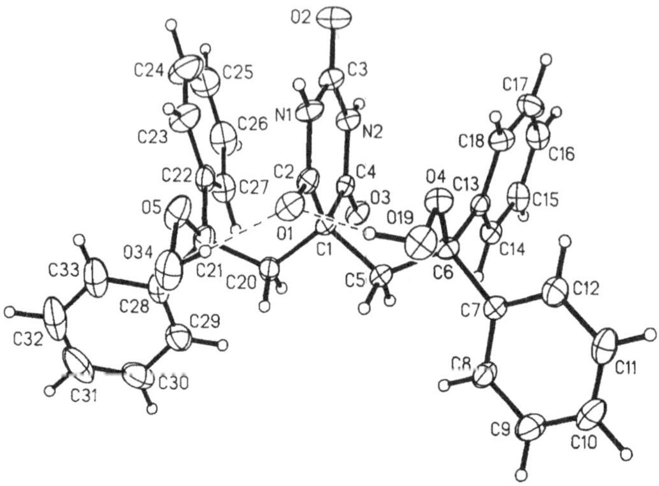

Fig. 6 ORTEP drawing of 5,5-bis(2-hydroperoxyethyl)barbituric acid ($R^1 = R^2 = Ph$, $R^3 = H$)

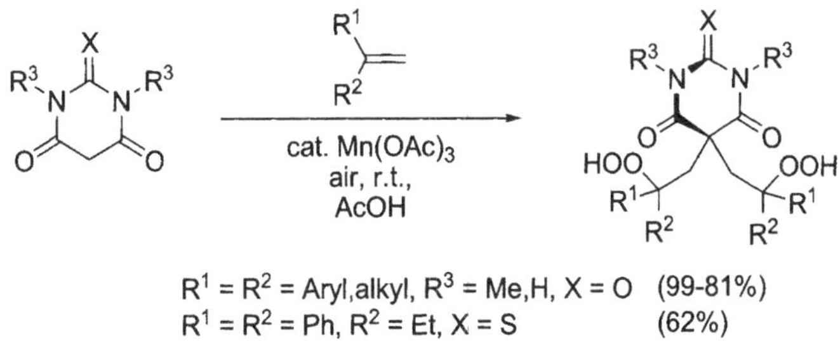

Scheme 23 Manganese(III)-catalyzed double 2-hydroperoxyalkylations of barbituric acids

separated into individual isomers, the structure was determined by spectroscopy, elemental analysis, and finally X-ray crystallography. As a result, it was found that the solid products were not 1,2-dioxanes, but 5,5-bis(2-hydroperoxyethyl)barbituric acids (Fig. 6) [137].

Similar reactions of other barbituric acid derivatives with alkenes gave the corresponding 5,5-bis(2-hydroperoxyethyl)barbituric acids in quantitative yields (Scheme 23). Using thiobarbituric acid also afforded a similar bis(hydroperoxide) in good yield.

5.2 Reaction Using Pyrazolidine-3,5-diones

Derivatives of the pyrazolidine-3,5-diones possess a wide variety of biological and pharmaceutical activities along with other uses (e.g., color agents, photographic light-sensitive and thermal printing materials [138, 139]). Especially, interest in the synthesis and pharmacological evaluation of numerous pyrazolidine-3,5-diones as AT_1 angiotensin II receptor antagonists and study of the inhibition of the enzyme activity for prostaglandin H synthase has increased [140, 141]. In order to synthesize functionalized pyrazolidine-3,5-dione derivatives with potent biological activities, the pyrazolidine-3,5-diones were selected as the second choice of heterocyclic diamides in the manganese(III)-catalyzed aerobic oxidation. Therefore, the reactions were investigated with particular attention being paid to the use of a combination of a catalytic amount of manganese(III) acetate and air. In this case, the oxidation also led to free double hydroperoxides in excellent yields, instead of cyclic peroxides as in the reaction of the barbituric acids (Scheme 24) [142, 143].

The structural assignment of the bis(hydroperoxyethyl)pyrazolidine-3,5-diones was based on their ^{1}H NMR, ^{13}C NMR, and IR spectra, as well as their elemental analyses. The exact structure of one of these was also confirmed by X-ray crystallography. The most characteristic feature of the bis(hydroperoxyethyl)pyrazolidine-3,5-diones was that two hydroperoxy groups are individually hydrogen-bonded to the two amide carbonyl oxygens since the interatomic distance between the carbonyl oxygen and the peroxy oxygen was found to be 2.688 Å [143]. The hydrogen bonding is stronger in the solid state than that in solution. A similar stabilization of the hydroperoxy group was also observed in 5,5-bis(hydroperoxy-2,2-diphenylethyl)barbituric acid in which the corresponding interatomic distances were 2.73 Å and 2.81 Å, respectively [137]. The hydroperoxy O – O bond length (1.465 Å) in the bis(hydroperoxyethyl)pyrazolidine-3,5-dione was analogous to those of the bis(hydroperoxyethyl)barbituric acid (1.450 Å and 1.464 Å). In addition, unlike many alkylhydroperoxides, the bis(hydroperoxyethyl)pyrazolidine-3,5-diones as well as the bis(hydroperoxyethyl)barbituric acids were found to be thermally stable at ambient temperature [144] and to prolonged ex-

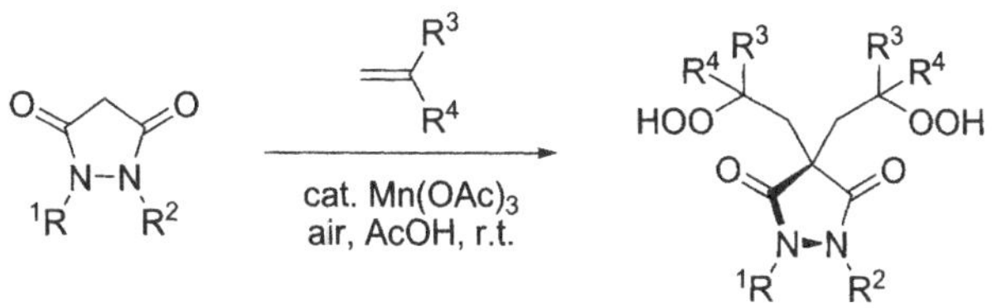

$R^1 = R^2$ = Ph,Bn, $R^3 = R^4$ = Aryl	(98-78%)
R^1 = Bn, R^2 = Ph,Me, $R^3 = R^4$ = Aryl	(91-74%)
$R^1 = R^2$ = Ph, $R^3 = R^4$ = Alkyl	(95-91%)
$R^1 = R^2 = R^3$ = Ph, R^4 = Me	(95%)
R^1, R^2 = $-(CH_2)_4-$, $R^3 = R^4$ = Ph	(98%)

Scheme 24 Manganese(III)-catalyzed double 2-hydroperoxyalkylations of 1,2-disubstituted pyrazolidine-3,5-diones

posure to sunlight or visible light [145], which could also be attributed to stabilization by strong hydrogen bonding.

5.3 Reaction Using 4-Hydroxy-1*H*-quinolin-2-ones

A variety of quinoline alkaloids exists in plants, and it is known that the quinoline alkaloids exhibit a wide range of biological activities [146]. Naturally occurring quinine and artificial chloroquine are the most well-known antimalarial agents using the quinoline skeleton (Fig. 1) [147–149]. Since the development of the reaction scheme of the quinoline-fused cyclic peroxides might also be significant in order to find a new class of artificial antimalarial reagents, the 4-hydroxy-1*H*-quinolin-2-ones were selected as the third choice of heterocyclic 1,3-dicarbonyl compounds in the manganese(III)-catalyzed aerobic oxidation.

Use of a 0.1 equivalent of manganese(III) acetate toward the substrate 4-hydroxy-1*H*-quinolin-2-one did not afford the expected cyclic peroxide, but the bis(hydroperoxide) analogous to the manganese(III)-catalyzed autoxidation of the barbituric acids and pyrazolidine-3,5-diones. Prolongation of the reaction period and the use of 0.5 equivalents of the catalyst resulted in increased yields (Scheme 25) [150].

X-ray crystallography indicated that the two hydroperoxy groups seemed to be stabilized by intramolecular hydrogen bonding with the quinolinedione carbonyls since the interatomic distance between the carbonyl oxygen and peroxy oxygen was 2.712 Å and 2.756 Å, respectively. A similar stabilization of the hydroperoxy group was observed in the 5,5-bis(2-hydroperoxyethyl)barbituric acid (2.73 Å and 2.81 Å) and 4,4-bis(2-hydroperoxyethyl)pyrazolidine-3,5-dione systems (2.688 Å) [137, 142, 143].

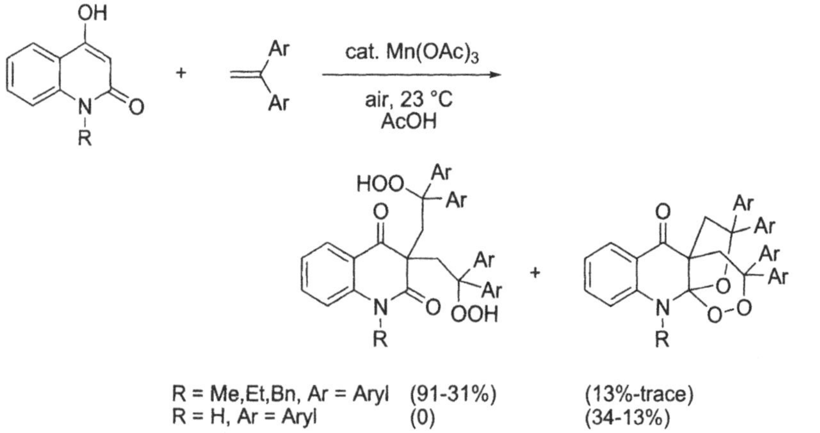

Scheme 25 Manganese(III)-catalyzed aerobic oxidation of 4-hydroxy-1*H*-quinolin-2-ones in the presence of 1,1-diarylethenes

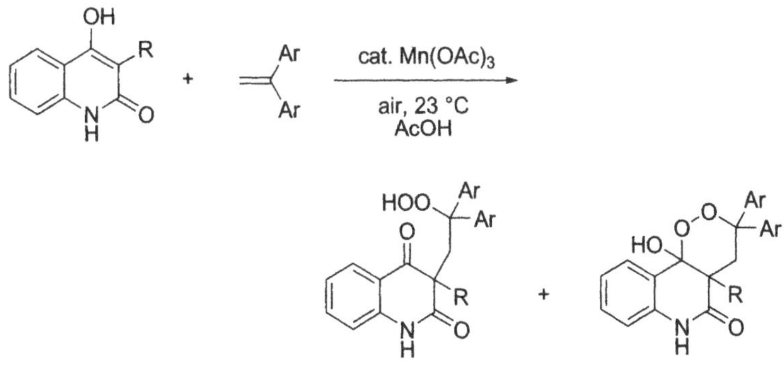

R = Me, Ar = Ph		(89%)
R = Me, Ar = 4-ClC_6H_4		(88%)
R = Me, Ar = 4-MeC_6H_4	(58%)	(38%)
R = Pr, Ar = Ph	(60%)	(22%)
R = Bu, Ar = Ph	(52%)	(24%)
R = Ph, Ar = Ph	(43%)	(32%)

Scheme 26 Manganese(III)-catalyzed aerobic oxidation of 3-substituted 4-hydroxy-1*H*-quinolin-2-ones in the presence of 1,1-diarylethenes

In order to find other products, the rest of the residue was scrutinized, and the quite unique [4.4.3]propellane-type cyclic peroxides were isolated (Scheme 25). The peroxy O – O bond length (1.464 Å) was analogous to that of the reported crystalline 1,2-dioxanes (normally 1.44–1.47 Å) [77, 128, 133]. Although the use of the *N*-nonprotected quinolinones (R = H) resulted in a complex mixture, only the [4.4.3]propellane-type cyclic peroxides were isolated (Scheme 25).

Since it was found that the α-carbon of the amide carbonyl in the 4-hydroxy-1*H*-quinolin-2-ones was very easily oxidized by manganese(III) to form double hydroperoxyalkylated quinolinediones, even when an excess amount of the quinolinones were employed under very mild aerobic oxidation conditions, the 3-substituted quinolinones were prepared. Aerobic oxidation with alkenes was investigated in an effort to prevent the double hydroperoxyalkylation from taking place and to produce the cyclic peroxide. This process was successful and the corresponding quinoline-fused endoperoxides were obtained in good yields (Scheme 26). However, the quinolinones substituted by a bulky group rather than by the methyl group preferentially gave the corresponding acyclic monohydroperoxides [150].

5.4 Manganese(III)-Catalyzed Direct Hydroperoxidation of Heterocyclic 1,3-Dicarbonyl Compounds

4-Butyl-1,2-diphenylpyrazolidine-3,5-dione (phenylbutazone), a nonsteroidal antiinflammatory drug, is an efficient reducing cofactor for the peroxidase activity of prostaglandin H synthase [141]. Phenylbutazone inhibits the production of lipid mediators causing inflammation, but paradoxically performs this via the intermediacy of the peroxy radical and hydroperoxide, which may themselves be proinflammatory. In the isolated heart preparations of guinea pigs and in rabbit hearts in vivo 4-butyl-4-hydroperoxy-1,2-diphenylpyrazolidine-3,5-dione shows a significantly stronger cardiodepressive and coronary-constricting effect than phenylbutazone itself, 4-butyl-4-hydroxy-1,2-diphenylpyrazolidine-3,5-dione, or the ring-opened decomposition product of the hydroperoxide [151]. These phenomena could shed light on the significance of the hydroperoxylated phenylbutazone regarding the antiinflammatory or other biological activities of the phenylbutazone (e.g., in rheumatoid arthritis [139, 152]) and could explain the side effects associated with phenylbutazone, such as gastric irritation and toxicity. A number of reagents have been utilized for the introduction of an oxygen functionality at the 2-position of the 1,3-dicarbonyl compounds [153–161]. However, in all of these cases, only the hydroxyl functionality could be introduced at the 2-position of the 1,3-dicarbonyl compounds. The direct peroxidation of the 4-monoalkylpyrazolidine-3,5-diones including phenylbutazone was then examined using the manganese(III)-catalyzed aerobic oxidation in the absence of alkenes. Stirring a 1 mmol solution of different 4-monoalkylpyrazolidine-3,5-diones in acetic acid in the presence of a catalytic amount of manganese(III) acetate under an aerobic atmosphere gave the direct hydroper-

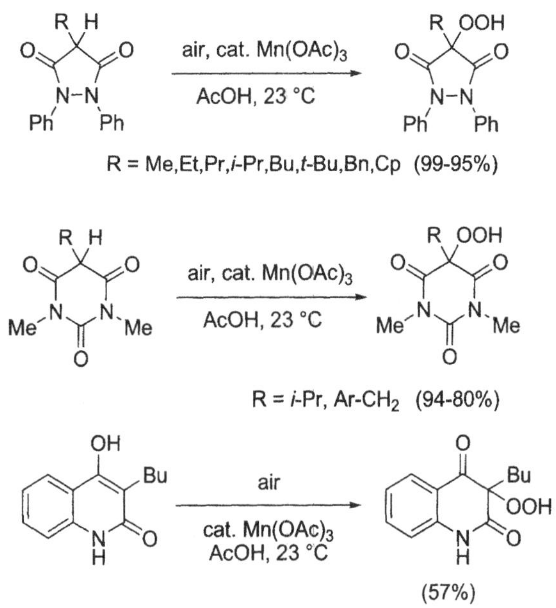

Scheme 27 Manganese(III)-catalyzed direct hydroperoxidation of heterocyclic 1,3-dicarbonyl compounds

oxylated pyrazolidine-3,5-diones in quantitative yields (Scheme 27) [162]. The structural assignment of the 4-hydroperoxypyrazolidine-3,5-diones was based on the ^1H NMR, ^{13}C NMR, and IR spectra, elemental analyses, and finally X-ray crystallography.

To examine the applicability of the manganese(III) acetate-catalyzed α-hydroperoxidation of other biologically important heterocyclic 1,3-dicarbonyl compounds, the reaction of the 5-monosubstituted barbituric acids and 3-butyl-4-hydroxy-2-quinolinone was carried out under similar aerobic conditions. Very similar autoxidation results were obtained, giving the corresponding hydroperoxides in excellent or moderate yields (Scheme 27).

6 Spectroscopic Feature and Stereochemistry of 1,2-Dioxan-3-ols

The cyclic peroxides and hydroperoxides obtained by the manganese(III)-based aerobic oxidation was characterized by ^1H NMR, ^{13}C NMR, and IR spectroscopies, FAB MS spectrometry, and combustion analysis. The absolute structure of the crystalline peroxides was determined by X-ray crystallography. The spectroscopic feature of the cyclic peroxides is that two quaternary carbons attached to the oxygen atom appear at 110–80 ppm in the ^{13}C NMR spectrum. On the other hand, in the case of the hydroperoxides, only one quaternary carbon attached to the oxygen atom appears at 88–85 ppm. The chemical shifts of the characteristic peaks in the ^{13}C NMR spectrum are summarized in Figs. 7 and 8.

The absolute structure of some cyclic peroxides and hydroperoxides were established by the X-ray crystallographic measurement. The O – O bond lengths of the cyclic peroxides and hydroperoxides were 1.43–1.47 Å, which were similar to those of the reported crystalline 1,2-dioxanes [3, 163–165]. Some O – O bond lengths are shown in Figs. 7 and 8.

The stereochemistry of the hydroxyl group in the synthesized cyclic peroxides is also characteristic. In the solid state, the hydroxyl group was arranged in an *axial*-like orientation due to the anomeric effect [166] based on the X-ray crystallographic analysis (Fig. 4) [77, 128, 133, 143].

Since the 1,2-dioxan-3-ol is a hemiketal structure, it is considered that it exists as an equilibrium mixture of an OH-*axial* and an OH-*equatorial* dioxane in solution (Scheme 28). In fact, the phenomenon was observed in the case of the 4-carbamoyl-1,2-dioxan-3-ols (Schemes 11 and 12) [83]. It seems that the equilibrium depends on the balance of the anomeric effect and the strength of the hydrogen bonding between the hydroxyl group and the carbonyl function.

Methylation of the 1,2-dioxan-3-ol in the presence of *p*-toluenesulfonic acid gave two stereoisomers in 95% yield, and the isomer ratio was 42 : 58

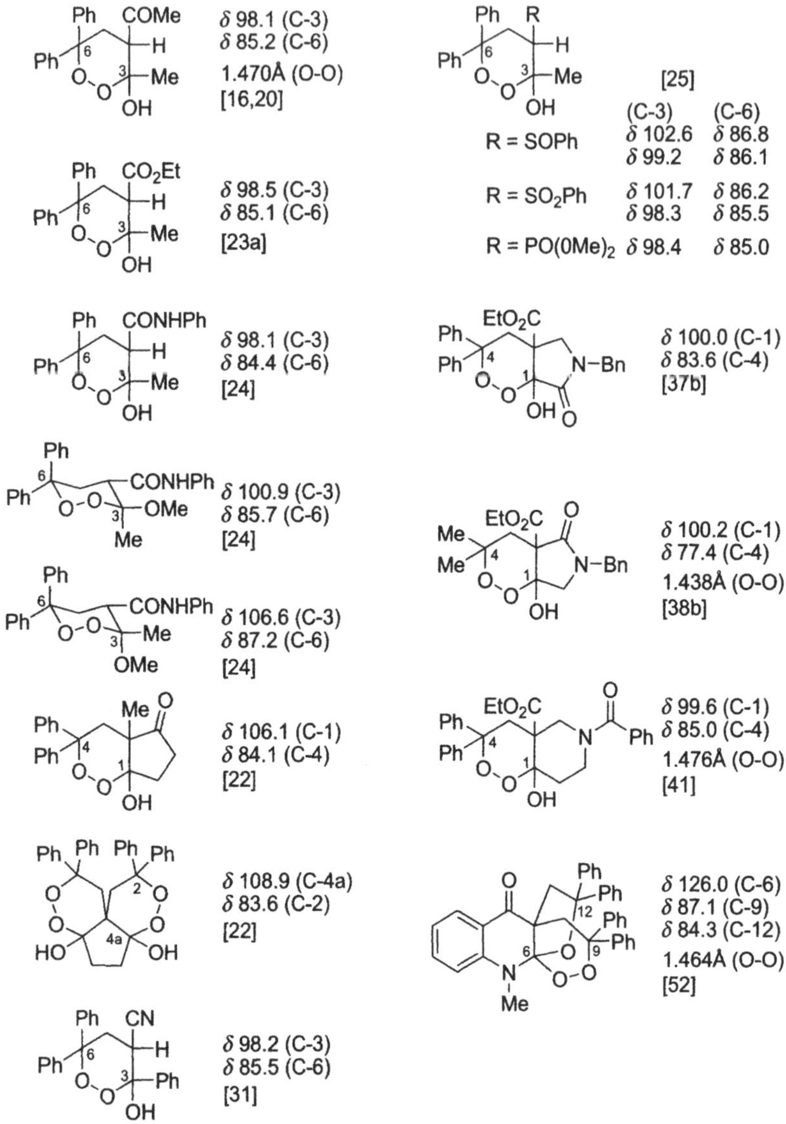

Fig. 7 Characteristic ^{13}C NMR chemical shifts and O – O bond length of some cyclic peroxide derivatives

(Scheme 29) [167]. The stereoisomers could be isolated by fractional recrystallization from ethanol. The major methyl ether was MeO-*equatorial* dioxane and the minor MeO-*axial* based on the X-ray crystallography.

In $CDCl_3$ solution of the minor dioxane, NMR analysis showed that it existed as an equilibrium mixture of twisted boat conformers. Adding shift reagent, $Eu(FOD)_3$, resulted in the MeO-*axial* chair conformer (Scheme 30).

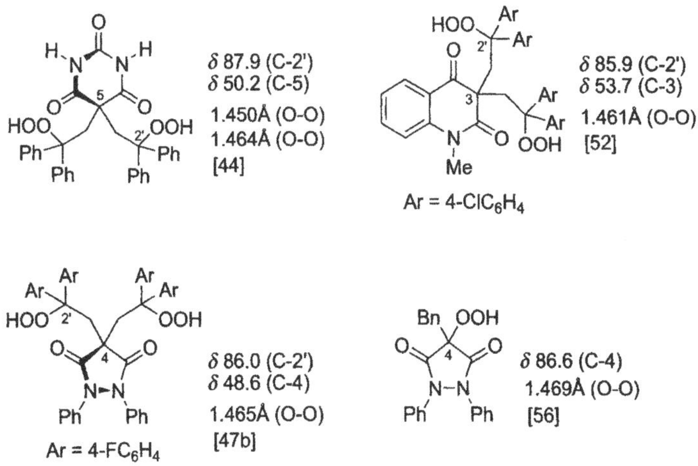

Fig. 8 Characteristic ^{13}C NMR chemical shifts and O – O bond length of some hydroperoxides

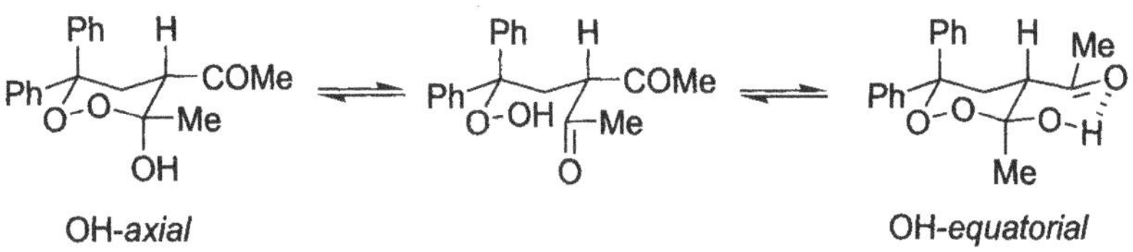

Scheme 28 An equilibrium mixture of 1,2-dioxan-3-ol in solution

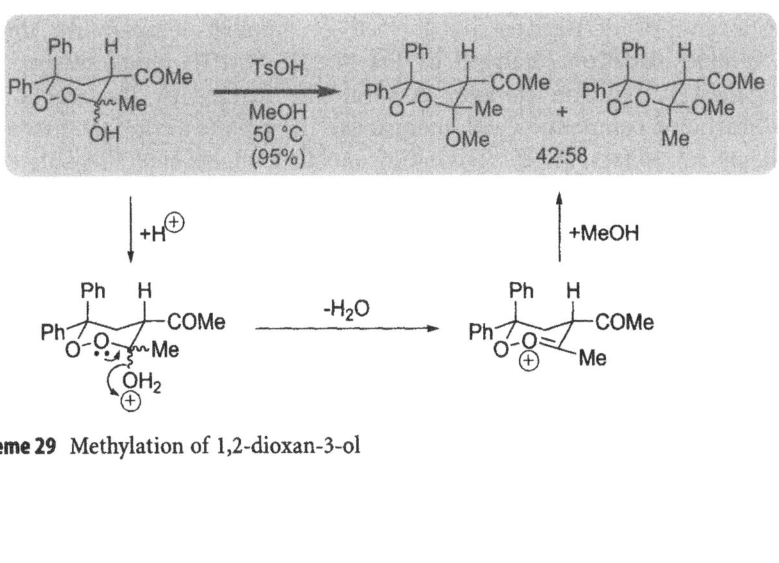

Scheme 29 Methylation of 1,2-dioxan-3-ol

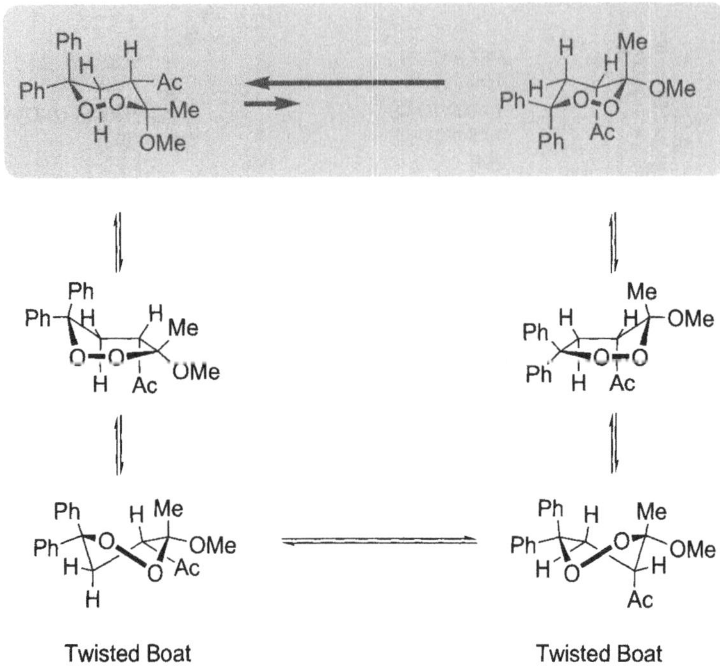

Scheme 30 Conformer equilibrium of the minor MeO-*axial* dioxane in $CDCl_3$

7
Mechanism for the Formation of 1,2-Dioxan-3-ols

1,2-Dioxan-3-ol derivatives are formed at ambient temperature during the aerobic oxidation catalyzed by the Mn(II)–Mn(III) redox system. The manganese(III)-enolate complex **A** would be formed by the reaction of 1,3-dicarbonyl compounds with manganese(III) acetate during the first stage (Scheme 31). In fact, the corresponding enolate complex could be isolated by the reaction of manganese(III) acetate with 4,4,4-trifluoro-1-phenylbutane-1,3-dione in acetic acid at 23 °C for 1 h [106]. The enolate complex formation is the rate-determining step [44, 124] and the key to the catalytic reaction. When alkenes are present in the reaction system, complexes between the alkenes (electron donor) and the manganese(III)-enolate complex (electron acceptor) are immediately formed in situ and a one-electron-transfer oxidation easily occurs to produce the corresponding carbon radicals **B** and release manganese(II) acetate [124]. The carbon radicals **B** take up dissolved molecular oxygen in the solvent to form the peroxy radicals **C**. The peroxy radicals **C** could be reduced by the manganese(II)-enolate complex **D**, which would be formed by the ligand-exchange reaction of the released manganese(II) acetate with 1,3-dicarbonyl compounds, followed by cyclization to finally produce the thermodynamically stable OH-*axial* 1,2-dioxanes [168].

The manganese(III) species **A** should be reproduced in the reaction system, and the catalytic cycle must be continued until the added alkenes and 1,3-dicarbonyl compounds are completely consumed [46, 77, 79, 80, 83, 106, 125–128, 134, 135, 142, 143]. The mechanism for the formation of peroxides is summarized in Scheme 31.

The formation of manganese(II)-enolate complex **D** is essential for the catalytic reaction since the complex **D** is lightly oxidized by the peroxy radicals **C**, dissolved oxygen, or other added metal oxidants (such as manganese(III) acetate, cobalt(III) acetate, potassium permanganate, lead(IV) acetate, chromium(IV) trioxide, thallium(III) acetate, ammonium cerium(IV) nitrate, copper(II) acetate, and iron(III) perchlorate) though the oxidation of manganese(II) acetate itself is very slow in air at ambient temperature [80]. Therefore, it is possible to use manganese(II) acetate instead of

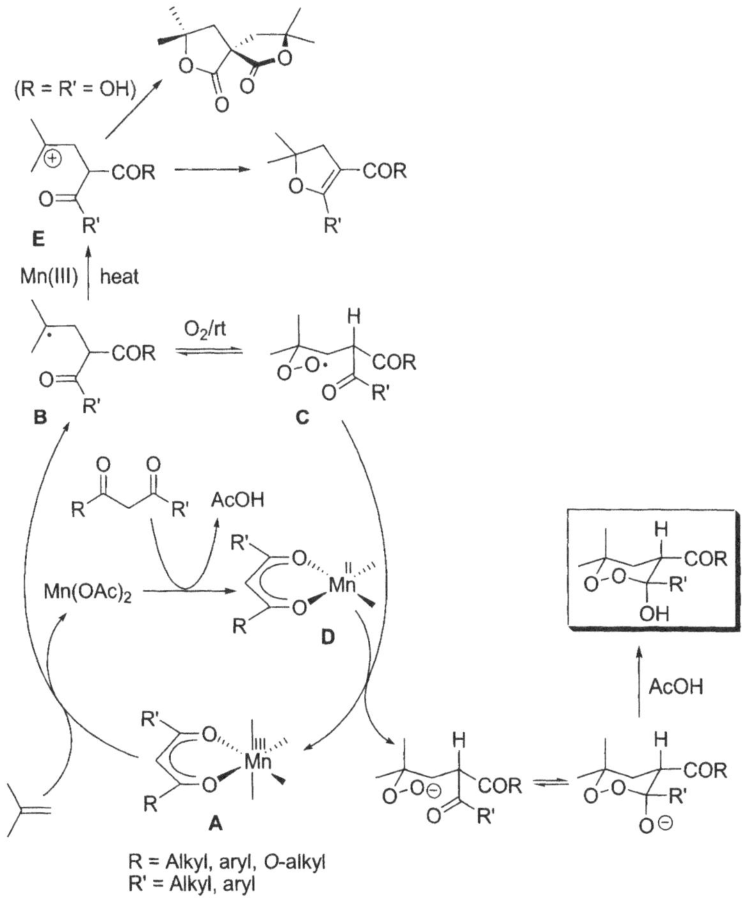

Scheme 31 Mechanism of the manganese(III)-catalyzed aerobic oxidation

manganese(III) species for the aerobic oxidation [79–82, 137]. The hydroperoxidation of cyclic amides could also be explained by a similar mechanism for the manganese(III)-catalyzed aerobic oxidation. The cyclization between the hydroperoxy group and amido carbonyl group might not occur because the electrophilicity of the amido carbonyl carbon is poor.

In contrast, when the reaction is carried out at elevated temperatures using a stoichiometric amount of manganese(III) acetate, the reaction is dramatically changed. The formed carbon radicals **B** would be quickly oxidized by manganese(III) species in the reaction at elevated temperature due to the absence of dissolved molecular oxygen and the presence of sufficient manganese(III) acetate or cooxidant such as copper(II) acetate [44]. As a result, carbocations **E** would be produced and would cyclize at the carbonyl oxygen to give furan or lactone derivatives [135, 169–173].

8
Chemical Transformation of the Synthesized 1,2-Dioxane Derivatives

1,2-Dioxan-3-ols can be converted into various functionalized heterocyclic compounds (Scheme 32). For example, the acid-catalyzed decomposition of the 6,6-disubstituted 1,2-dioxan-3-ols quantitatively gave the 2,3,5-trisubstituted furans [78–82]. The reaction could be accounted for as the consequence of an oxygen–oxygen bond cleavage by acid and the migration of a phenyl group at the C-6 position, followed by cyclization and elimination of a phenol. The migratory aptitude was in the order $4\text{-MeOC}_6H_4 > 4\text{-MeC}_6H_4 > Ph = 4\text{-FC}_6H_4 > 4\text{-ClC}_6H_4 = 4\text{-BrC}_6H_4$, which was found from the competitive phenyl migration in the reaction of 1,2-dioxan-3-ols bearing two different substituents at the C-6 position.

Treatment of the 4-acetyl-1,2-dioxan-3-ols with alkali resulted in the deacetylated 1,2-dioxan-3-ols via the retro-Claisen condensation by hydroxide ion followed by recyclization [79].

The palladium-catalyzed hydrogenolysis of azabicyclic peroxides led to the formal extrusion of one of the peroxide oxygens and produced the corresponding tetrahydrofuranols in quantitative chemical yields [83, 129, 174]. The reduction of the 1,2-dioxan-3-ols using triphenylphosphine also afforded tetrahydrofuranols, which were further dehydrated to yield the corresponding dihydrofurans [77].

The hydroxyl group of the 1,2-dioxan-3-ols could be substituted by a methoxyl, acetoxyl, and chloro group [73, 167]. The Grignard reaction of the 3-methoxy-1,2-dioxane-4-carboxylates afforded the corresponding alcohols [80]. Further, the 4-bromoacetyl-3-methoxy-1,2-dioxanes could also be converted by the reaction of acetamide, hydrazines, thioureas, thioamides, 4-amino-5-mercapto-1,2,4-triazole, and 2-aminothiophenol into heterocycle-substituted 1,2-dioxane derivatives [167].

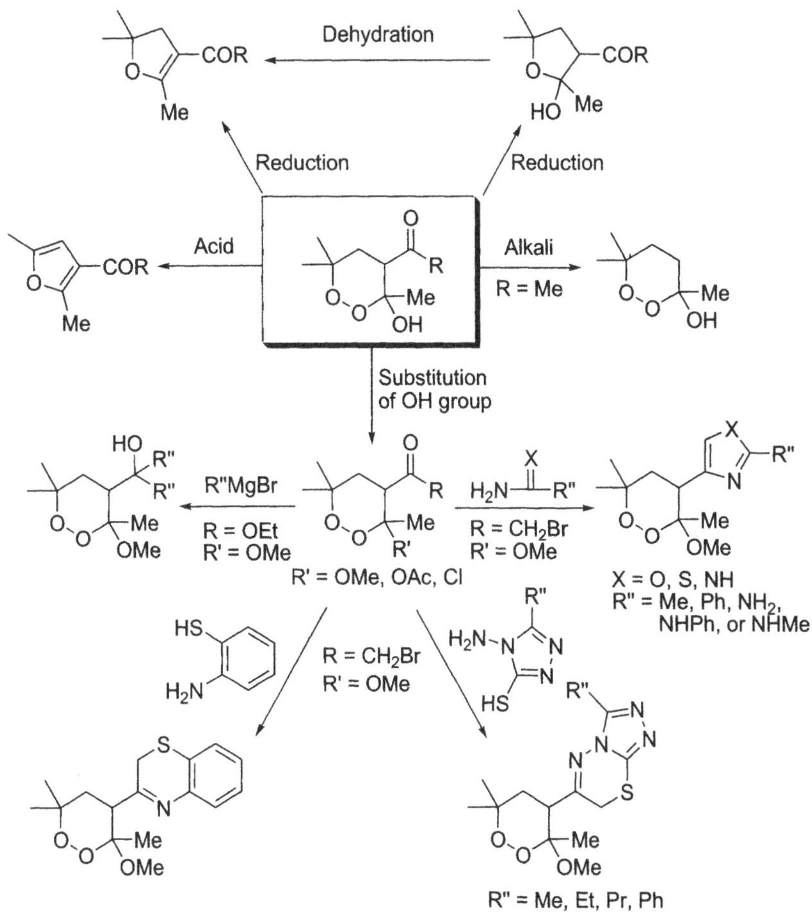

Scheme 32 Chemical transformation of the synthesized 1,2-dioxane derivatives

9 Summary

The synthesis of functionalized 1,2-dioxane derivatives was developed using the tris(2,4-pentanedionato)manganese(III)-alkene system and the manganese(III) acetate-1,3-dicarbonyl compound-alkene system. The endoperoxidation catalytically proceeded in air under very mild reaction conditions and the excellent yield of the endoperoxides was achieved. In addition, heterocycle-fused or -substituted dioxanes could be synthesized according to the manganese(III)-catalyzed endoperoxidation. The hydroperoxidation also occurred in the reaction of the cyclic 1,3-diamides with alkenes. Furthermore, the direct hydroperoxidation of the cyclic 1,3-diamides was effective in the absence of alkenes.

However, all types of alkenes are not always used in the peroxidation. The most efficient alkene is the 1,1-disubstituted alkene due to the stability of the intermediate tertiary radicals **B** in Scheme 31. Aliphatic terminal alkenes such as *n*-hexene are a poor substrate. In addition, ambient air is the best atmosphere for the catalytic autoxidation since overoxidation occurs under an oxygen atmosphere.

Antimalarial testing was performed for the five synthesized 1-hydroxy-8-aza-2,3-dioxabicyclo[4.4.0]decane-6-carboxylates (R^1 = Et, NEt_2, cyclohexyl, or Ph; R^2 = Et; R^3 = R^4 = 4-ClC_6H_4 in Scheme 21). Unfortunately, the azadioxabicyclo[4.4.0]decane-6-carboxylates did not show any activity against *Plasmodium falciparum*, but a cytotoxicity toward FM3A [175]. However, it was found that both 4,4-bis(2-hydroperoxy-2,2-diphenylethyl)-1,2-diphenylpyrazolidine-3,5-dione (Scheme 24) and 4-benzyl-4-hydroperoxy-1,2-diphenylpyrazolidine-3,5-dione (Scheme 27) showed a weak antimalarial activity [162].

Acknowledgements I gratefully acknowledge Professor Shoji Eguchi, Editor of Japan, Heterocyclic Communications, for giving me the chance to write this review. The research was supported by Grants-in-Aid for Scientific Research from the Ministry of Education, Culture, Sports, Science and Technology, Japan, and also by Grants-in-Aid for Scientific Research from the Japan Society for the Promotion of Science.

References

1. Casteel DA (1992) Nat Prod Rep 9:289
2. Casteel DA (1999) Nat Prod Rep 16:55
3. Cremer D (1983) General and theoretical aspects of the peroxide group. In: Patai S (ed) The chemistry of peroxides. Wiley, New York, chap 1, p 1
4. Haynes RY, Vonwiller SC (1997) Acc Chem Res 30:73
5. Mekonnen B, Ziffer H (1997) Tetrahedron Lett 38:731
6. Posner GH, O'Dowd H, Caferro T, Cumming JN, Ploypradith P, Xie S, Shapiro TA (1998) Tetrahedron Lett 39:2273
7. O'Neill PM, Searle NL, Raynes KJ, Maggs JL, Ward SA, Storr RC, Park BK, Posner GH (1998) Tetrahedron Lett 39:6065
8. Provot O, Camuzat-Dedenis B, Hamzaoui M, Moskowitz H, Mayrargue J, Robert A, Cazelles J, Meunier B, Zouhiri F, Desmaële D, d'Angelo J, Mahuteau J, Gay F, Cicéron L (1999) Eur J Org Chem 1935
9. Oh CH, Kim HJ, Wu SH, Won HS (1999) Tetrahedron Lett 40:8391
10. Wu W-M, Wu Y-L (2000) J Chem Soc Perkin Trans 1, 4279
11. Borstnik K, Paik I, Shapiro TA, Posner GH (2002) Int J Parasitology 32:1661
12. Zhang L, Zhou W-S, Xu X-X (1988) J Chem Soc Chem Commun 523
13. Zhou W-S, Xu X-X (1994) Acc Chem Res 27:211
14. Bachi MD, Korshin EE, Hoos R, Szpilman AM, Ploypradith P, Xie S, Shapiro TA, Posner GH (2003) J Med Chem 46:2516
15. Tokuyasu T, Kunikawa S, Abe M, Masuyama A, Nojima M, Kim H-S, Begum K, Wataya Y (2003) J Org Chem 68:7361

16. Tang Y, Dong Y, Vennerstrom JL (2004) Med Res Rev 24:425
17. Robert A, Dechy-Cabaret O, Cazelles J, Meunier B (2002) Acc Chem Res 35:167
18. Crow WD, Nicholls W, Sterns M (1971) Tetrahedron Lett 1353
19. Bolte ML, Crow WD, Yoshida S (1982) Aust J Chem 35:1411
20. Bolte ML, Crow WD, Yoshida S (1982) Aust J Chem 35:1421
21. Bolte ML, Crow WD, Yoshida S (1982) Aust J Chem 35:1431
22. Baltas M, Benbakkar M, Gorrichon L (1991) J Chem Soc Chem Commun 1044
23. Wells RJ (1976) Tetrahedron Lett 2637
24. Higgs M, Faulkner DJ (1978) J Org Chem 43:3454
25. Quiñoá E, Kho E, Manes LV, Crews P (1986) J Org Chem 51:4260
26. Sakemi S, Higa T, Anthoni U, Christophersen C (1987) Tetrahedron 43:263
27. Capon RJ, MacLeod JK, Willis AC (1987) J Org Chem 52:339
28. Murayama T, Ohizumi Y, Nakamura H, Sasaki T, Kobayashi J (1989) Experientia 45:898
29. Snider BB, Shi Z (1990) J Org Chem 55:5669
30. De Guzman FS, Schmitz FJ (1990) J Nat Prod 53:926
31. Snider BB, Shi Z (1992) J Am Chem Soc 114:1790
32. Murakami N, Kawanishi M, Itagaki S, Horii T, Kabayashi M (2001) Tetrahedron Lett 42:7281
33. Murakami N, Kawanishi M, Itagaki S, Horii T, Kobayashi M (2002) Bioorg Med Chem Lett 12:69
34. Davidson BS (1991) J Org Chem 56:6722
35. Phillipson DW, Rinehardt KL Jr (1983) J Am Chem Soc 105:7735
36. Hernández R, Velázquez SM, Suárez E, Prangé T (1996) Tetrahedron Lett 37:6409
37. Bush JB Jr, Finkbeiner H (1968) J Am Chem Soc 90:5903
38. Heiba EI, Dessau RM, Koehl WJ Jr (1968) J Am Chem Soc 90:5905
39. Arndt D (1981) Manganese compounds as oxidizing agents in organic chemistry. Open Court, La Salle
40. De Klein WJ (1986) Reactions with manganese(III) acetate. In: Mijs WJ, De Jonge CRHI (eds) Organic syntheses by oxidation with metal compounds. Prenum, New York, chap 4, p 261
41. Demir AS, Jeganathan A (1992) Synthesis 235
42. Melikyan GG (1993) Synthesis 833
43. Iqbal J, Bhatia B, Nayyar NK (1994) Chem Rev 94:519
44. Snider BB (1996) Chem Rev 96:339
45. Melikyan GG (1997) Carbon–carbon bond-forming reactions promoted by trivalent manganese. In: Paquette LA (ed) Organic Reactions. Wiley, New York, chap 3, p 427
46. Nishino H (2001) Oleoscience 1:491
47. Snider BB, Merritt JE, Dombroski MA, Buckman BO (1991) J Org Chem 56:5544
48. Bar G, Parsons AW, Thomas CB (2001) Chem Commun 1350
49. van Helden R, Kooyman EC (1961) Recl Trav Chim Pays-Bas 80:57
50. Andrulis PJ Jr, Dewar MJS, Dietz R, Hunt RL (1966) J Am Chem Soc 88:5473
51. Aratani T, Dewar MJS (1966) J Am Chem Soc 88:5479
52. Andrulis PJ Jr, Dewar MJS (1966) J Am Chem Soc 88:5483
53. Kurosawa K (1969) Bull Chem Soc Jpn 42:119
54. Kurosawa K, Higuchi J (1972) Bull Chem Soc Jpn 45:1132
55. Kurosawa K, Sasaki Y, Ikeda M (1973) Bull Chem Soc Jpn 46:1498
56. Ueda S, Kurosawa K (1977) Bull Chem Soc Jpn 50:193
57. Kurosawa K, Araki F (1979) Bull Chem Soc Jpn 52:529
58. Nishino H, Kurosawa K (1983) Bull Chem Soc Jpn 56:1982

59. Nishino H, Itoh N, Nagashima M, Kurosawa K (1992) Bull Chem Soc Jpn 65:620
60. Heiba EI, Dessau RM, Koehl WJ Jr (1969) J Am Chem Soc 91:138
61. Anderson JA, Kochi JK (1970) J Am Chem Soc 92:2450
62. Heiba EI, Dessau RM (1972) J Am Chem Soc 94:2888
63. Okano M (1973) Chem Lett 165
64. Nikishin GI, Vinogradov MG, Fedorova TM (1973) J Chem Soc Chem Commun 693
65. Heiba EI, Dessau RM, Rodewald PG (1974) J Am Chem Soc 96:7977
66. Kasahara A, Izumi T, Suzuki A, Takeda T (1976) Bull Chem Soc Jpn 49:3711
67. Kurosawa K, Harada H (1979) Bull Chem Soc Jpn 52:2386
68. Kurosawa K, McOmie JFW (1981) Bull Chem Soc Jpn 54:3877
69. Nishino H, Kurosawa K (1983) Bull Chem Soc Jpn 56:474
70. Midgley G, Thomas CB (1984) J Chem Soc Perkin Trans 1, 1537
71. Fristad WE, Peterson JR (1985) J Org Chem 50:10
72. Firstad WE, Peterson JR, Ernst AB, Urbi GB (1986) Tetrahedron 42:3429
73. Nishino H (1985) Bull Chem Soc Jpn 58:1922
74. Nishino H (1986) Bull Chem Soc Jpn 59:1733
75. Tategami S, Yamada T, Nishino H, Korp JD, Kurosawa K (1990) Tetrahedron Lett 31:6371
76. Bhattacharjee MN, Chaudhuri MK, Khathing DT (1982) J Chem Soc Dalton Trans 669
77. Nishino H, Tategami S, Yamada T, Korp JD, Kurosawa K (1991) Bull Chem Soc Jpn 64:1800
78. Qian C-Y, Hirose J, Nishino H, Kurosawa K (1994) J Heterocycl Chem 31:1219
79. Qian C-Y, Yamada T, Nishino H, Kurosawa K (1992) Bull Chem Soc Jpn 65:1371
80. Yamada T, Iwahara Y, Nishino H, Kurosawa K (1993) J Chem Soc Perkin Trans 1, 5:609
81. Kajikawa S, Noiri Y, Shudo H, Nishino H, Kuroswa K (1998) Synthesis 1457
82. Ouyang J, Nishino H, Kurosawa K (1995) J Heterocycl Chem 32:1783
83. Qian C-Y, Nishino H, Kurosawa K (1991) Bull Chem Soc Jpn 64:3557
84. Qian C-Y, Nishino H, Kurosawa K (1993) J Heterocycl Chem 30:209
85. Quinterio M, Seoane C, Soto JL (1977) Tetrahedron Lett 18:1835
86. Quinterio M, Soto JL (1978) J Heterocycl Chem 15:57
87. Soto JL, Seoane C, Martin N, Quinterio M (1984) Heterocycles 22:1
88. Seoane C, Soto JL, Zamorano MP (1980) Heterocycles 14:639
89. Martin N, Seoane C, Soto JL (1985) Heterocycles 23:2013
90. Soto JL, Seoane C, Zamorano P, Rubio MJ, Monforte A, Quinterio M (1985) J Chem Soc Perkin Trans 1, 1681
91. Aran VJ, Soto JL (1982) Synthesis 513
92. Ciller JA, Martin N, Seoane C, Soto JL (1985) J Chem Soc Perkin Trans 1, 2581
93. Elnagdi MH, Fleita DH, Ml-Moghayar MRH (1975) Tetrahedron 31:63
94. Augustin M, Rudorf W-D, Schmidt U (1975) Tetrahedron 32:3055
95. Weber K-H, Langbein A, Dantiel H (1978) Justus Liebigs Ann Chem 1241
96. Fristad WE, Peterson JR (1985) J Org Chem 50:10
97. Fristad WE, Peterson JR, Ernst AB (1985) J Org Chem 50:3143
98. Ernst AB, Fristad WE (1985) Tetrahedron Lett 26:3761
99. Corey EJ, Gross AW (1985) Tetrahedron Lett 26:4291
100. Nishino H, Kamachi H, Baba H, Kurosawa K (1992) J Org Chem 57:3551
101. Nishino H, Yoshida T, Kurosawa K (1991) Bull Chem Soc Jpn 64:1097
102. Sato H, Nishino H, Kurosawa K (1987) Bull Chem Soc Jpn 60:1753
103. Nishino H, Hashimoto H, Korp JD, Kurosawa K (1995) Bull Chem Soc Jpn 68:1999

104. Nishino H, Ishida K, Hashimoto H, Kurosawa K (1996) Synthesis 888
105. Nguyen V-H, Nishino H, Kurosawa K (1996) Tetrahedron Lett 37:4949
106. Nguyen V-H, Nishino H, Kurosawa K (1997) Synthesis 899
107. Royles BJL (1995) Chem Rev 95:1981
108. Stickings CE (1959) Biochem J 72:332
109. Ito S, Hirata Y (1972) Tetrahedron Lett 1181
110. Whitesell JK, Minton MA (1987) J Am Chem Soc 109:6403
111. Paquette LA, Romine JL, Lin H-S (1987) Tetrahedron Lett 28:31
112. Rinehart KL Jr, Beck JR, Borders DB, Kinstle TH, Krauss D (1963) J Am Chem Soc 85:4038
113. Kohl H, Bhat SV, Patell JR, Gandhi NM, Nazareth J, Divekar PV, de Souza NJ, Berscheid HG, Fehlhaber H-W (1974) Tetrahedron Lett 983
114. van der Baan JL, Barnick JWFK, Bickelhaupt F (1978) Tetrahedron 34:223
115. Jones RCF, Tankard M (1990) J Chem Soc Chem Commun 765
116. MacKellar FA, Grostic MF, Olson EC, Wnuk RJ, Branfman AR, Rinehart KL Jr (1971) J Am Chem Soc 93:4943
117. Warner TD (1994) Cardiovasc Drug Rev 12:105
118. Bhagwat SS, Gude C, Chan K (1996) Tetrahedron Lett 37:4627
119. Samuelsson B (1983) Angew Chem Int Ed Engl 22:805
120. Snider BB, Shi Z (1992) J Am Chem Soc 114:1790
121. Meconnen B, Ziffer H (1997) Tetrahedron Lett 38:731
122. Xu X-X, Zhu J, Huand D-Z, Zhou W-S (1986) Tetrahedron 42:819
123. Liang X-T (1985) In: Chang H-M, Yeung H-W, Tso W-W, Koo A (eds) Advances in Chinese Medicinal Materials Research. World Scientific, p 427
124. Nishino H, Nguyen V-H, Yoshinaga S, Kurosawa K (1996) J Org Chem 61:8264
125. Nguyen V-H, Nishino H, Kurosawa K (1997) Tetrahedron Lett 38:1773
126. Nguyen V-H, Nishino H, Kurosawa K (1998) Heterocycles 48:465
127. Chowdhury FA, Nishino H, Kurosawa K (1998) Tetrahedron Lett 39:7931
128. Chowdhury FA, Nishino H, Kurosawa K (1999) Heterocycles 51:575
129. Chowdhury FA, Nishino H, Kurosawa K (2001) Heterocycl Commun 4:17
130. Rindone B, Scolastico C (1974) Tetrahedron Lett 3379
131. Galliani G, Rindone B, Scolastico C (1975) Tetrahedron Lett 1285
132. Nishino H, Kurosawa K (1983) Bull Chem Soc Jpn 56:1682
133. Kumabe R, Nishino H, Yasutake M, Nguyen V-H, Kurosawa K (2001) Tetrahedron Lett 42:69
134. Asahi K, Nishino H (2005) Heterocycl Commun 11:379
135. Asahi K, Nishino H (2005) Tetrahedron 61:11107
136. Guillen SR, Guzman CM (1988) Pharmazie 43:827
137. Qian C-Y, Nishino H, Kurosawa K, Korp JD (1993) J Org Chem 58:4448
138. Elguero J (1989) Comprehensive Heterocyclic Chemistry, vol 5. Pergamon, New York, p 167
139. Chauhan SMS, Srinivas KA, Mohapatra PP (1999) Ind J Chem 31B:724
140. Bourdonnec BL, Meulon E, Yous S, Goossens J-F, Houssin R, Hénichart J-P (2000) J Med Chem 43:2685
141. Vennerstorm JL, Holmes TJ Jr (1987) J Med Chem 30:563
142. Rahman MT, Nishino H, Qian C-Y (2003) Tetrahedron Lett 44:5225
143. Rahman MT, Nishino H (2003) Tetrahedron 59:8383
144. Hiatt R, Irwin KC (1968) J Org Chem 33:1436
145. Boyaci FG, Takaç S, Özdamar TH (2000) Rev Chem Eng 16:249
146. Michael JP (1999) Nat Prod Rep 16:697

147. Nontpresert A, Pukrittayakamee S, Kyle DE, Vanijanonta S, White NJ (1996) Trans Royal Soc Trop Med Hygiene 90:553
148. Epperson MT, Hadden CE, Waddell TG (1995) J Org Chem 60:8113
149. Shibuya H, Kitamura C, Maehara S, Magahata M, Winarno H, Simanjuntak P, Kim H-S, Wataya Y, Ohashi K (2003) Chem Pharm Bull 51:71
150. Kumabe R, Nishino H (2004) Tetrahedron Lett 45:703
151. Mentz VP, Schulz M, Kluge R (1987) Arzneim-Forsch 37:1229
152. Rechenberg HKV (1962) Phenylbutazone, 2nd edn. Edward Arnold, London
153. Demir AS, Jeganathan A (1992) Synthesis 235
154. Vedejs E, Engler DA, Telschow JE (1978) J Org Chem 43:188
155. Hubert AJ, Starcher PS (1968) J Chem Soc C 2500
156. Adam W, Smerz AK (1996) Tetrahedron 52:5799
157. Christoffers J (1999) J Org Chem 64:7668
158. Christoffers J, Werner T, Unger S, Frey W (2003) Eur J Org Chem 425
159. Baucherel X, Levoirier E, Uziel J, Juge S (2000) Tetrahedron Lett 41:1385
160. Nair V, Nair LG, Mathew J (1998) Tetrahedron Lett 39:2801
161. Watanabe T, Ishikawa T (1999) Tetrahedron Lett 40:7795
162. Rahman MT, Nishino H (2003) Org Lett 5:2887
163. Suzuki M, Ohtake H, Kameya Y, Hamanaka M, Noyori R (1989) J Org Chem 54:5292
164. Sterns M (1971) J Cryst Mol Struct 1:373
165. Takenaka A, Sasada Y, Yamamoto H (1977) Acta Crystallogr Sect B 33:3566
166. Edward JT (1955) Chem Ind 1102
167. Sakata M, Shirakawa Y, Kamata N, Sakaguchi Y, Nishino H, Ouyang J, Kurosawa K (2000) J Heterocycl Chem 37:269
168. Yoshida J, Nakatani S, Sakaguchi K, Isoe S (1989) J Org Chem 54:3383
169. Heiba EI, Dessau DM (1974) J Org Chem 23:3456
170. Kumabe R, Nishino H (2004) Heterocycl Commun 10:135
171. Nguyen V-H, Nishino H (2004) Tetrahedron Lett 45:3373
172. Chowdhury FA, Nishino H (2005) J Heterocycl Chem 42:1337
173. Fujino R, Nishino H (2005) Synthesis 731
174. Chowdhury FA, Kajikawa S, Nishino H, Kurosawa K (1999) Tetrahedron Lett 40:3765
175. Kumabe R, Nishino H (2003) Cryo Rep Kumamoto Univ 14:1

Top Heterocycl Chem (2006) 6: 77–111
DOI 10.1007/7081_037

Published online: 29 April 2006

A Frontier in Indole Chemistry: 1-Hydroxyindoles, 1-Hydroxytryptamines, and 1-Hydroxytryptophans

Masanori Somei

Division of Pharmaceutical Sciences, Graduate School of Natural Science and Technology, Kanazawa University, Kakuma-machi, 920-1192 Kanazawa, Japan
somei@mail.p.kanazawa-u.ac.jp

Abstract Many new members of the 1-hydroxyindole, -tryptamine, and -tryptophan families are prepared. They demonstrate totally different reactivities from those of the corresponding indoles upon electrophilic substitution reactions. They also undergo nucleophilic substitution reactions that are unprecedented in indole chemistry. Doors to various kinds of new reactions and novel products open depending on the substrate structures and reaction conditions. 1-Hydroxyindoles, -tryptamines, and -tryptophans show a variety of biological activities, which makes this chemistry worth exploring for the development of new drugs.

Keywords Indole · Electrophilic substitution · Nucleophilic substitution · DFT calculations · Biologically active compound

1 Introduction

It is said "Necessity is the mother of invention". We believe that "Imagination is also the mother of invention". About 30 years ago when we conceived the "1-hydroxyindole hypothesis" [1–7], it was an imaginary story because it involved two impossible problems set against common sense in indole chemistry at that time [8]. The first problem was the lack of both synthetic and natural products having 1-hydroxyindole skeletons. In addition, no one knew how to synthesize the unknown 1-hydroxytryptamines and -tryptophans. The second was the prediction that they would undergo unprecedented nucleophilic substitution reactions.

Presently, the chemistry of 1-hydroxyindoles, -tryptamines, and -tryptophans seems to have become a gold mine in which we can find new facts and reactions everywhere. In this review, we would like to report briefly on the progress we have made since our previous reviews (up to the end of August, 2005) avoiding any overlap with their contents [5–7].

2 Synthesis of 1-Hydroxyindoles, -Tryptamines, and -Tryptophans

In 1989, we created a simple and general synthetic method (Somei's Method) for 1-hydroxyindoles [5–7, 9] consisting of the following two steps: (1) reduction of indoles to 2,3-dihydroindoles; (2) their oxidation with 30% H_2O_2 (or urea-hydrogen peroxide addition compound) in the presence of a catalytic amount of sodium tungstate dihydrate ($Na_2WO_4 \cdot 2H_2O$) or sodium phosphotungstate ($2Na_2O \cdot P_2O_5 \cdot 12WO_3 \cdot 18H_2O$). Throughout the following text, we abbreviate the oxidation step as "the tungstate method". Employing this method, many new members of the 1-hydroxyindole family have been discovered since our previous reviews [5–7].

2.1 1-Hydroxy- and 1-Methoxy-2,3-dimethylindoles

1-Hydroxyindoles having an electron-donating group on the indole ring are generally unstable [5]. 1-Hydroxy-2,3-dimethylindole (**1a**) is not easily prepared and is even more difficult to store (Scheme 1) [10]. Although the application of the tungstate method to 2,3-dihydro-2,3-dimethylindole (**3**), obtained from 2,3-dimethylindole (**2**), generates **1a** in the reaction mixture, its isolation is cumbersome. During the work-up, **1a** is oxidized rapidly to 3-hydroxy-2,3-dimethyl-3*H*-indole *N*-oxide (**4**) under air. The structure (**4**) is determined by X-ray single-crystal analysis.

Once the unstable **1a** is converted to 1-acetoxy-2,3-dimethylindole (**1b**) in situ, its isolation and purification is easy because of its stability. 1-Methoxy-2,3-dimethylindole (**1c**) is also a stable compound. Mild hydrolysis of **1b** gives rise to pure **1a**. As a result, the half-life of **1a** in $CHCl_3$ at room temperature under air is shown to be about 4 h. In the pure state in the refrigerator, the half-life of **1a** is determined to be 24 h [10].

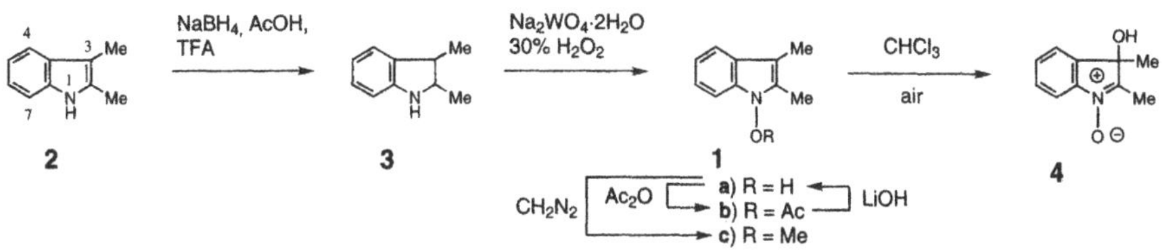

Scheme 1 Synthesis of 1-hydroxy- and 1-methoxy-2,3-dimethylindoles

2.2 1-Hydroxy-5- and -6-Nitroindole-3-carbaldehydes, and Analogs

Compounds (**5a** and **6a**, Scheme 2) are useful synthetic intermediates as shown later in Sect. 3.3.2. In order to obtain **5a**, three routes are explored starting from indoline (**7**) [11]. Since 1-hydroxy-6-nitroindole (**8a**) is readily available in two steps from **7** according to our method [12], the Vilsmeier–Haack reaction of **8a** is first attempted in vain and gives disappointing results with the formation of many products.

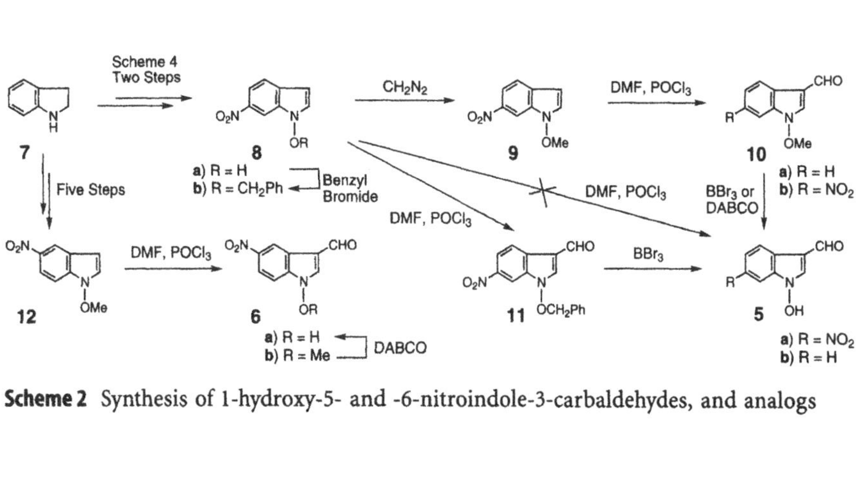

Scheme 2 Synthesis of 1-hydroxy-5- and -6-nitroindole-3-carbaldehydes, and analogs

As a second route, we have tried the ether cleavage of the 1-methoxy group in 1-methoxy-6-nitroindole-3-carbaldehyde (**10b**), which is obtained from **8a** through 1-methoxy-6-nitroindole (**9**) [13]. Although the treatment of **10b** with trimethylsilyl iodide is unsuccessful, BBr_3 is found to generate **5a**. The yield is, however, not improved to more than 40% under various examined reaction conditions. With this point in mind, we have decided to change the alkyl group on the 1-hydroxy oxygen from a methyl to a benzyl group. Benzylation of **8a** with benzyl bromide/K_2CO_3 affords 1-benzyloxy-6-nitroindole (**8b**, 95%). Subsequent Vilsmeier–Haack reaction of **8b** affords 1-benzyloxy-6-nitroindole-3-carbaldehyde (**11**, 96%). Benzyl ether cleavage of **11** with BBr_3 proceeds successfully as expected, culminating in the formation of **5a** (97%).

As a third route, we conceived the idea to attack the methyl carbon of the 1-methoxy group in **10b** with a base relying on the acidic nature of 1-hydroxyindole compounds [14]. The idea has actually been realized by employing 1,4-diazabicyclo[2.2.2]octane (DABCO) as explained in detail in Sect. 3.3.2.

This new reaction is successfully applied for the preparation of 1-hydroxy-5-nitroindole-3-carbaldehyde (**6a**, 96%) from **6b** that is obtained in 97% yield by the Vilsmeier–Haack reaction of readily available 1-methoxy-5-nitroindole (**12**) [5–7, 9].

2.3
5- and 6-Acetylamino-1-methoxytryptamines

*N*b-Acetyl-5-acetylamino-1-methoxytryptamine (**13**) is prepared with an overall yield of 32% in eight steps from tryptamine (**14**, Scheme 3) [15]. *N*b,1-Diacetyl-2,3-dihydrotryptamine (**15**) is obtained from **14** through *N*b-acetylation (97%), reduction with Et_3SiH/CF_3COOH (98%) [16], and subsequent acetylation of N(1) – H (78%). 5-Nitro compound (**16**, 81%), ob-

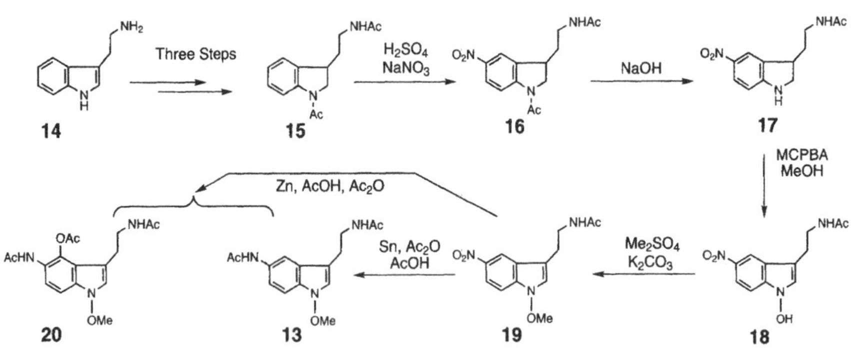

Scheme 3 Synthesis of *N*b-acetyl-5-acetylamino-1-methoxytryptamine

tained by nitration of **15**, is then converted to *N*b-acetyl-2,3-dihydro-5-nitrotryptamine (**17**, 93%).

Oxidation of **17** with *m*-chloroperbenzoic acid (MCPBA) gives rise to *N*b-acetyl-1-hydroxy-5-nitrotryptamine (**18**, 70%). In this oxidation step, MCPBA is superior to the tungstate method. Methylation of the 1-hydroxy group with Me_2SO_4 and subsequent reduction of the resultant **19** (93%) with $Sn/Ac_2O/AcOH$ affords **13** (89%). Interestingly, reduction of **19** with $Zn/Ac_2O/AcOH$ generates **13** (34%) and 4-acetoxy-*N*b-acetyl-5-acetylamino-1-methoxytryptamine (**20**, 16%), a Bamberger rearrangement product.

For the preparation of *N*b-acetyl-6-acetylamino-1-methoxytryptamine (**21**) two routes have been developed [15]. The first is seven steps from indoline (**5**) with an overall yield of 38%. The second is eight steps with a 34% overall yield starting from tryptamine (**14**).

In the first route (Scheme 4), indoline (**5**) is nitrated to 6-nitro-2,3-dihydroindole (**22**, 92%). Application of the tungstate method to **22** and subsequent methylation provide 1-methoxy-6-nitroindole (**9**, 77%) via **8a**. A Vilsmeier–Haack reaction (94%), followed by nitroaldol reaction (85%), leads **9** to 1-methoxy-6-nitro-3-(2-nitrovinyl)indole (**23**) through 1-methoxy-6-nitroindole-3-carbaldehyde (**10b**). After selective reduction of the nitrovinyl part of **23** with $NaBH_4$, the resultant 1-methoxy-6-nitro-3-(2-nitroethyl)indole (**24**, 84%) is treated with Zn/HCl and then Ac_2O to give **21** (81%). Reduction of **24** with Zn/AcOH produces 6-amino-1-methoxytryptamine (**25**, 30%).

In the second route (Scheme 5), tryptamine (**14**) is transformed to *N*b-trifluoroacetyl-2,3-dihydro-6-nitrotryptamine (**28**) by the successive trifluoroacetylation (99%), reduction (99%), and nitration (75%) through **26** and **27**. Application of the tungstate method to **28** and subsequent methylation afford *N*b-trifluoroacetyl-1-methoxy-6-nitrotryptamine (**30**, 73%) via *N*b-trifluoroacetyl-1-hydroxy-6-nitrotryptamine (**29**). After alkaline hydrolysis, acetylation of the resultant **31** (99%) affords **32** (90%). Reduction of **32**

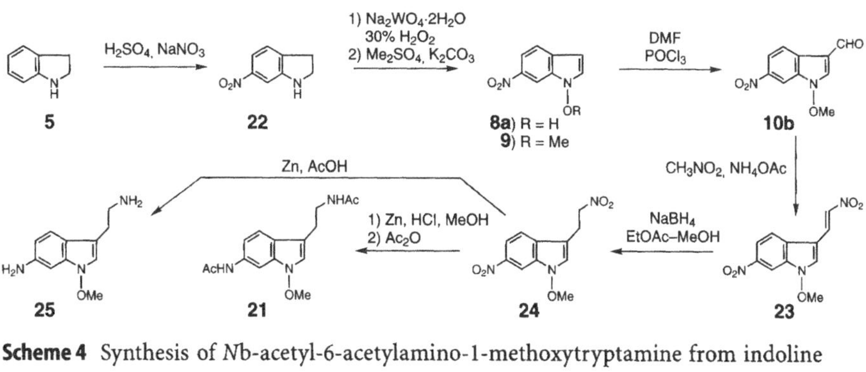

Scheme 4 Synthesis of *N*b-acetyl-6-acetylamino-1-methoxytryptamine from indoline

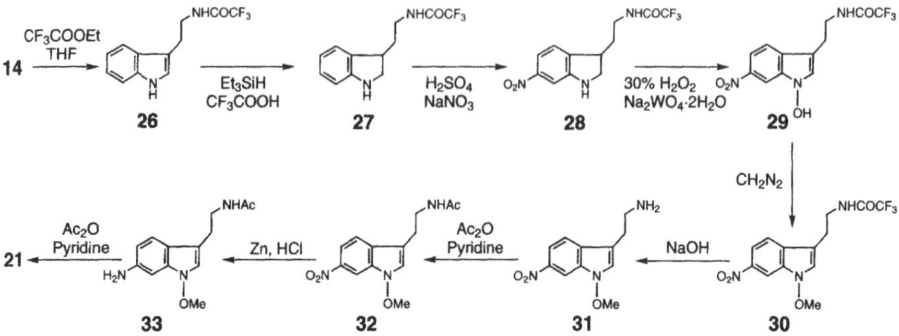

Scheme 5 Synthesis of *N*b-acetyl-6-acetylamino-1-methoxytryptamine from tryptamine

with Zn/HCl provides *N*b-acetyl-6-amino-1-methoxytryptamine (**33**, 91%). Finally, acetylation of the 6-amino group of **33** provides the desired product (**21**, 72%). Compounds (**25** and **33**) are useful building blocks for the future preparation of a variety of 6-substituted 1-methoxytryptamines.

2.4
5-Bromo-1-hydroxytryptamines

5-Bromo-1-hydroxy-*N*b-methoxycarbonyltryptamine (**34**) is obtained in seven steps with a 17% overall yield from tryptamine (**14**) (Scheme 6) [17]. 1-Methoxy-*N*b-methoxycarbonyltryptamine (**35**), available from **14** in four steps [5–7], is converted to 5-bromo-*N*b-methoxycarbonyltryptamine (**36**, 50%) employing our nucleophilic substitution reaction with 47% HBr [18]. Subsequent reduction (93%) with Et_3SiH/CF_3COOH, followed by oxidation with 30% H_2O_2/sodium phosphotungstate, gives rise to **34** (66%) through a 2,3-dihydroindole (**37**).

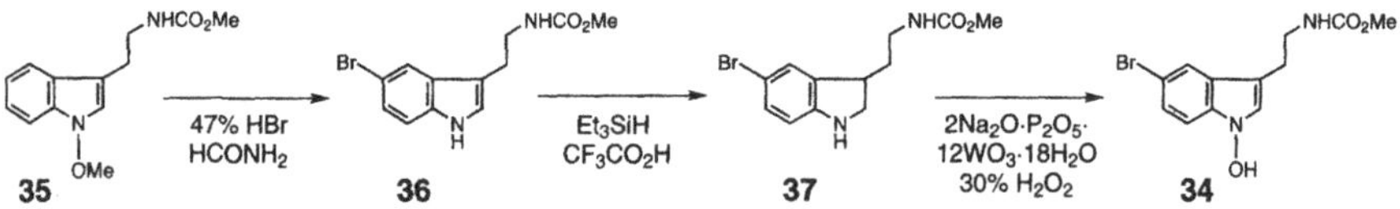

Scheme 6 Synthesis of 5-bromo-*N*b-methoxycarbonyl-1-hydroxytryptamine

2.5
1-Hydroxytryptophans

Reduction of *N*b-acetyl-L-tryptophan methyl ester ((*S*)-**38**, Scheme 7) with $NaBH_3CN/AcOH$ [19] provides *N*b-acetyl-2,3-dihydro-L-tryptophan methyl esters (**39a** and **40a**, 1.4 : 1) in 68% yield as a mixture of diastereomers. These diastereomers (**39a** and **40a**) are easily separated with high performance liquid chromatography (HPLC). Their stereochemistries are de-

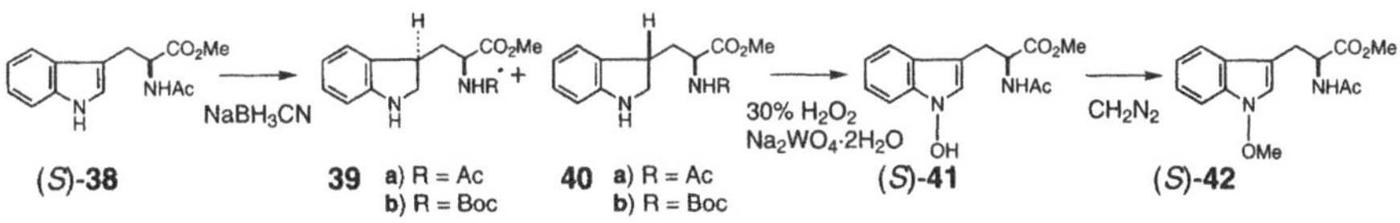

Scheme 7 Synthesis of (*S*)-*N*b-acetyl-1-hydroxy- and -1-methoxytryptophan methyl esters

termined as shown in Scheme 7 comparing each ^1H-NMR spectrum with the known set of diastereomers of *N*b-*tert*-butoxycarbonyl-2,3-dihydro-L-tryptophan methyl esters (**39b** and **40b**), whose absolute stereochemistries were determined by the Vranken group [20].

The tungstate method is successfully applied to both **39a** and **40a** resulting in the generation of the same product, *N*b-acetyl-1-hydroxy-L-tryptophan methyl ester ((*S*)-**41**), in 69 and 67% yields, respectively [21]. The application to the mixture of diastereomers (**39a** and **40a**) without separation gives (*S*)-**41** (69%) as reported previously [22]. Subsequent treatment of (*S*)-**41** with excess ethereal CH_2N_2 yields *N*b-acetyl-1-methoxy-L-tryptophan methyl ester ((*S*)-**42**, 94%) [22]. Optical purity of (*S*)-**42** is established to be more than 99% ee by its analysis using chiral column chromatography.

2.6 *N*-[2-(1-Alkoxyindol-3-yl)ethyl]-1-alkoxyindole-3-acetamides

N-[2-(1-Methoxyindol-3-yl)ethyl]-1-methoxyindole-3-acetamide (**43a**), its *N*-methyl analog (**43b**), and *N*-[2-(1-methoxyindol-3-yl)ethyl]indole-3-acetamide (**51**) are prepared by the following route (Scheme 8) [23].

Condensation of tryptamines (**14**,**44**) with indole-3-acetic acid (**45**) provides *N*-[2-(indol-3-yl)ethyl]indole-3-acetamides (**46a**, 81%; **46b**, 98%). Reduction of **46a** with Et_3SiH/TFA produces bis-2,3-dihydroindole derivative (**47a**, 90%), while the reduction with $NaBH_3CN$ gives rise to **47a** (42%) and **48** (45%), a product reduced selectively at the tryptamine pyrrole part.

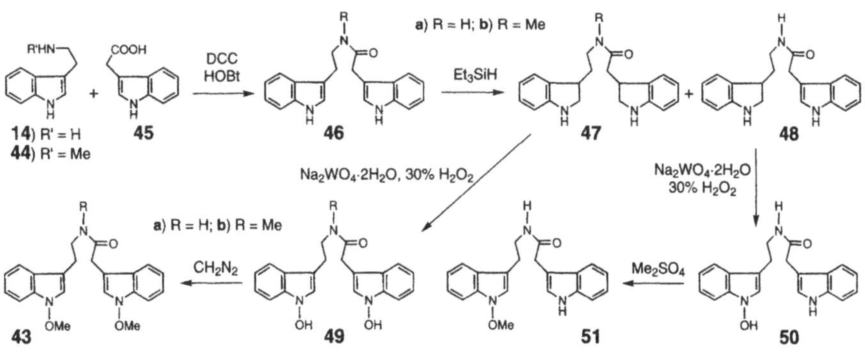

Scheme 8 Synthesis of *N*-[2-(1-alkoxyindol-3-yl)ethyl]-1-alkoxyindole-3-acetamides

Reduction of **46b** with Et_3SiH/TFA produces similarly *N*-methyl-bis-2,3-dihydroindole (**47b**, 85%). Employing the tungstate method, **47a,b** and **48** are converted to *N*-[2-(1-hydroxyindol-3-yl)ethyl]-1-hydroxyindole-3-acetamides (**49a**, 51%; **49b**, 39%) and *N*-[2-(1-hydroxyindol-3-yl)ethyl]indole-3-acetamide (**50**, 55%), respectively. Subsequent methylation of **49a,b** and **50** affords **43a** (91%), **43b** (72%), and **51** (100%).

3
Chemical Reactions of 1-Hydroxyindoles, -Tryptamines, and -Tryptophans

3.1
Alkylation with *sec*-Alkyl Halides and *tert*-Butyldimethylsilyl Chloride

Alkylation of 1-hydroxyindoles, -tryptamines, and -tryptophans is readily achieved with various *prim*-alkyl halides in the presence of bases due to the acidic nature [14] of these 1-hydroxyindole compounds. Diazoalkanes are also useful alkylating reagents [5–7].

As for the introduction of a secondary alkyl moiety to the 1-hydroxy oxygen, the usual alkyl halide method has given poor results. However, employing preformed cesium salt of 1-hydroxyindoles in DMF, this difficulty has been overcome (Scheme 9) [24]. Thus, the reaction of *N*b-methoxycarbonyl-1-hydroxytryptamine (**52**) with methyl 2-bromopropionate, 2-bromopropane, cyclohexyl bromide, and 1-phenethylbromide gives birth to the corresponding 1-(1-methoxycarbonyl)ethoxy (**53a**, 88%), 1-(2-propyl)oxy (**53b**, 87%), 1-cyclohexyloxy (**53c**, 19%), and 1-(1-methyl)benzyloxy derivatives (**53d**, 80%), respectively. 1-*t*-Butyldimethylsilyloxy-*N*b-methoxycarbonyltryptamine (**54**) can be obtained by the conventional *t*-butyldimethylsilyl chloride and imidazole method [24].

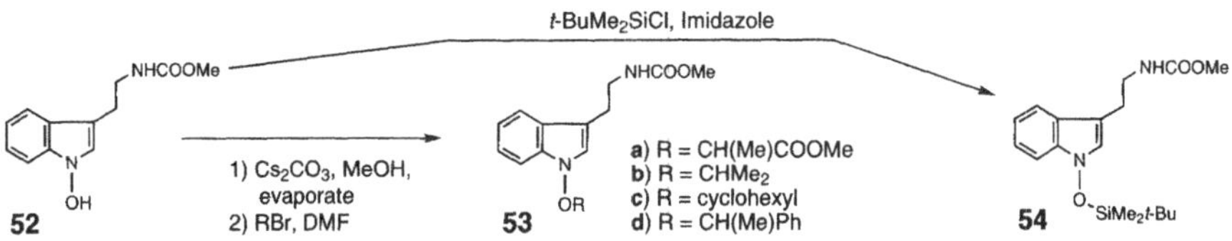

Scheme 9 Alkylation with *sec*-alkyl halides and *tert*-butyldimethylsilyl chloride

3.2
Electrophilic Substitution Reactions

Upon electrophilic substitution reactions, the introduction of either a methoxy or a hydroxy group onto the 1-position of the indole nucleus causes alteration of its positional reactivity [5–7]. Halogenation and Friedel–Crafts acylation are presented as additional examples.

3.2.1
Halogenation and Acylation

Electrophilic substitution reactions of 1-methoxyindoles proceed in a quite different way from that of the corresponding N(1) – H indoles [25]. For example, bromination of *N*b-methoxycarbonyltryptamine (**55**) with 1.2-mol eq. of Br_2/NaOAc/AcOH affords 2-oxindole (**56**, 44%), **36** (3%), 5,7-dibromo- (**57**, 3%), and 2,5-dibromotryptamines (**58**, 4%) (Table 1).

Under similar reaction conditions, **35** generates many products such as 7-bromo- (**59**), 2,7-dibromotryptamines (**60**), **57**, 7-bromo- (**61**), 5-bromo-2-oxindoles (**62**), **56**, 2-bromotryptamine (**63**), and **58**, depending on the bromination conditions (entries 1–3). 1-Hydroxy-*N*b-methoxycarbonyltryptamine (**52**) shows almost the same results as **35**. It should be noted that the ratio of all of the 7-brominated indoles to the total products, observed in the bromination of 1-hydroxyindole derivatives (**35** and **52**), is much higher than that of the N(1)–H compound (**55**).

Melatonin (**64**) shows characteristic positional selectivity upon bromination depending on the quantity of Br_2 (Table 2) [26]. With an increase in the quantity of Br_2 from 1 to 2-mol eq., the product distribution changes from 4-bromo- (**65**), 2-bromo- (**66**), and 2,4-dibromomelatonins (**68**) to **65**, 2,6-dibromo- (**67**), and **68** (entries 1 and 2). When 3-mol eq. of Br_2 is employed, 2,4,6-tribromomelatonin (**69**) is obtained as a sole product in 94% yield (entry 3).

Once a methoxy group is introduced onto the N(1) of melatonin, namely in the case of 1-methoxymelatonin (**70**), remarkable differences in the po-

Table 1 Bromination of *N*b-methoxycarbonyltryptamine and its 1-hydroxy analogs

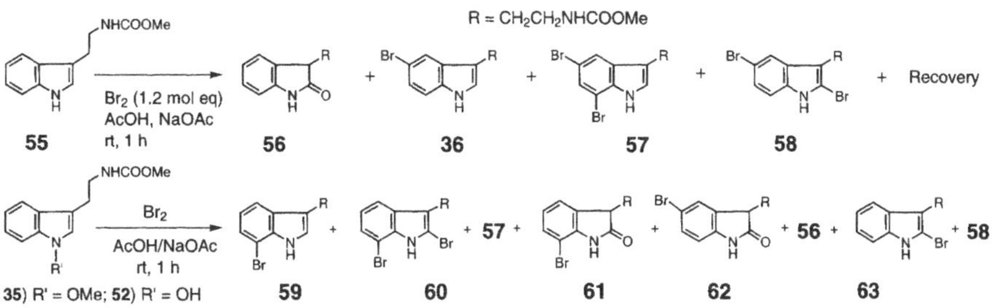

Entry	Bromine (mol eq.)	Yield (%) of 59	60	57	61	62	56	63	58	7-Br (total)	Total yield
1	0.7	23	9	0	1	0	9	8	7	33	57
2	1.2	0	22	12	3	5	5	0	10	37	57
3	3.0	0	0	0	0	7	0	0	0	many spots	

sitional reactivity are observed upon bromination. As can be seen from Table 3, 7-bromo- (**71**), 4,7-dibromo- (**72**), 2,4,7-tribromo- (**73**), 2,4,6,7-tetrabromomelatonins (**74**), **68**, **69**, 3,4,7-tribromo- (**75**), and 4,7-dibromo-2-oxindoles (**76**) are produced depending on the quantity of Br_2 [25].

The results shown in Tables 2 and 3 clearly demonstrate that the 1-methoxy group promotes selective bromination at the 7-position. This phenomenon can be explained by the following working hypothesis, either electrophilic (**A**) or nucleophilic mechanisms (**B**), as illustrated in Fig. 1 [25].

In the former mechanism, the bromonium ion initially approaches the indole ring and forms a π-complex (**C**). Then a coordination bond develops between the bromonium ion and the lone pair of 1-methoxy oxygen that is

Table 2 Bromination of melatonin

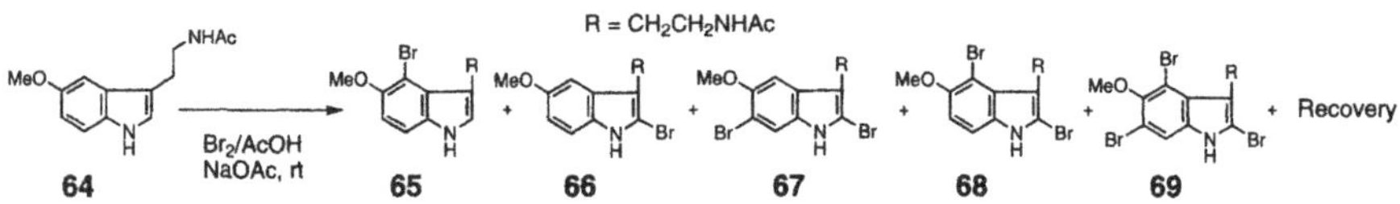

Entry	Bromine (mol eq.)	Reaction time (h)	Yield (%) of 65	66	67	68	69	Recovery	7-Br (total)	Total yield
1	0.95	5	8	28	0	15	0	34	0	85
2	1.9	2.5	10	0	34	49	0	0	0	93
3	3	1.5	0	0	0	0	94	0	0	94

Table 3 Bromination of 1-methoxymelatonin

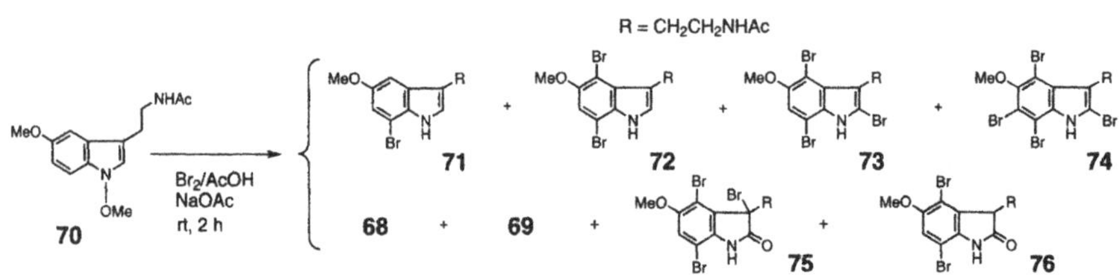

Entry	Bromine (mol eq.)	Yield (%) of 71	72	73	74	68	69	75	76	Recovery	7-Br (total)	Total yield
1	0.95	11	10	0	0	2	0	0	2	4	23	29
2	1.9	0	11	37	0	0	0	0	16	0	64	64
3	3.0	0	0	13	15	0	6	9	5	0	42	48

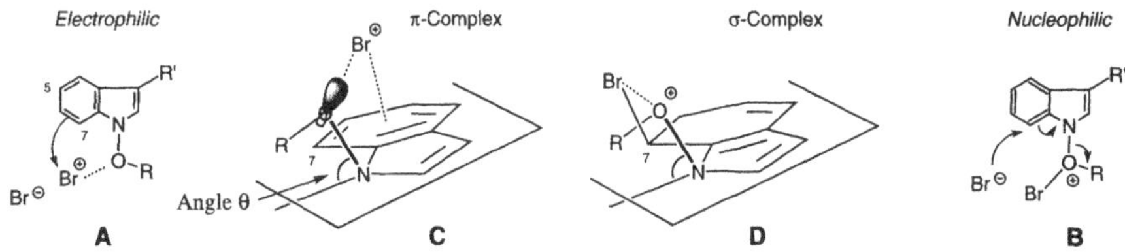

Fig. 1 A possible mechanism for the regio-selective bromination at the 7-position

deviated from the indole molecular plane with an angle θ as shown in Sect. 4. As a result, the bromonium ion is attracted close to the 7-position and forms a σ-complex (**D**) at the 7-position. In the latter mechanism, a nucleophilic substitution reaction takes place through transition state (**B**), where the bromide ion attacks the nearby 7-position with the concomitant liberation of the oxonium group from the 1-position [25].

Iodination of methyl indole-3-carboxylate (**77a**) with $KI/NaIO_4$ afforded 5-iodo- (**78**, 58%), 6-iodo- (**79**, 25%)-, and 7-iodoindole-3-carboxylates (**80**, 2%) as reported previously [27], while the same reaction of methyl 1-methoxyindole-3-carboxylate (**77b**) provided methyl 5-iodo-1-methoxy-indole-3-carboxylate (**81**, 72%) as a sole product (Scheme 10).

Additional halogenation of **78** and **81** exhibits interesting results [28]. Bromination (1.3-mol eq.) of the former in AcOH provides methyl 6-bromo-5-iodoindole-3-carboxylate (**82**, 56%) and the corresponding carboxylic acid (**83**, 18%). On the other hand, bromination of the latter gives complex mixtures of products under various reaction conditions. Relatively clean reaction takes place with Br_2 (1.3-mol eq.) in AcOH at 100 °C providing methyl 6-bromo- (**84**, 17%), methyl 2-bromo-5-iodo-1-methoxyindole-3-carboxylates (**85**, 33%), 5-iodo-1-methoxyindole-3-carboxylic acid (**86**, 7%),

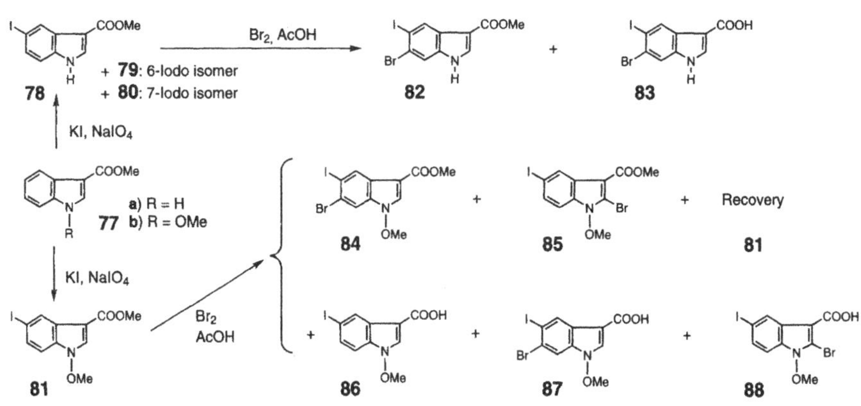

Scheme 10 Halogenation of methyl indole-3-carboxylates

6-bromo- (**87**, 6%), and 2-bromo-5-iodo-1-methoxyindole-3-carboxylic acids (**88**, 5%) together with unreacted **81** (16%).

The above results show that the bromination of **78** proceeds regioselectively at the 6-position, while the introduction of a methoxy group onto the 1-position changes the positional reactivity and tends to favor the 2-position.

Friedel–Crafts acylation of **77b** with chloroacetyl chloride and $AlCl_3$ in nitrobenzene provides cleanly methyl 5-chloroacetyl- (**90a**, 52%) and 6-chloroacetyl-1-methoxyindole-3-carboxylates (**91a**, 20%) with the ratio of 2.5 : 1 (Scheme 11) [28]. Under similar reaction conditions, Nakatsuka and co-workers reported the chloroacetylation of methyl 1-methylindole-3-carboxylate (**89**) to give **90b** and **91b** in the ratio of 3 : 1 [29]. Comparing these results with ours, it would be safe to conclude that the introduction of a methoxy group onto the N(1) prefers the 6- to 5-substitution in Friedel–Crafts acylation.

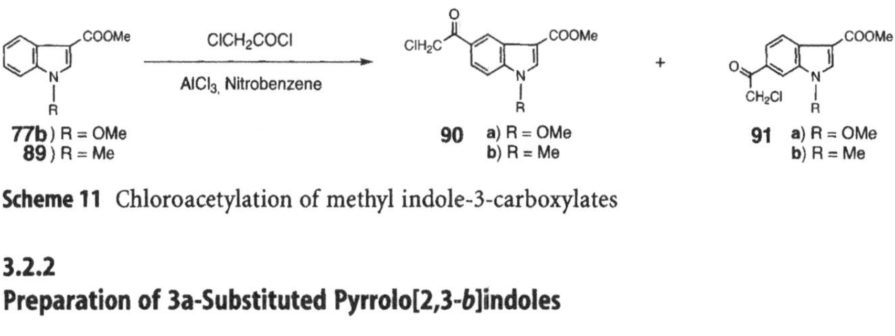

77b) R = OMe
89) R = Me

90 a) R = OMe
b) R = Me

91 a) R = OMe
b) R = Me

Scheme 11 Chloroacetylation of methyl indole-3-carboxylates

3.2.2
Preparation of 3a-Substituted Pyrrolo[2,3-*b*]indoles

Utilizing the unique features of 1-hydroxy- and 1-methoxyindoles, three kinds of useful routes for preparing 3a-alkoxy-1,2,3,3a,8,8a-hexahydropyrrolo[2,3-*b*]indoles have been developed [30].

The first route is the treatment of 1-hydroxy-*N*b-methoxycarbonyltryptamine (**52**) in refluxing Ac_2O giving rise to 3a-acetoxy-8-acetyl-1-methoxycarbonyl-1,2,3,3a,8,8a-hexahydropyrrolo[2,3-*b*]indole (**92**, 72%, Scheme 12), whose structure is proved by X-ray single-crystal analysis [30]. The key step is a [3,3] sigmatropic rearrangement of the 1-acetoxy group of **93** to the 3-position providing the imine (**94**).

Mild hydrolysis of **92** produces 8-acetyl-3a-hydroxy-1-methoxycarbonyl-1,2,3,3a,8,8a-hexahydropyrrolo[2,3-*b*]indole (**95a**, 96%). Various 3a-alkoxy derivatives, for example **95b** (95%), can be prepared by the alkylation of the 3a-hydroxy group with alkyl halides and NaH.

A second route involving either chlorination or bromination and subsequent substitution with water has been developed [30]. The reaction of 1-methoxy-*N*b-methoxycarbonyltryptamine (**35**) with NCS (1-mol eq.) in MeOH affords 3a-chloro-1,2,3,3a,8,8a-hexahydro-8-methoxy-1-methoxycar-

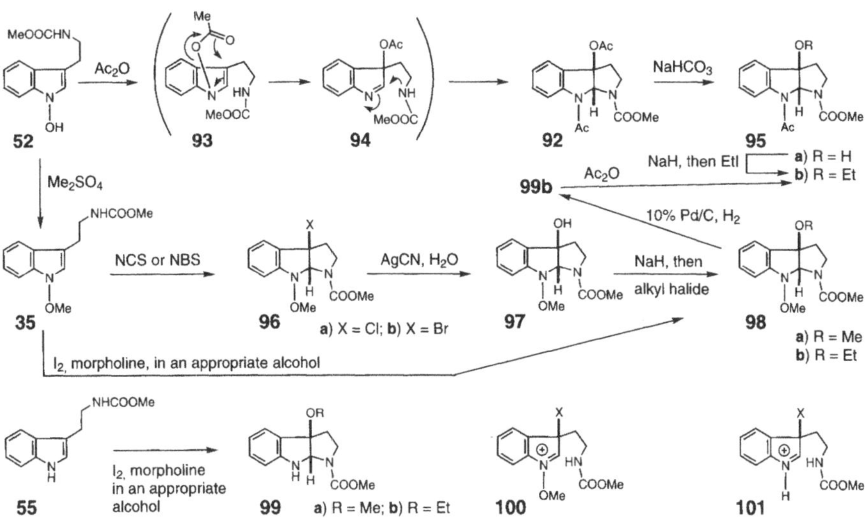

Scheme 12 Preparation of 3a-substituted pyrrolo[2,3-*b*]indoles from 1-hydroxytryptamines

bonylpyrrolo[2,3-*b*]indole (**96a**, 92%) as a sole product. Similar reaction of **35** with NBS (0.9-mol eq.) in MeCN provides 3a-bromo-1,2,3,3a,8,8a-hexahydro-8-methoxy-1-methoxycarbonylpyrrolo[2,3-*b*]indole (**96b**, 85%) cleanly. Treatment of **96a** or **96b** with silver cyanide in MeCN/H_2O yields the same product, 8-acetyl-3a-hydroxy-1-methoxycarbonyl-1,2,3,3a,8,8a-hexahydropyrrolo[2,3-*b*]indole (**97**, 85%, 94%, respectively). Alkylation of **97** with alkyl halides and NaH affords **98**.

A simple one-step synthetic method from **35** for the desired product (**98**) has been created as the third route [30]. Addition of iodine to the solution of **35** in morpholine and an appropriate alcohol as a solvent at room temperature generated **98a–i** in excellent yields as shown in Table 4. In this reaction, use of 10-mol eq. of iodine is recommended to achieve high product yields (compare entry 2 with 1). The reagent system is selected among various trials including Br_2, Br_2/NaOAc, 4-dimethylaminopyridinium tribromide, NIS, iodine/triethylamine, iodine/K_2CO_3, iodine/$NaHCO_3$, iodine/pyridine, iodine/NaI, iodine/NH_4Cl, and iodine only in various solvents.

Although the N(1) – H compound (**55**) provides the 3a-alkoxypyrrolo[2,3-*b*]indoles such as **99a** and **99b** upon the reaction with iodine/morpholine in either MeOH or EtOH, their yields are always inferior, accompanied by tar formation, to those of the 1-methoxy compound (**35**). Compound (**99b**, 96%) is also obtained by the catalytic reduction of **98b**. Since the acetylation of **99b** provides **95b** (95%), all of the structures are correlated and determined unequivocally. These findings suggest that the presence of the methoxy oxygen at the 1-position on the indole nucleus makes the intermediate (**100**) more stable than the corresponding immonium salt (**101**).

The above iodine-morpholine reaction has been successfully applied for the preparation of optically active methyl 1,2,3,3a,8,8a-hexahydropyrrolo[2,3-*b*]indole-2-carboxylates having a halogen or an oxygen functional group at the 3a-position (Scheme 13) [31].

Treatment of (*S*)-**42** with iodine and morpholine in MeOH results in the formations of (2*S*,3a*S*,8a*S*)- (**102a**, 6%) and (2*S*,3a*R*,8a*R*)-methyl 1-acetyl-1,2,3,3a,8,8a-hexahydro-3a,8-dimethoxypyrrolo[2,3-*b*]indole-2-carboxylates (**103a**, 48%). When isopropyl alcohol is employed as a solvent, **102b** (6%) and **103b** (34%) are obtained. The reaction of (*S*)-**42** with NCS in MeCN generates (2*S*,3a*S*,8a*S*)- (**102c**, 42%) and (2*S*,3a*R*,8a*R*)-methyl 1-acetyl-3a-chloro-1,2,3,3a,8,8a-hexahydro-8-methoxypyrrolo[2,3-*b*]indole-2-carboxylates (**103c**, 42%). Upon the reaction with NBS, (2*S*,3a*S*,8a*S*)- (**102d**,

Table 4 A novel synthetic method for 3a-alkoxy-1,2,3,3a,8,8a-hexahydropyrrolo[2,3-*b*]indoles

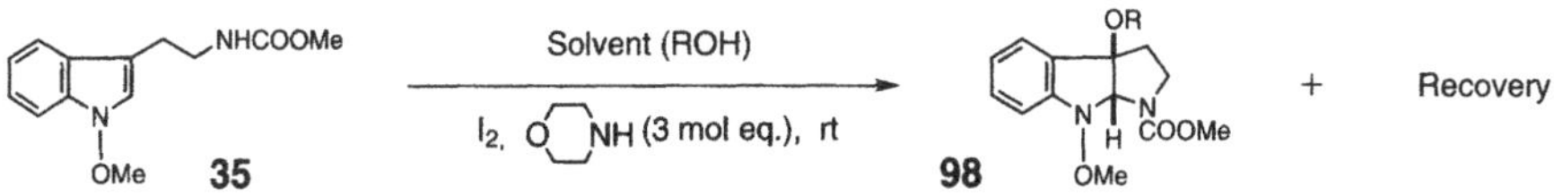

Entry	Iodine (mol eq.)	Solvent (ROH)	Reaction time (h)	Product **98**	R	Yield (%) of **98**	Recovery (%)	Total yield (%)
1	1.5	MeOH	21	a)	Me	36	45	81
2	10	MeOH	1/6		Me	98	0	98
3	10	EtOH	1/6	b)	Et	97	0	97
4	10	Me_2CHOH	1/3	c)	$CH(Me)_2$	90	0	90
5	10	Me_3COH	1/3	d)	CMe_3	80	0	90
6	10	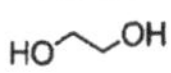	3/2	e)	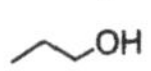	92	0	92
7	10	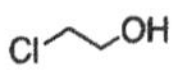	1/3	f)	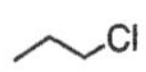	90	0	90
8	10	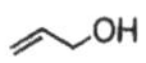	1/2	g)	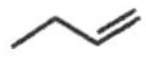	96	0	96
9	10	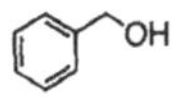	1/6	h)	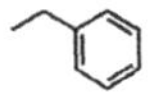	97	0	97
10	10	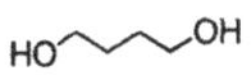	3/2	i)	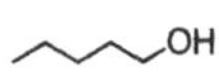	89	0	89

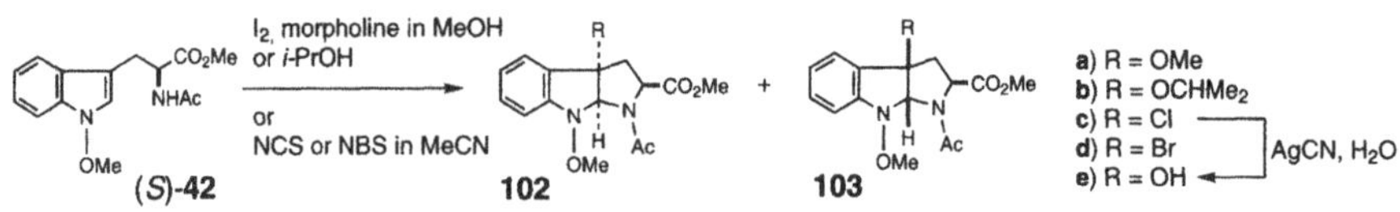

Scheme 13 Preparation of 3a-substituted pyrrolo[2,3-*b*]indoles from (*S*)-1-methoxytryptophan derivative

8%) and (2*S*,3a*R*,8a*R*)-methyl 1-acetyl-3a-bromo-8-methoxy-1,2,3,3a,8,8a-hexahydropyrrolo[2,3-*b*]indole-2-carboxylates (**103d**, 81%) are obtained. Productions of (2*S*,3a*S*,8a*S*)- (**102e**, 52%) and (2*S*,3a*R*,8a*R*)-methyl 1-acetyl-3a-hydroxy-8-methoxy-1,2,3,3a,8,8a-hexahydropyrrolo[2,3-*b*]indole-2-carboxylates (**103e**, 51%) are achieved by reacting **102c** and **103c** with AgCN in MeCN/H_2O.

3.3 Nucleophilic Substitution Reactions

Nucleophilic substitution reactions are rarely observed in indole chemistry [32]. We have disclosed that 1-hydroxy- and 1-methoxyindoles can undergo nucleophilic substitution reactions that are impossible in the corresponding indole compounds [1–8].

Depending on the structures, the reactivity of 1-methoxyindoles in the nucleophilic substitution reactions changes dramatically. The approximate order of increasing reactivity is summarized in Fig. 2 with respect to typical compounds. From these data, we can conclude that the introduction of an electron-withdrawing group or a halogen onto the indole ring increases the reactivity significantly.

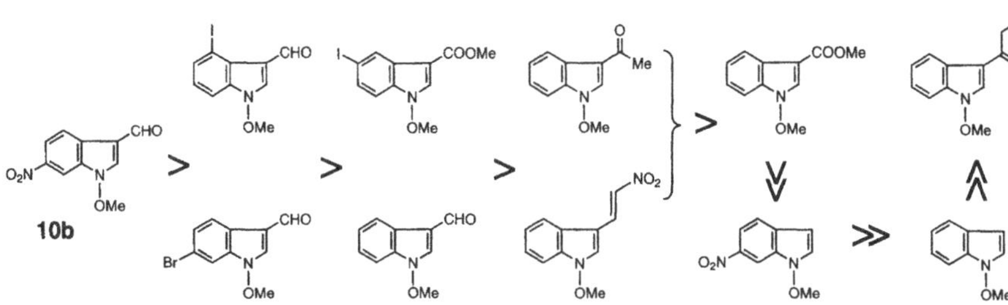

Fig. 2 The reactivity order of 1-methoxyindoles with nucleophiles

3.3.1 Effect of an Acid on the Reaction of 1-Hydroxytryptamines

Various acids promote the substitution reaction of 1-hydroxyindoles, 1-hydroxytryptamines, and 1-hydroxytryptophans with nucleophiles. We have demonstrated that p*K*a of the organic acids has a remarkable effect on the reaction employing 1-hydroxy-*N*b-methoxycarbonyltryptamine (**52**) as a substrate [33]. As can be seen from Table 5, neither propionic nor acetic acids have sufficient acidity for inducing the substitution reaction. They yield a slight quantity of dehydroxylated tryptamine (**55**) (entries 1–2). As the acidity increases, the yield of 5-hydroxytryptamine (**104a**), a product of nucleo-

Table 5 Effect of an acid on the reaction of *N*b-methoxycarbonyl-1-hydroxytryptamine

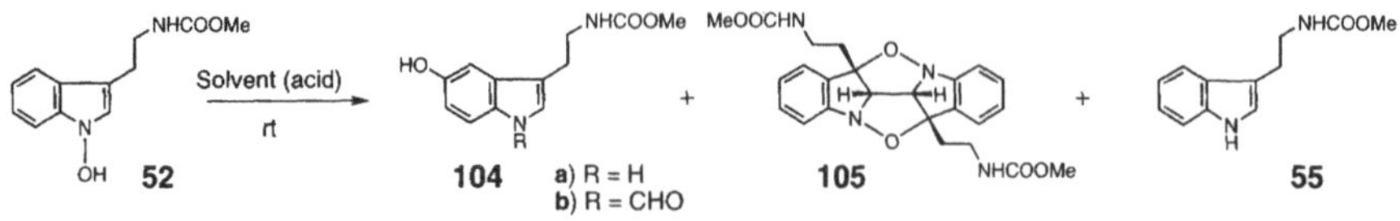

Entry	Solvent (Acid)	p*K*a	Reaction Time (h)	Yield (%) of 104a	104b	105	55	Recovery	Total Yield (%)
1	$MeCH_2COOH$	4.9	24	0	0	0	11	40	51
2	MeCOOH	4.8	24	0	0	0	5	91	96
3	HCOOH	3.2	14	8	54	0	0	0	62
4	Citric Acid*	3.1	1	34	0	0	13	6	53
5	$ClCH_2COOH$	2.9	4	12	0	0	20	0	32
6	Taurine**	1.5	12	23	0	0	18	38	79
7	$Cl_2CHCOOH$	1.3	0.5	12	0	6	0	0	18
8	CF_3COOH	0.2	1/12	59	0	5	5	0	69

* Citric Acid: $HOC(COOH)(CH_2COOH)_2$. ** Taurine: $H_2NCH_2CH_2SO_3H$.

(*dl*)-**41** $\xrightarrow{\text{Citric Acid*}}$ (*dl*)-**187** + (*dl*)-**38**

philic substitution reaction, increases. Strong acids tend to generate kabutane (**105**) and **55** as by-products (entries 7–8).

Although the p*K*a is 3.2, HCOOH has an interesting feature. The yield of **104a** is seemingly low, however, a large quantity of **104a** is present as its 1-formyl derivative (**104b**). Consequently, the nucleophilic substitution reaction of **52** is attained most successfully with HCOOH (entry 3).

In our 1-hydroxyindole hypothesis, we had predicted that 5-hydroxytryptamines (serotonins) and 5-hydroxytryptophans were supplied from the 1-hydroxytryptophan residues in the tryptophans-containing peptides by the mediation of acids that are linked with the TCA (Kreb's) cycle [2, 5–7].

Now, we have examined citric acid (p*K*a, 3.1), which is widespread in living organisms. To our delight, in the presence of citric acid, 1-hydroxy-*N*b-methoxycarbonyltryptamine (**52**) and (*dl*)-1-hydroxytryptophan derivative ((*dl*)-**41**) actually react with H_2O giving **104a** (entry 4), and (*dl*)-**187** (39%) and (*dl*)-**38** (9%), respectively. Taurine (pKa$_1$, 1.5) affords **104a** as well (entry 6). Pyruvic acid, oxaloacetic acid, 2-oxoglutaric acid, and phosphoric acid are other predicted candidates as mediators in vivo. The study whether or not they mediate the nucleophilic substitution reaction of 1-hydroxytryptophans is now under investigation.

3.3.2
Reaction of 1-Methoxy-6-nitroindole-3-carbaldehyde

1-Methoxyindole-3-carbaldehyde (**10a**) undergoes nucleophilic substitution reactions in sharp contrast with indole-3-carbaldehyde that does not react with nucleophiles under forcing reaction conditions. The most reactive reactant among the thus-far obtained 1-hydroxy and 1-methoxy derivatives is 1-methoxy-6-nitroindole-3-carbaldehyde (**10b**) as shown in Fig. 2. So, the nucleophilic substitution reactions of **10b** in DMF are examined for the synthesis of new 2,3,6-trisubstituted indoles. The representative results are shown in Scheme 14 [13].

In the presence of a strong base (NaH, KH, KO*t*-Bu), the nitrogen-, sulfur-, carbon-, and oxygen-centered nucleophiles attack **10b** regioselectively on the 2-position in excellent yields giving rise to 2-substituted 6-nitroindole-3-carbaldehydes (**106–116**). Better still, the nitro group at the 6-position can be transformed into various functional groups.

The reaction of **10b** with *p*-chlorophenoxyacetonitrile (**117**) produces 4-amino-3-*p*-chlorophenoxy-2-*p*-chlorophenoxymethyl-7-nitropyrimido[1,2-*a*]indole-10-carbaldehyde (**118**, 71%). X-ray single-crystal analysis of its derivative proves the structure [13].

In the reaction of **10b** with piperidine in DMF, a novel methyl ether cleavage reaction of the 1-methoxy group occurs in addition to the nucleophilic substitution reaction, resulting in the formation of 1-hydroxy-6-nitroindole-3-carbaldehyde (**5a**, 10%) and 2-piperidinyl product (**109**, 59%, Scheme 15) [11].

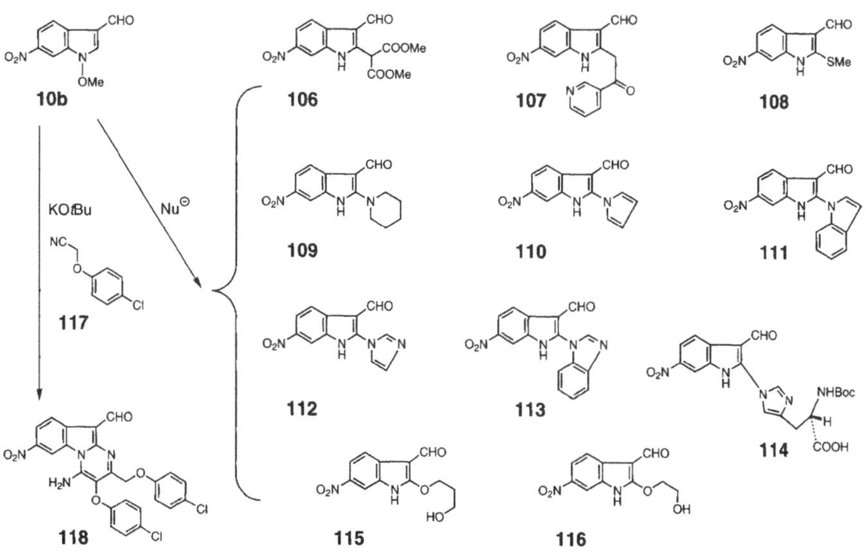

Scheme 14 Nucleophilic substitution reaction of 1-methoxy-6-nitroindole-3-carbaldehyde

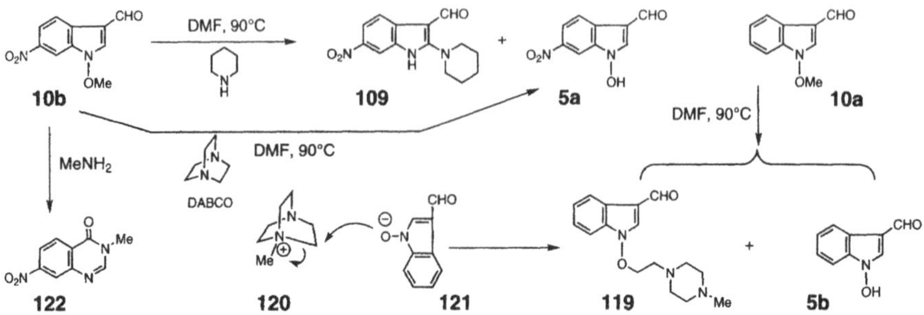

Scheme 15 Nucleophilic substitution reaction of 1-methoxy-6-nitroindole-3-carbaldehyde with amines

When 1,4-diazabicyclo[2,2,2]octane (DABCO) is employed as a reactant, the predominant ether cleavage reaction takes place and only **5a** (90%) is obtained as noted in 2.2. Application of the reaction to daikon-phytoalexin (**10a**) produces 1-hydroxyindole-3-carbaldehyde (**5b**, 42%) [5] and 1-[2-(4-methylpiperazin-1-yl)ethoxy]indole-3-carbaldehyde (**119**, 49%). The isolation of **119** suggests a mechanism where the nitrogen of DABCO first attacks the methyl carbon of the 1-methoxy group in **10a** affording *N*-methylammonium salt (**120**) and the oxide (**121**) of **5b**. Then **121** attacks the carbon adjacent to the positively charged nitrogen of **120** resulting in the formation of **119** [11].

The reaction of **10b** with methylamine in DMF produces 3-methyl-7-nitroquinazoline-4-one (**122**, 25%) as a sole isolable product with tar, which is formed by the initial attack of methylamine on the 3-position of **10b**. The structure is proved by X-ray single-crystal analysis [34].

The above results demonstrate that the reaction site of **10b** changes depending on the species of nitrogen-centered nucleophiles. Nitrogen anions favor the 2-position. The methyl carbon of the 1-methoxy group is favored by *tert*-amines. On the other hand, *sec*-amines attack both sites, while *prim*-amines attack the 3-position.

3.3.3
Solvent Effect on the Reaction Site

Solvents are known to exert large effects on reaction rates, equilibrium constants, chemical reactions, and so on [35].

We have discovered that, in the reaction of 1-methoxy-3-(2-nitrovinyl)indole (**123a**) with nucleophiles (Scheme 16), the choice of the solvent also governs the reaction site [36]. The compound (**123a**) has two reaction sites. One is the β-carbon of the nitrovinyl side chain at the 3-position giving the Michael addition product. The other is the 2-position with concomitant liberation of the 1-methoxy group producing 2-substituted indoles.

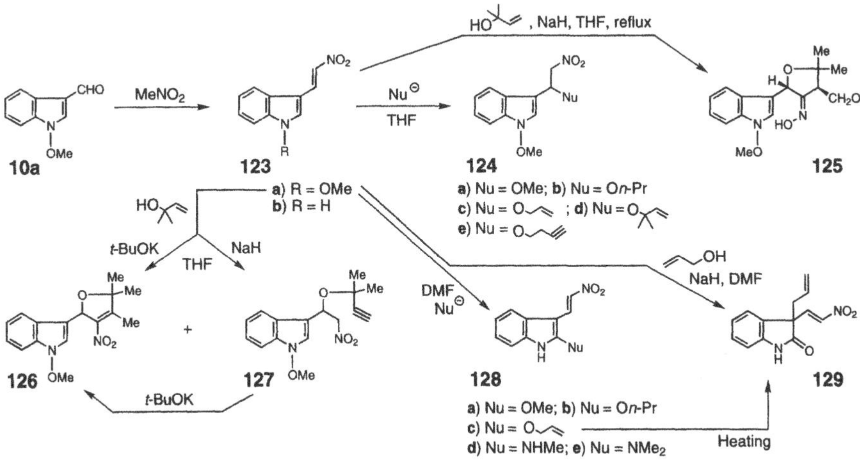

Scheme 16 Solvent effect on the reaction site of 1-methoxy-3-(2-nitrovinyl)indole

The substrate (**123a**) is available by the nitroaldol reaction of 1-methoxyindole-3-carbaldehyde (**10a**). Employing THF as a solvent, the reaction of **123a** with NaOMe or NaO*n*-Pr at 0 °C provides the Michael addition products, 1-methoxy-3-(1-methoxy-2-nitroethyl)indole (**124a**, 90%) and 1-methoxy-3-(2-nitro-1-*n*-propyloxyethyl)indole (**124b**, 92%), respectively. Similar reactions of **123a** with sodium allyl oxide, potassium 1,1-dimethylallyl oxide, and sodium 3-butyn-1-oxide in THF produce **124c** (58%), **124d** (81%), and **124e** (95%), respectively. It is interesting to note that the reaction of **123a** with potassium 1,1-dimethylallyl oxide at reflux instead of 0 °C generates a novel cyclic product (**125**, 55%) probably through the expected intermediate (**124d**). The structure of **125** is proved by X-ray single-crystal analysis of its derivative [36].

We know in some reactions the employment of a different base results in a different product from the original one. The reaction of **123a** with 1,1-dimethylpropargyl alcohol in THF, employing KO*t*-Bu as the base, forms a novel cyclic product (**126**, 35%). A similar reaction using NaH as the base in THF affords a conjugate addition product (**127**, 75%). The fact that the treatment of **127** with KO*t*-Bu provides **126** (60%) clearly shows **126** is formed through **127** [36].

In dipolar aprotic solvents, **123a** reacts with nucleophiles regioselectively at the 2-position with concomitant liberation of the 1-methoxy group. Treatment of **123a** with either NaOMe or NaO*n*-Pr in DMF generates **128a** (85%) and **128b** (45%), respectively. In these reactions, formation of conjugate addition products to the β-carbon atom of the nitrovinyl side chain is not observed at all. In addition, it should be noted that the N(1)–H compound (**123b**) does not undergo a nucleophilic substitution reaction even under forcing reaction conditions [36].

The reaction of **123a** with allyl alcohol/NaH in DMF produces 2-allyloxy-3-(2-nitrovinyl)indole (**128c**, 65%) and (*dl*)-3-allyl-3-(2-nitrovinyl)-2-oxindole (**129**, 6%) as expected. Alternatively, compound (**129**) can be obtained by heating **128c** at 144 °C on celite by the Claisen-type rearrangement (95%). Unexpectedly, the reaction of **123a** with sodium allyl oxide in *N*-methylformamide or DMF at around 105 °C gives rise to **128d** (58%) and **128e** (49%), respectively. In both cases, formation of **129** is not observed at all [36].

3.3.4 Reaction of 1-*t*-Butyldimethylsilyloxytryptamine

Generally speaking, a Si − O bond (370 KJ/mol) is stronger than a N − O bond (163 KJ/mol). We can therefore expect that 1-*t*-butyldimethylsilyloxytryptamine (**54**) would function as a good substrate for the preparation of 5-substituted tryptamines (**130**) with a concomitant liberation of *t*-butyldimethylsilanol from N(1) upon reaction with nucleophiles.

Contrary to expectation, both nucleophiles, NaCN and NaN_3, provide only 1-hydroxytryptamine (**52**) as shown in Table 6 (entries 1 and 2) [24]. In $MeOH/H_2O$ (2 : 1) NaI affords **52** as well (entry 3), while in $MeOH/H_2O$ (1 : 10) formation of **52** is not observed at all (entry 4) instead the starting material is recovered. In every case, a product with a nucleophile in the indole nucleus is not detected. The reaction of **54** with TBAF provides **52** in quanti-

Table 6 Reaction of 1-*t*-butyldimethylsilyloxytryptamine with nucleophiles

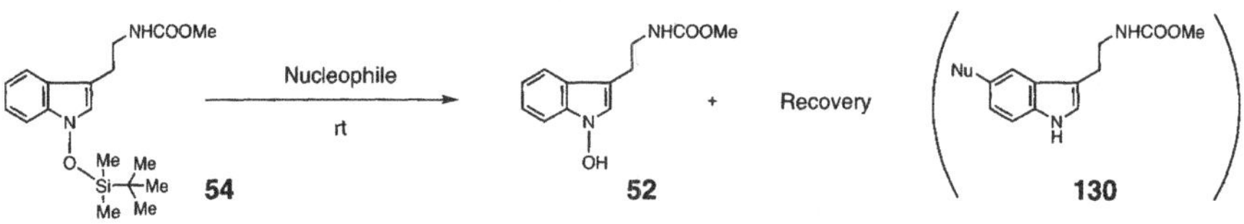

Entry	Nucleophile (mol eq.)	Solvent system	Reaction time (h)	Yield (%) of 52	Recovery	Total yield (%)
1	NaCN (10)	MeCN − H_2O (1 : 1, v/v)	1.5	98	0	98
2	NaN_3 (10)	MeCN − H_2O (1 : 1, v/v)	2	82	16	98
3	NaI (10)	MeOH − H_2O (2 : 1, v/v)	24	65	33	88
4	NaI (10)	MeOH − H_2O (1 : 10, v/v)	216	0	82	82
5	TBAF · $3H_2O$ (2)	THF	1	97	0	97

Table 7 Reaction of 1-*t*-butyldimethylsilyloxytryptamine with acids

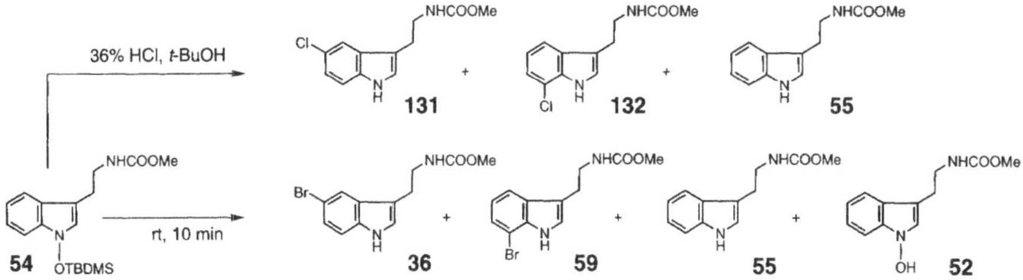

Entry	Reaction conditions	Yield (%) of 36	59	55	52	Total yield (%)
1	47% HBr –*t*-BuOH (1 : 1, v/v)	31	15	30	0	76
2	47% HBr –*t*-BuOH – $HCONH_2$ (1 : 1 : 1, v/v)	0	0	0	91	91

tative yield. These results suggest that the Si – O bond of 1-hydroxyindole is remarkably weaker than we expected. However, the TBDMS group is shown to be useable for the protection of the 1-hydroxy moiety.

When **54** is treated with hydrogen halides, rapid nucleophilic substitution reactions take place within 10 min at room temperature (Table 7). In the reaction with HCl/*t*-BuOH, 5-chloro- (**131**, 60%), 7-chlorotryptamines (**132**, 4%), and tryptamine (**55**, 3%) are produced. The reaction of **54** with HBr/*t*-BuOH affords 5-bromo- (**36**), 7-bromotryptamines (**59**), and tryptamine (**55**) (entry 1). In the same reaction, the employment of formamide as a solvent dramatically changes the reaction pathway to yield 1-hydroxytryptamine (**52**) as a sole product (entry 2) [24].

3.3.5
Acid-Promoted Nucleophilic Substitution Reaction

The unprecedented nucleophilic substitution reaction on the indole nitrogen N(1) was discovered when 1-hydroxyindoles were reacted with good nucleophiles in 85% HCOOH [1–7]. Since then, this type of reaction has been widely observed among the reactions of 1-hydroxytryptamine derivatives and many examples have been accumulated.

Upon reaction with indole as a good nucleophile, (*dl*)-1-hydroxytryptophan methyl ester (*dl*-**41**) and 1-propyloxy-*N*b-methoxycarbonyltryptamine (**53b**) provide the corresponding 1-(indol-3-yl)indoles, **133** (51%) and **134** (38%), respectively (Scheme 17) [37]. Neither the introduction of a sterically hindered group on the *N*b-position nor the existence of a substituent on the

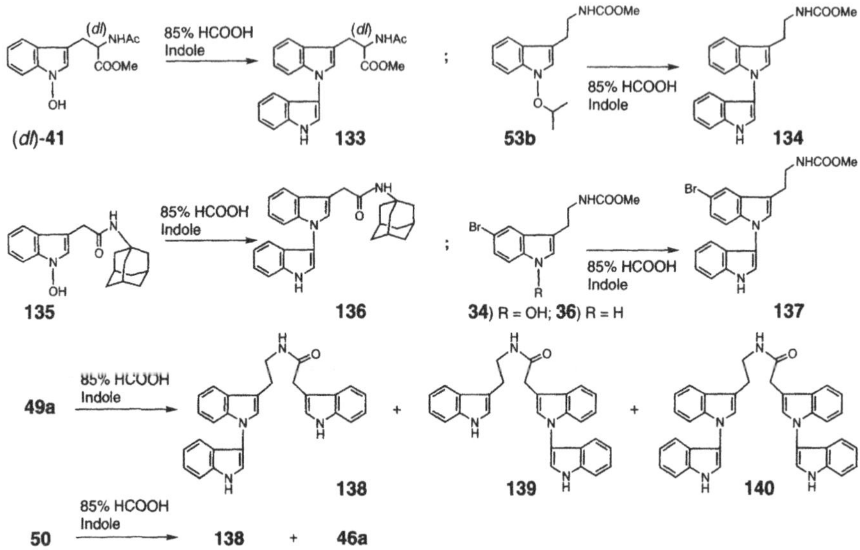

Scheme 17 Nucleophilic substitution on the N(1) position of indole

indole benzene part alters the reaction pathway. Thus, **135** and **34** give rise to the expected products, **136** (73%) and **137** (34%), respectively.

The compound (**49a**) reacts with indole in 85% HCOOH to produce *N*-{2-[1-(indol-3-yl)indol-3-yl]ethyl}indole-3-acetamide (**138**, 7%), *N*-[2-(indol-3-yl)ethyl]- (**139**, 3%), and *N*-{2-[1-(indol-3-yl)indol-3-yl]ethyl}-1-(indol-3-yl)indole-3-acetamide (**140**, 39%). In a similar reaction, the compound (**50**) provides **138** (52%) and a dehydroxylated product (**46a**, 8%).

The structure of the nucleophile is an important factor for governing the reaction pathway [38]. Phenol (**141a**) can react with **52** as a nucleophile in 85% HCOOH to give **142a** (20%) and **104** (11%) (Scheme 18). In a similar reaction with anisole (**141b**), **52** affords **142b** (15%) and **104** (26%).

When veratrole is employed in the reaction with **52**, interesting results are observed [38]. In addition to the expected products, **143** (10%) and **55** (13%), *N*b-methoxycarbonyl-5-(3,4-dimethoxyphenyl)tryptamine (**144**, 9%) is isolated. This type of C – C bond formation on the indole nucleus becomes a major one when **52** is reacted with 2-naphthol culminating in the formations of **145** (16%), **55** (13%), **146** (5%), and **147** (4%).

The reaction site is influenced by the substituent of the benzene part as well (Scheme 19). In the reaction of *N*b-methoxycarbonyl-5-bromo-1-hydroxytryptamine (**34**) with phenol in 100% HCOOH, **148** (37%), **149** (6%), and **36** (11%) are produced.

*N*b-Acetyl-5-acetylamino-1-methoxytryptamine (**13**) is of special interest because of its nature to change the reaction site depending on the nucleophile [33]. When **13** reacts with 2-naphthol in 85% HCOOH, *N*b-acetyl-5-acetylamino-1-(2-hydroxynapht-1-yl)- (**150**, 19%), *N*b-acetyl-5-acetylamino-

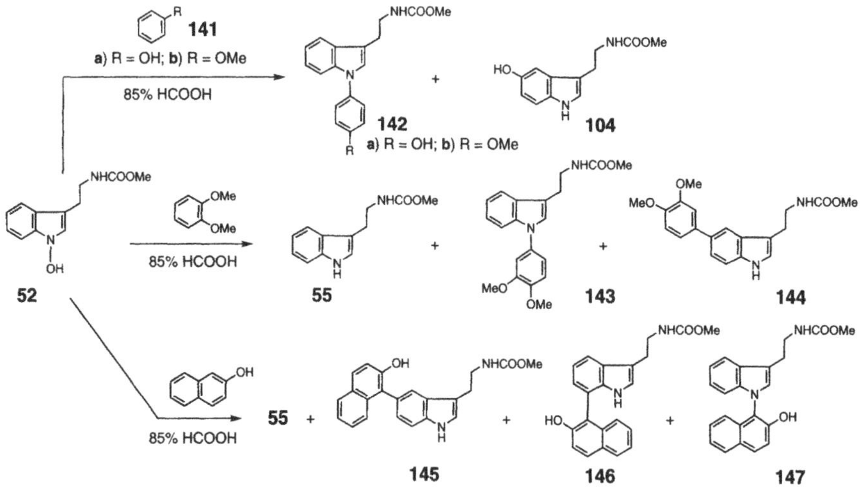

Scheme 18 Nucleophilic substitution reaction of *N*b-methoxycarbonyl-1-hydroxytryptamine

7-(2-hydroxynapht-1-yl)- (**151**, 68%), and *N*b-acetyl-5-acetylaminotryptamines (**152**, 13%) are isolated. In the reaction with indole, **13** produces **152** (62%) as a major product together with *N*b-acetyl-5-acetylamino-6-(indol-3-yl)tryptamine (**153**, 13%). Its structure is determined by leading it to the *tri*-Boc derivative (**154**, 45%) and by observing the deshielded C(7) proton compared to the $^{1}H-NMR$ spectrum of **153**.

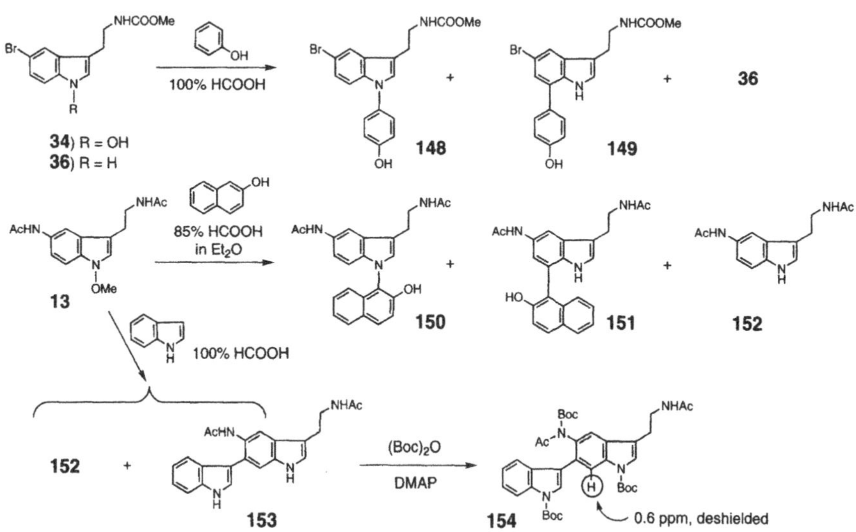

Scheme 19 Nucleophilic substitution reaction of 1-hydroxy- and 1-methoxytryptamines having a substituent on the indole nucleus

The reaction conditions are another factor for altering the reaction pathway. When **52** is treated with mesyl chloride (1.3-mol eq.) and triethyl amine in the presence of a nucleophile, the products are totally different from those of the above-mentioned reactions (Scheme 20) [39]. In the presence of pyrrole, 1-methoxycarbonyl-1,2,3,3a,8,8a-hexahydro-3a-(pyrrol-2-yl)-8*H*-pyrrolo[2,3-*b*]indole (**155**, 40%), -3a-(pyrrol-3-yl)-8*H*-pyrrolo[2,3-*b*]indole (**156**, 15%), and *N*b-methoxycarbonyl-5-(pyrrol-2-yl)tryptamine (**157**, 8%) are formed. Although the reaction of **52** with thiophene, a less-effective nucleophile than pyrrole, is not a clean one with concomitant tar formation, the expected 1-methoxycarbonyl-1,2,3,3a,8,8a-hexahydro-3a-(thiophen-2-yl)-8*H*-pyrrolo[2,3-*b*]indole (**158**, 10%) and **55** (10%) are isolated. When furan is employed as a nucleophile, 1-methoxycarbonyl-1,2,3,3a,8,8a-hexahydro-3a-(furan-2-yl)-8*H*-pyrrolo[2,3-*b*]indole (**159**, 11%) and **55** (10%) are obtained.

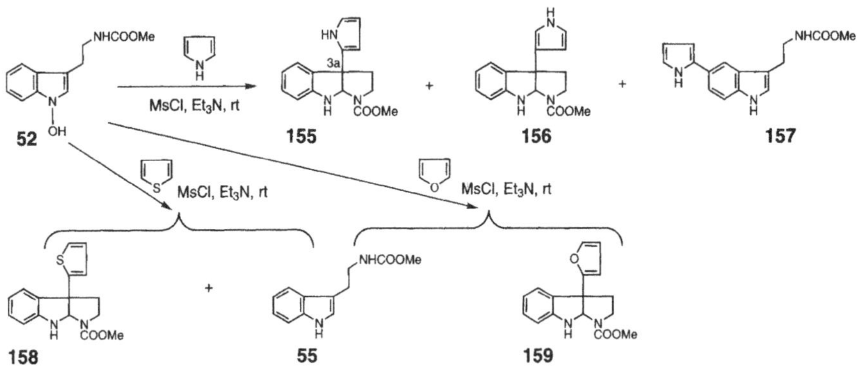

Scheme 20 Preparation of 3a-substituted pyrrolo[2,3-*b*]indoles with an heteroaryl group

3.3.6
Reaction with Enamines

With an attempt to develop a synthetic method for pyrrolo[2,3-*b*]indoles having a C – C bond at the 3a-position like **160**, TsCl (or MsCl) is added to a mixture of 1-hydroxymelatonin (**161**), 1-(4-morpholinyl)cyclohexene (**162**), and $Et_3NinCHCl_3$. An unusual reaction has occurred, though it's based on the nature of 1-hydroxyindole skeletons: products are *N*b-acetyl-3-hydroxy-5-methoxy-2-(4-morpholinyl)-3*H*-indole-3-ethanamine (**163a**, 37%), 2-[3-(2-acetylaminoethyl)-5-methoxy-3*H*-indol-3-yl]cyclohexanone (**164**, 4%), and melatonin (**64**, 8%) (Scheme 21) [40]. The same reaction of 1-hydroxy-*N*b-methoxycarbonylindole-3-ethanamine (**52**) provides 3-hydroxy-*N*b-methoxycarbonyl-2-(4-morpholinyl)-3*H*-indole-3-ethanamine (**163b**, 44%) and *N*b-methoxycarbonyl-6-tosyloxyindole-3-ethanamine (**165**, 11%).

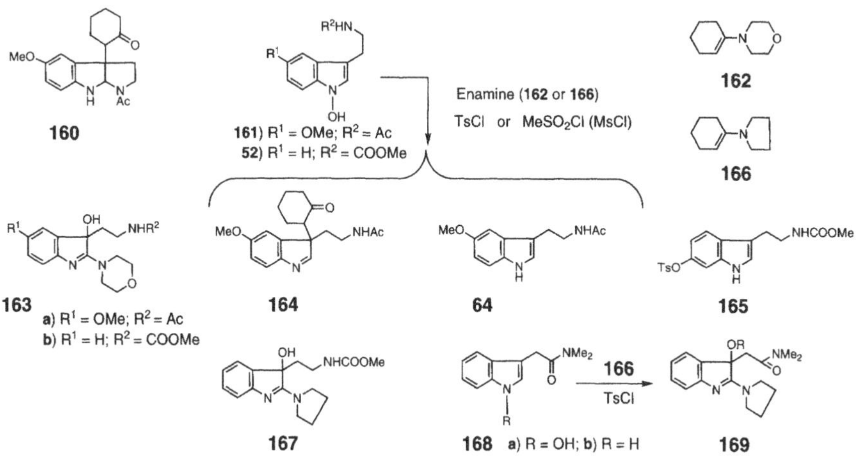

Scheme 21 Reaction of 1-hydroxytryptamines with enamines

When the enamine component is changed from **162** to 1-(1-pyrrolidinyl)-cyclohexene (**166**), **52** affords 3-hydroxy-*N*b-methoxycarbonyl-2-(1-pyrrolidinyl)-3*H*-indole-3-ethanamine (**167**, 38%) as a sole isolable product. Similarly, 1-hydroxy-*N*,*N*-dimethylindole-3-acetamide (**168a**) provides 3-hydroxy-*N*,*N*-dimethyl-2-(1-pyrrolidinyl)-3*H*-indole-3-acetamide (**169**, 21%) and *N*,*N*-dimethylindole-3-acetamide (**168b**, 26%). The structure of **167** is determined by X-ray single-crystal analysis.

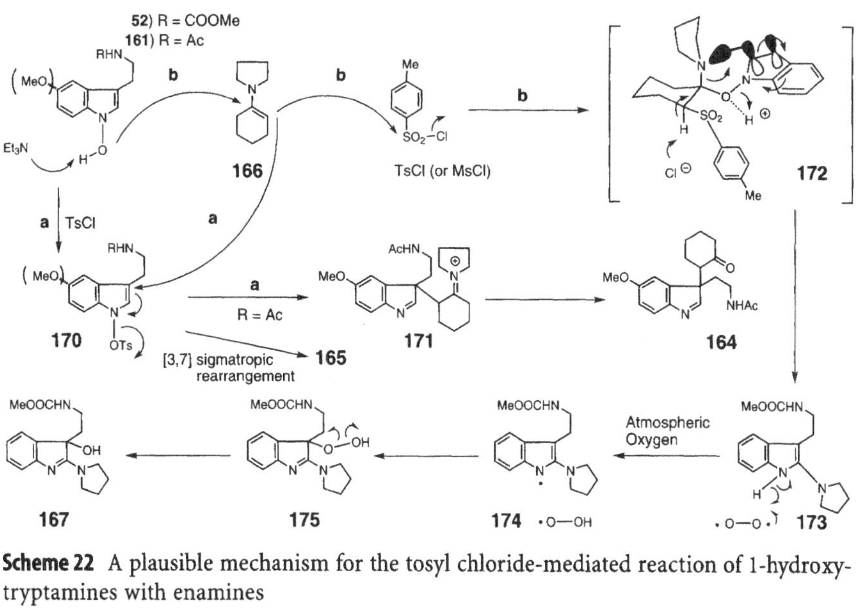

Scheme 22 A plausible mechanism for the tosyl chloride-mediated reaction of 1-hydroxytryptamines with enamines

A plausible mechanism for the formations of **164**, **165**, and **167** is shown in Scheme 22 [40]. Following the route **a**, an initially attempted reaction pathway, the tosylation of **52** occurs. Subsequent nucleophilic attack of the enamine (**166**) on the resultant tosylate (**170**) at the 3-position can afford **164** through **171** after work-up. Compound (**165**) is the product of [3,7] sigmatropic rearrangement of **170**.

On the other hand, following the route **b**, the tosylation of the enamine (**166**) would occur simultaneously with the attack by the 1-hydroxy oxygen of **52** to generate the intermediate (**172**), in which the nitrogen lone pair orbital of pyrrolidine can interact with the π-orbital (2,3-bond) of indole. Subsequent acid catalyzed O – N bond cleavage results in the formation of **173** following the electron movement as shown by the curved arrows. The highly electron-rich compound (**173**) would be oxidized to the final product (**167**) through intermediates (**174**) and (**175**) under atmospheric oxygen.

3.3.7 Preparation of 5*H*-pyrrolo[2,3-*f*]benzoxazoles, -[2,3-*f*]benzothiazole, and 8*H*-pyrrolo[3,2-*g*]benzothiazole

Novel compounds having either 5*H*-pyrrolo[2,3-*f*]benzoxazole, 5*H*-pyrrolo[2,3-*f*]benzothiazole, or 8*H*-pyrrolo[3,2-*g*]benzothiazole skeletons are now available utilizing 1-hydroxyindole chemistry (Scheme 23) [39].

A simple treatment of *N*b-acetyl-5-acetylamino-1-methoxytryptamine (**21**) with 85% HCOOH affords 7-(2-acetylamino)ethyl-2-methyl-5*H*-pyrrolo[2,3-*f*]benzoxazole (**176a**, 32%), *N*b-acetyl-6-acetylamino-5-hydroxytryptamine (**177a**, 16%), and *N*b-acetyl-6-acetylaminotryptamine (**178a**, 15%). A failure of the attempt to cyclize **177a** to **176a** suggests that the oxygen in the molecule (**176a**) comes from the acetylamino oxygen of **21** [39].

Interestingly, 7-(2-acetylamino)ethyl-5*H*-pyrrolo[2,3-*f*]benzoxazole (**176c**, 16%) is isolated together with **176b** (23%), **177b** (18%), and **178b** (17%)

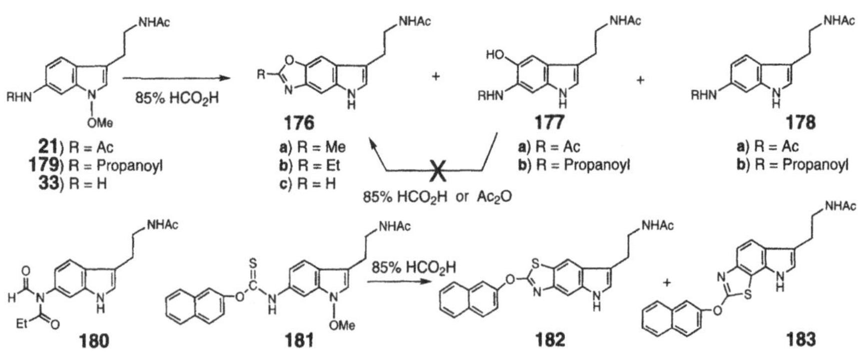

Scheme 23 Preparation of 5*H*-pyrrolo[2,3-*f*]benzoxazoles, -[2,3-*f*]benzothiazole, and 8*H*-pyrrolo[3,2-*g*]benzothiazole

upon treatment of *N*b-acetyl-5-propanoylamino-1-methoxytryptamine (**179**) with 85% HCOOH. Its production could be explained by the prior formation of the intermediate, *N*b-acetyl-5-(*N*-formyl-*N*-propanoyl)amino-1-methoxytryptamine (**180**). The structure of **176c** is confirmed by direct comparison with the product obtained by the reaction of **33** with 85% HCOOH [39].

A novel intramolecular nucleophilic attack of the thioamide sulfur on the 7-position of the indole nucleus is realized to provide 7-(2-acetylamino)ethyl-2-(2-naphtyl)oxy-5*H*-pyrrolo[2,3-*f*]benzothiazole (**182**, 32%) and 6-(2-acetylamino)ethyl-2-(2-naphtyl)oxy-8*H*-pyrrolo[3,2-*g*]benzothiazole (**183**, 25%), when *N*b-acetyl-1-methoxy-6-*N*-(2-naphthyloxy)thiocarbonylaminotryptamine (**181**) is treated with 85% HCOOH [39].

4
X-ray Analysis, Theoretical Calculation, and Working Hypothesis

At the very beginning of the study, we had predicted in our 1-hydroxyindole hypothesis that 1-hydroxyindole compounds should undergo unprecedented nucleophilic substitution reaction and the reaction must be responsible for the generation of serotonin (melatonin, etc.) in our central nervous system [1–7].

Theoretically, however, why could 1-hydroxyindole compounds undergo nucleophilic substitution reactions? Let's consider the well-known reaction of diazonium salt (**E**, Fig. 3). Liberation of the nitrogen molecule leaves positive charge on the carbon initially diazonium group attached as shown in **F**. Because the positively charged vacant sp^2 orbital is orthogonal to the *p*-orbitals forming a 6π-electron aromatic system, electron delocalization does not occur. As a result, resonance structures such as **G** and **H** cannot be generated.

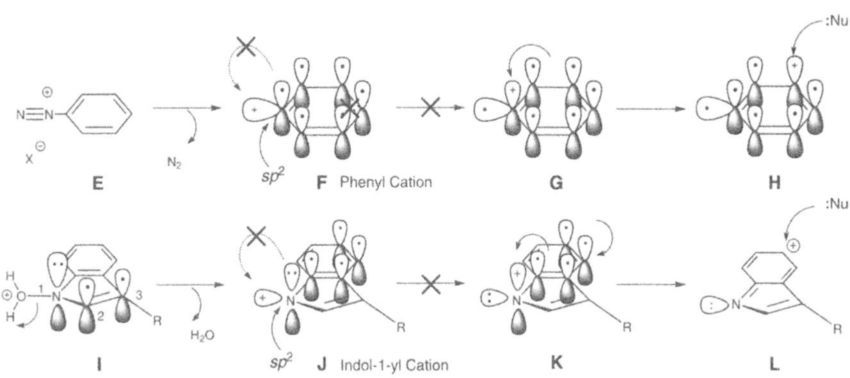

Fig. 3 Theoretical consideration of indol-1-yl cation

This is the reason why diazonium salts do not undergo nucleophilic substitution reactions at the carbon to which the diazonium group is not attached initially.

Similar discussions are applicable for the acid-mediated reaction of 1-hydroxyindole compounds (**I**, Fig. 3). A proton adds to the 1-hydroxy oxygen. Subsequent liberation of H_2O leaves the indol-1-yl cation (**J**), where the positively charged sp^2 orbital of N(1) is orthogonal to the *p*-orbitals consisting of the 10π-electron indole ring. On the basis of the above theory, the generation of resonance structures such as **K** and **L**, and the nucleophilic substitution reaction at the 5-position (**L**) are nonsense [41].

We have thus far demonstrated that only 1-hydroxyindoles having a C – C – N side chain at the 3-position can undergo nucleophilic substitution reactions [1–7]. To explain the paradox between the above theory and experimental results, we have conceived the following factors as a working hypothesis.

The first factor is named "bishomoallylic conjugation" [41]. In the conformation (**M**) as shown in Fig. 4, the lone pair on the bishomoallylic *N*b-nitrogen could interact with the π^* orbital of the C(2) – C(3). The *N*b lends a little electron density to the π^* orbital, thereby making the π-bond weaken and the N(1) free from the C(2) – C(3) π-bond resulting in the formation of the sp^3-like nitrogen. Now the 6π-electrons of the isolated benzene can delocalize to the positively charged sp^3 nitrogen. Consequently, the nucleophilic substitution reaction at the 5- and/or other positions of the indole nucleus becomes possible.

The second factor is the following: either the lone pair on the N(1) or the O – N(1) bonding electrons are able to interact with both the 6π-benzene orbital and the π^* orbital of the C(2) – C(3) if the O – N(1) bond deviates from the plane of the indole (**N**, Fig. 4). In addition, the resulting sp^3-like lone pair orbital of the N(1) can form a hydrogen bond with the 1-hydroxy hydrogen.

In 1-hydroxyindole molecules where the *N*b-nitrogen stretches away from the indole nucleus like **O**, they cannot assume a conformation similar to **M**. At the same time, when the second factor of the molecule becomes weak by chance, indole nitrogen becomes sp^2 hybridized and naturally a nucleophilic

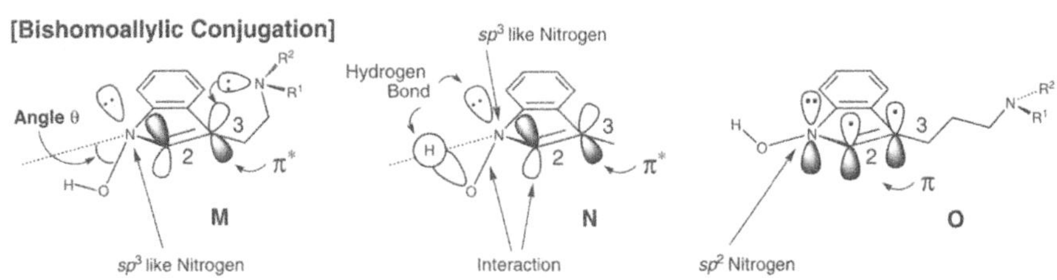

Fig. 4 Working hypothesis for explaining the deviation of the O – N(1) bond from the indole molecular plane

substitution reaction on the indole N(1) does not take place. The balance of the above factors would govern the reaction pathway.

In order to determine whether or not the 1-hydroxy oxygen deviates with an angle θ from the indole molecular plane, X-ray single-crystal analyses of four stable crystalline compounds ((*dl*)-**41** [22], **184**, **185**, and **19**) are carried out [41]. The results shown in Fig. 5 demonstrate that the N(1) – O bonds in *N*b-acetyl-1-hydroxytryptophan methyl ester ((*dl*)-**41**) and *N*,*N*-dimethyl-1-hydroxyindole-3-acetate (**184**) have angle θs of 15.2° [22] and 10.9°, respectively. On the other hand, the corresponding angle θs of **185** and **19** are 2.3° and 4.1°, respectively, showing their N(1) – O bonds are almost on the same plane of the indole [41] (Acheson and co-workers have reported that 1-benzoyloxyindole also has a pyramidal nitrogen atom [42]).

Molecular orbital calculations based on B3LYP/6-31G(d,p) are made to confirm the angle θ [41]. The angle θs of (*dl*)-**41** and **184** are calculated to be in the range of 7–9° in the most stable conformation, while those of both **185** and **19** are about 4°. These results are in accord with those obtained by X-ray analysis.

The difference in angle θ observed among the four compounds is small but real. Consequently 1-hydroxyindoles (**186**, Fig. 6) having a C – C – Nb side chain are classified into two subtypes. Subtype A compounds have angle θs near 10°, while subtype B compounds have no substantial angle θs.

On the basis of comparison of the above results with the actual reaction profiles of the four compounds ((*dl*)-**41**, **184**, **185**, and **19**), it would be natural to conclude that only subtype A compounds undergo acid-promoted

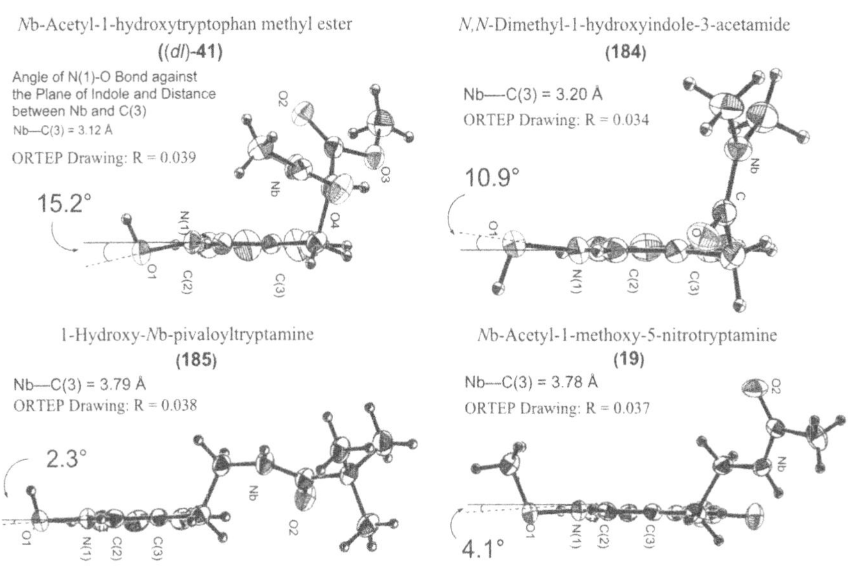

Fig. 5 X-ray analyses of four compounds and their angle θ

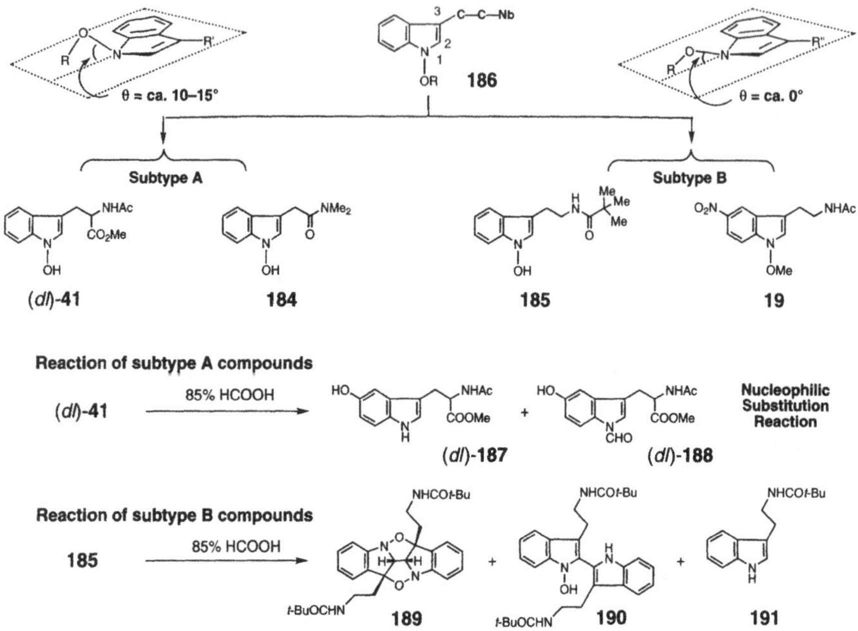

Fig. 6 Classification of 1-hydroxyindoles into two subtypes and their reaction profiles

nucleophilic substitution reactions [41]. The typical example is (*dl*)-**41** which gives rise to **187** (60%) and **188** (16%). Under the same reaction conditions, subtype B compounds provide kabutanes, dimeric indoles, and dehydroxylated products with no detectable formation of compounds originating from nucleophilic substitution. For example, **185** generates **189** (19%), **190** (3%), and **191** (10%).

In conclusion, the deviation of the N(1) – O bond from the indole molecular plane is responsible for the nucleophilic substitution reactions of the 1-hydroxytryptamine and 1-hydroxytryptophan derivatives.

5 Biologically Active Compounds

We had predicted in our 1-hydroxyindole hypothesis that the introduction of a hydroxy or an alkoxy group onto the 1-position of biologically inactive indole compounds makes them become biologically active compounds [1–7]. In addition, the application of the idea to the known biologically active indoles has their original activities improved or modified for the discovery of new drugs [1–7].

With these basic concepts, we have given birth to various biologically active compounds thus far and some examples are shown in our previous reviews [1–7]. Notably, we have shown that simple 1-hydroxytryptamines are fundamentally inhibitors of blood platelet aggregation (for our new lead compounds for cerebral infarction and myocardial infarction see [43, 44]). New compounds with potent activity are in preparation for patent application. Excluding these, we would like to introduce some of our biologically active compounds in this section.

Yohimbine (**192a**, Fig. 7) is a widely known α_2-blocker effective in treating erectile dysfunction (ED). They have been used all over the world. A slight modification of its mother skeleton and the introduction of an alkyl group onto its N(1) position are shown to change its biological activity [45]. Therefore, we are interested in the preparation of 1-hydroxyyohimbine (**192b**) and its derivatives. As expected, **192b** and some derivatives (**192c–f**) are α_2-blockers as potent as yohimbine [46, 47].

Compounds (**193a,b–199a,b**) have also been discovered to be α_2-blockers [46, 47]. When the activity of yohimbine is made a standard for 100, those of **193a** and **194a** are 87.7 ± 8.53 and 17.8, respectively. The activity of compound (**195a**, 100.0 ± 0.00) is found to be equal to or more potent than yohimbine (**192a**) [46, 47]. The 1-hydroxy compounds (**193a** and **195a**) are stronger than the corresponding N(1)H compounds (**193b** and **195b**), while **194a** is weaker than **194b**. In a series of *N*b-acyltryptamines (**196–199**), differences in activities are small between 1-hydroxy and N(1)H compounds.

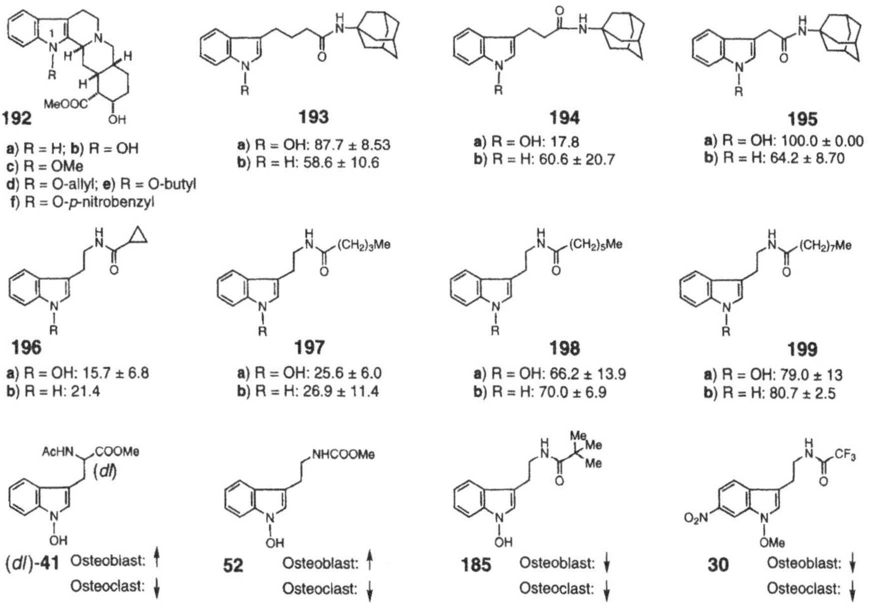

Fig. 7 Biologically active 1-hydroxyindoles and analogs (1)

It is important to note that in these compounds the activity depends on the length of the *N*b-alkyl side chain: the longer the side chain the stronger the activity [46, 47].

In our project to develop a drug for osteoporosis, we have created 1-hydroxytryptophan ((*dl*)-**41**) and 1-hydroxytryptamine derivatives (**52**, **185**, and **30**). Among them, (*dl*)-**41** and **52** have the same trend not only to stimulate osteoblasts, but also to suppress osteoclast activity in the cultured scales of goldfish [48, 49]. On the other hand, compounds (**185** and **30** suppress both osteoblasts and osteoclasts. During the study, we have succeeded in creating potent and promising compounds for osteoporosis [50, 51].

From the point of view of developing an anti-cancer drug, some potent 1-hydroxyindole derivatives have been discovered through the test using human tumor cells. Further study is in progress (Somei and co-workers, personal communication, 2006).

An attempt at developing an antifungal drug leads us to find 1-hydroxy- (**200**) and 1-methoxyindole derivatives (**201a,b**, **202**, and **203**) as shown in Fig. 8 (Somei and co-workers, personal communication, 2006). These have activity against gram-negative bacteria; *Erwinia carotovora* and *Xanthomonas oryza*. They also have mild activity against filamentous fungi; *Pythium debaryanum*, *Phytophthora infestans*, *Pyricularia oryzae*, *Botrytis cinerea*, and *Rhizoctonia solani*. Compounds (**204a,b**) are active against *Erwinia carotovora* and **204a** is more potent than **204b**. As for the inhibition of IAA-induced ethylene formation with the rice seedling test, compound (**205**) shows weak activity.

Phytoalexin is also an interesting compound. The plant family, *Cruciferae*, utilizes 1-methoxyindole compounds as phytoalexins. We expected there was a chance to find new agrochemicals among analogs of 1-methoxyindole-3-carbaldehyde [52] (**10a**, daikon-phytoalexin [53]). So, we have thus far prepared a lot of related compounds. Some of them are **5a,b**, **6a–f**, **10b**, **119**, **206a–d**, **207a,b**, and **208a,b**. These are found to have mild activities (Somei and co-workers, personal communication, 2006).

Melatonin (**64**) is a hormone secreted from the pineal gland and is well known to control the circadian rhythms. Its multimodality of biological activities [49, 54–59] has led us to synthesize unprecedented 1-hydroxy- (**161**) and 1-methoxymelatonin (**209a**) from melatonin [26] as lead compounds for new drugs [46, 47, 50, 51]. Quite recently, Tsotinis and co-workers reported its analogs (**209b,c**) as melatonin receptor agonists [60].

We have been engaged in a project for a long time aimed at increasing the harvest of farm products. During this study, we have discovered that 4-iodoindole-3-carbaldehydes (**210a–c**), 2-bromoindole-3-carbaldehydes (**211a–c**), 2-(3-hydroxy-3-methylbuten-1-yl)indole-3-carbaldehydes (**212a–c**), and 2-(3-hydroxy-3-methylbuten-1-yl)-3-(2-nitrovinyl)indoles (**213a–c**) are potent plant-growth regulators and promote root growth of various plants (Somei and co-workers, personal communication, 2006). 1-Methoxy analogs

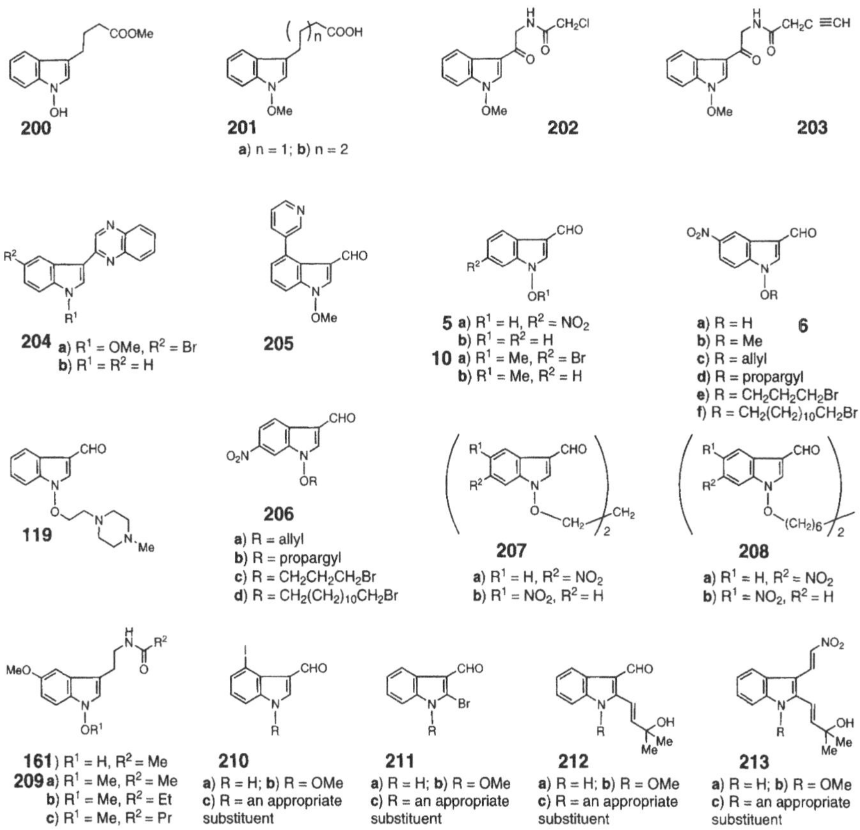

Fig. 8 Biologically active 1-hydroxyindoles and analogs (2)

(**210b, 211b**) are more potent than the N(1)–H compounds (**210a, 211a**). In a test using seeds of rice and cucumber, for example, **211a** elongates their roots 1.8 times longer than that of the control. With these and related derivatives in hand, we have begun a challenge this summer to make the Gobi desert greener in Neimenggu, Inner Mongolia, by making the native plants' roots longer and closer to the ground water level.

In conclusion, the initially imaginary world of 1-hydroxytryptamines and -tryptophans has become a real frontier in indole chemistry. Since this undeveloped world is rich in treasures, we hope many investigators come together to hunt down and own some of its valuable jewels. Our discoveries are only a part of the infinite abundance of treasures.

Acknowledgements I am deeply grateful to a number of graduate students and undergraduates for their hard work and dedication. For biological evaluations, I also express cordial gratitude to Dr. Suzuki N (Noto Marine Lab., Kanazawa Univ.), Prof. Hattori A (Tokyo Medical and Dental Univ.), Prof. Shigenobu H, Assoc. Prof. Tanaka Y (Toho Univ.), Dr. Ekimoto H (Nippon Kayaku), and Dr. Kamuro Y (Nissan Chemical Ind. Ltd.).

References

1. 1-Hydroxyindole Hypotheses was first presented orally: Somei M, Karasawa Y, Tokutake S, Shoda S, Yamada F, Kaneko C (1980) 13th Congress of Heterocyclic Chemistry, Shizuoka, Japan. Abstracts Papers, p 33
2. Somei M, Kawasaki T, Fukui Y, Yamada F, Kobayashi T, Aoyama H, Shinmyo D (1992) Heterocycles 34:1877
3. Somei M, Fukui Y (1993) Heterocycles 36:1859
4. Hasegawa M, Tabata M, Satoh K, Yamada F, Somei M (1996) Heterocycles 43:2333
5. Somei M (1991) J Synth Org Chem 49:205
6. Somei M (1999) Heterocycles 50:1157
7. Somei M (2002) Recent Advances in the Chemistry of 1-Hydroxyindoles, 1-Hydroxytryptophans, and 1-Hydroxytryptamines. In: Katritzky AR (ed) Advances in Heterocyclic Chemistry, vol 82. Academic Press, London, 305 pp
8. Sundberg RJ (1970) The Chemistry of Indoles. Academic Press, New York; Sundberg RJ (1996) Indoles. Academic Press, New York
9. Somei M, Kawasaki T (1989) Heterocycles 29:1251
10. Yamada F, Kawanishi A, Tomita A, Somei M (2003) ARKIVOC viii:102
11. Yamada K, Tomioka S, Tanizawa N, Somei M (2004) Heterocycles 63:1601
12. Yamada K, Kawasaki T, Fujita T, Somei M (2001) Heterocycles 55:1151
13. Yamada K, Yamada F, Shiraishi T, Tomioka S, Somei M (2002) Heterocycles 58:53
14. Yamada F, Hayashi T, Yamada K, Somei M (2000) Heterocycles 53:1881
15. Nakai Y, Godou T, Yamada F, Somei M (2005) 125th Annual Meeting of Pharmaceutical Society of Japan, Tokyo, Japan. Abstract Papers, p 31
16. Lanzilotti AE, Littell R, Fanshawe WJ, McKenzie TC, Lovell FM (1979) J Org Chem 44:4809
17. Hayashi T, Yamada F, Somei M (2002) 107th Meeting of Pharmaceutical Society of Japan, Hokuriku Branch, Fukui, Japan. Abstract Papers, p 18
18. Hasegawa M, Yamada K, Nagahama Y, Somei M (1999) Heterocycles 51:2815
19. Gribble GW, Hoffman JH (1977) Synthesis, p 859
20. Dinh TD, Vranken DLV (1999) J Peptide Res 53:465
21. Yamada A, Fukui Y, Iwaki T, Ogasawara S, Okigawa M, Tanaka S, Somei M (2006) Heterocycles 67:129
22. Somei M, Kawasaki T, Shimizu K, Fukui Y, Ohta T (1991) Chem Pharm Bull 39:1905
23. Nakai Y, Goto A, Yamada F, Somei M (2003) Heterocycles 60:1589
24. Nakai Y, Asada M, Santo R, Hayashi T, Somei M (2004) 111st Meeting of Pharmaceutical Society of Japan, Hokuriku branch, Kanazawa, Japan. Abstract Papers, p 6
25. Hayashi T, Iwaki T, Yamada K, Somei M (2003) 33rd Congress of Heterocyclic Chemistry, Sapporo, Japan. Abstract Papers, p 60
26. Somei M, Fukui Y, Hasegawa M, Oshikiri N, Hayashi T (2000) Heterocycles 53:1725
27. Somei M, Tanimoto A, Orita H, Yamada F, Ohta T (2001) Heterocycles 54:425
28. Yamada K, Kanbayashi Y, Tomioka S, Somei M (2002) Heterocycles 57:1627

29. Nakatsuka S, Asano O, Ueda K, Goto T (1987) Heterocycles 26:1471
30. Iwaki T, Yamada F, Funaki S, Somei M (2005) Heterocycles 65:1811
31. Yamada K, Izumi T, Yamada F, Somei M (2005) Heterocycles 66:583
32. Yamada F, Shinmyo D, Somei M (1994) Heterocycles 38:273
33. Nakai Y, Hayashi T, Yamada F, Somei M (2004) 34th Congress of Heterocyclic Chemistry. Kanazawa, Japan. Abstract Papers, p 253
34. Tomioka S, Yamada K, Yamada F, Tanizawa N, Shiraishi K, Somei (2003) 29th Symposium on Progress in Organic Reactions and Syntheses, Gifu, Japan. Abstracts, p 136
35. Reichardt C (1979) Solvent Effects in Organic Chemistry. Verlag Chemie GmbH, Weinheim
36. Yamada K, Yamada F, Somei M (2002) Heterocycles 57:1231
37. Yamada F, Goto A, Peng W, Hayashi T, Saga Y, Somei M (2003) Heterocycles 61:163
38. Hayashi T, Peng W, Nakai Y, Yamada K, Somei M (2002) Heterocycles 57:421
39. Nakai Y, Imai K, Tokumura K, Yamada F, Somei M (2005) 35th Congress of Heterocyclic Chemistry, Osaka, Japan. Abstract Papers, p 163
40. Hayashi T, Nakai Y, Yamada F, Somei M (2004) Heterocycles 62:437
41. Somei M, Imai K, Yamada F, Hayashi T, Nakai Y, Tokumura K (2004) 34th Congress of Heterocyclic Chemistry, Kanazawa, Japan. Abstract Papers, p 255
42. Acheson RM, Benn MH, Jacyno J, Wallis JD (1983) J Chem Soc Perkin Trans 2:497
43. Somei M (1996) JP Patent 157475 [(1996) Chem Abstr 125:195426y]
44. Somei M (1991) JP Patent 31257 [(1991) Chem Abstr 114:247138a]
45. Muramatsu I (1996) Nihon Yakuzaishikai Zasshi 48:1987
46. Somei M, Shigenobu H, Tanaka Y JP Patent 2004-280104, applied Sept. 2004
47. Somei M, Shigenobu H (2001) JP Patent 2001-255963 [(2002) Chem Abstr 137:217126c]
48. Suzuki N, Hattori A (2002) J Pineal Res 33:253
49. Suzuki N, Hattori A (2003) Life Sciences 73:2237
50. Somei M, Hattori A, Suzuki N JP Patent 2004-64408, applied March 2004
51. Somei M, Hattori A, Suzuki N JP Patent 2005-209753, applied July 2005
52. Somei M, Ohnishi H, Shoken Y (1986) Chem Pharm Bull 34:677
53. Takasugi M, Monde K, Katsui N, Shirota A (1987) The 29th Symposium on The Chemistry of Natural Products, Sapporo, Japan. Symposium Papers, p 629
54. Pappolla M, Bozner P, Soto C, Shao H, Robakis NK, Zagorski M, Frangione B, Ghiso J (1998) J Biol Chem 273:7185
55. Loeffler M (1996) Exp Clin Endocrinol Diabetes 104:308
56. Ying SW, Niles LP, Crocker C (1993) Eur J Pharmacol Mol Pharmacol Sect 246:89
57. Schmid H (1993) Gerontology 39:189
58. Waldhauser F, Ehrart B, Forster E (1993) Experientia 49:671
59. Pierpaoli W (1991) Aging 3:9
60. Tsotinis A, Eleutheriades A, Hough K, Sugden D (2003) Chem Commun, p 382

Top Heterocycl Chem (2006) 6: 113–156
DOI 10.1007/7081_022

Published online: 7 April 2006

Quinazoline Alkaloids and Related Chemistry

Shoji Eguchi

Department of Molecular Design and Engineering, Graduate School of Engineering, Nagoya University, Furo-cho, Chikusa-ku, 468-0863 Nagoya, Japan
eguchi@lilac.ocn.ne.jp

Abstract Recent progress in quinazoline alkaloids and related chemistry was reviewed focusing on developments of the synthetic methodologies and their synthetic applications. A brief historical background, aza-Wittig methodology, microwave assisted synthesis, solid-phase synthesis, and a variety of new syntheses of quinazoline compounds by organometallic reagents, metal-catalyzed reactions, heterocyclizations, pericyclic reactions, etc. are briefly reviewed. Selected topics of total synthesis of various types of quinazoline alkaloids including substituted type like febrifugine and heterocycle-fused type such as pyrroloquinazolines, indolopyridoquinazolines, pyrazinoquinazolines, and pyrroloquinazolinoquinolines, etc. by these methodologies are discussed.

Keywords Aza-Wittig reaction · Quinazolinone annelation · Quinazoline alkaloids · Microwave-assisted synthesis · Solid-phase synthesis

Abbreviations

aq	Aqueous
BOP	Benzotriazol-1-yloxytris(dimethylamino)phosphoniumhexafluorophosphate (peptide coupling agent)
i-Bu	*Iso*-butyl
CCK	Cholecystokinin
Cy_3P	Tricyclohexylphosphine
1,2-DCE	1,2-Dichloroethane
DCI	1,3-Diisopropylcarbodiimide
DCM	Dichloromethane
Dppp	Bis(diphenylphosphino)propane
DS	Dodecyl sulfate
MCRs	Multi-component reactions
MDR	Multi-drug resistance
μw(MW)	Microwave
refl	Reflux
SAR	Structure activity relationship
TBP	Tributylphosphine
TPP	Triphenylphosphine

1 Introduction

1.1 Quinazoline Pharmaceuticals

Quinazoline (1) is 1,3-diazanaphthalene. It is also known as 5,6-benzopyrimidine or benzo[*a*]pyrimidine, or phenmiazine [1], and its 4-oxo derivative is called 4(3*H*)-quinazolinone (2) [2–4] (Fig. 1). Quinazoline and quinazolinone derivatives have continued to attract a widespread interest for a long time due to their diverse pharmacological activities such as antibacterial [5, 6], antitubecular [7], antifungal [8], antihyperglysemic [9], anti-inflammatory [10], bronchodilatory [11], cholineesterase inhibitor [12], antifolate [13, 14], antitumor [15–17], protein kinase inhibitor [18] and many others [2, 17]. A vast number of quinazoline derivatives have been synthesized to provide synthetic drugs and to design more effective medicines. There are a number of reviews [2–4, 19–29] and monographs [30–32] on quinazoline and quinazoline alkaloids.

As the actual example, methaqualone (3) [33–35] is perhaps the most well-known synthetic quinazoline drug, which was first synthesized in

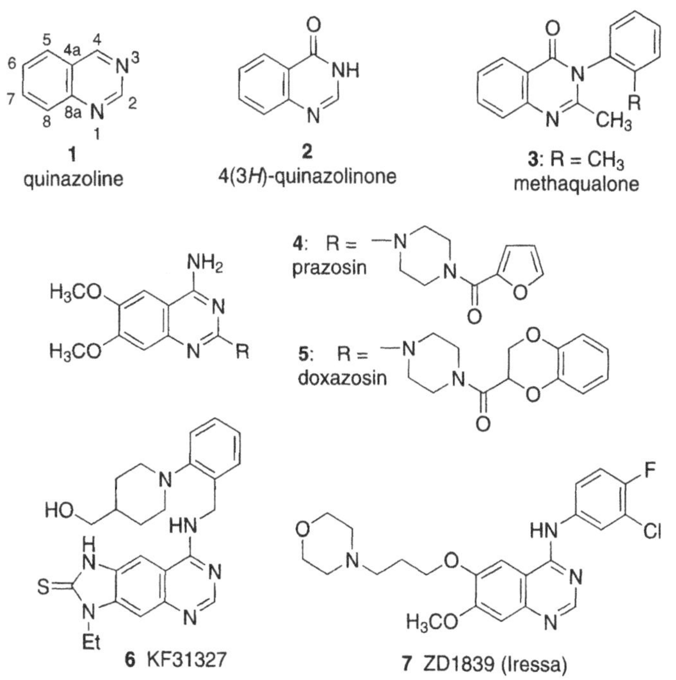

Fig. 1 Quinazoline, 4(3*H*)-quinazolinone, and examples of quinazoline-based medicines

1951 [34] and was used as sedatives and hypnotics as somewhat modified etaqualone [36] and mecloqualone [37, 38] (Fig. 1). The effective α-adrenergic blocker, prazosin (**4**) [39], bunazosin [40], and doxazosin (**5**) [41] (Fig. 1) are useful medicines for antihypertensives. KF31327 (**6**) was developed as a heart disease remedy and an impotence medicine [42, 43]. ZD1839 (iressa) (**7**) is known as a potent EGFR (epidermal growth factor receptor) kinase inhibitor with excellent bioavailability and is clinically used for the treatment of nonsmall cell lung carcinoma [18].

1.2 Natural Quinazoline Compounds

The occurrence of the quinazoline skeleton in various natural sources as alkaloids, fungal metabolites, and marine natural products has generated interest of many groups on account of their useful biological properties and often very intriguing structures. For example, the unique novel structure of tetrodotoxin known as one of the most powerful non-protein neurotoxins from the Japanese fugu (puffer fish), *Sphoerdes rubripes* and *S. phyreus* was finally elucidated as the labile complicated quinazoline derivative (**8**) (Fig. 2) in 1964 by Hirata et al., Tsuda et al., and Woodward et al., respectively [44–47]. It is of interest that the same toxin was also found in the Californian newt or salamander, Taricha torosa [48] and others [49, 50]. Total synthesis of (±)-**8** was achieved by Kishi in 1972 [51], and asymmetric synthesis of (–)-**8** has been achieved by Isobe et al. [52], and also by Du Bois et al. [53] in 2003. Quinazoline alkaloids are continuously updated in Natural Product Reports [21–28], and comprehensive reviews are available [19, 20, 29–32]. Quinazoline alkaloids have been isolated from several families in the plant kingdom, as well as from bacteria and animal species, and many are biogenetically derived from anthranilic acid [54].

The first quinazoline alkaloid to be isolated was vasicine (peganine) (**9**) in 1888, produced by Indian medicinal tree *Adhatoda vasica*, and later isolated from other species along with the related pyrrolo[2,1-*b*]quinazoline alkaloids, vasicinone (**10**) and deoxyvasicinone (**11**), etc. [21, 55, 56]. A variety of other quinazoline natural products have been isolated, characterized and synthesized thereafter. Some of these examples including simple 2-substituted 4(3*H*)-quinazolinones such as **12** [57] and **13** (–)-chrysogenine [58, 59] are shown in Fig. 2. *N*(3)-Substituted derivatives are also included as antimalarial (+)-febrifugine **14** and (+)-isofebrifugine **15** isolated as the active ingredient of Chinese medicinal plant *Dichora febrifuga* [24, 25, 60–63]. Unfortunately these are also toxic to man. This has led to extensive synthetic studies and biological screening of many quinazoline derivatives (see Sect. 3.6). Familiar quinazolin alkaloid rutaecarpine (**16**) [64, 65] and evodiamine are known as the active compounds of *Evodia rutaecarpa*, used in

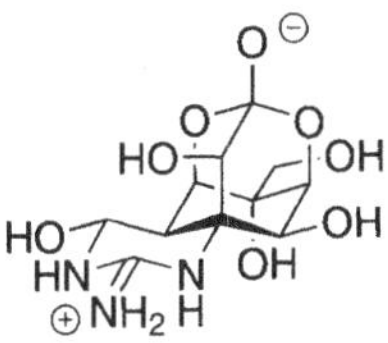

8
(-)-tetrodotoxin

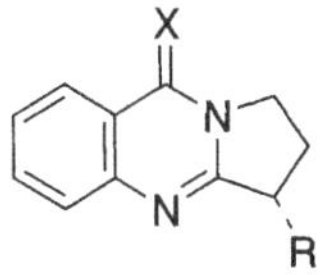

9: X = H_2, R = OH
(-)-vasicine
10: X = O, R = OH
(-)-vasicinone
11: X = O, R = H
deoxyvasicinone

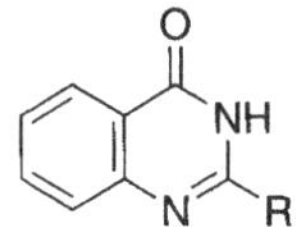

12: R = CH_3
13: R = (*S*)-CH(OH)CH_3
(-)-chrysogenine

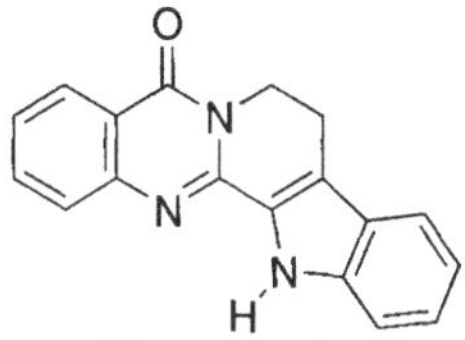

16: rutaecarpine

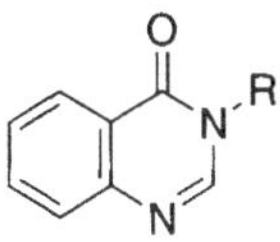

14: R =
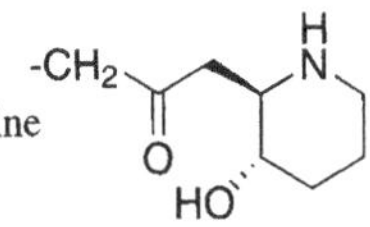
(+)-febrifugine

15: R =
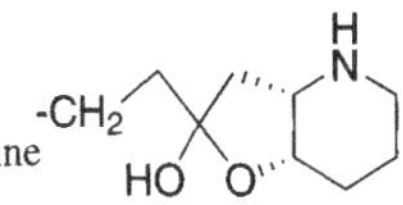
(+)-isofebrifugine

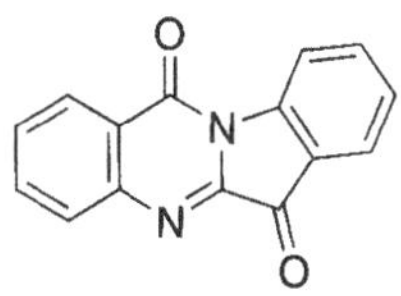

17: tryptanthrin

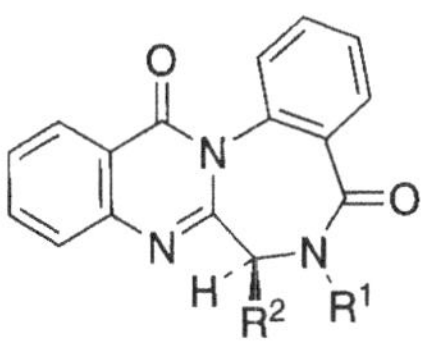

18: R^1 = Me, R^2 = Bn
(-)-benzomalvin A
19: R^1 = R^2 = H
sclerotigenin
20: R^1 = H, R^2 =
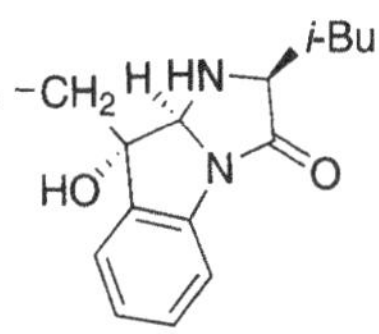
(-)-asperlicin

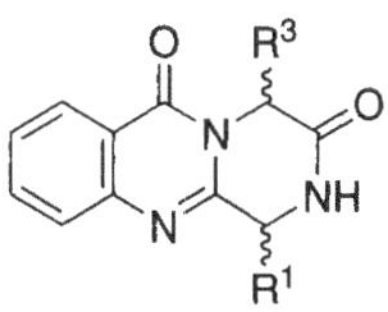

21: R^1 = H, R^2 = (*S*)-indol-3-ylmethylen
glyantrypine
22: R^1 = (*S*)-Me, R^2 = (*S*)-indol-3-ylmethylen
fumiquinazoline F
23: R^1 = (*R*)-Me, R^2 = (*S*)-indol-3-ylmethylene
fumiquinazoline G

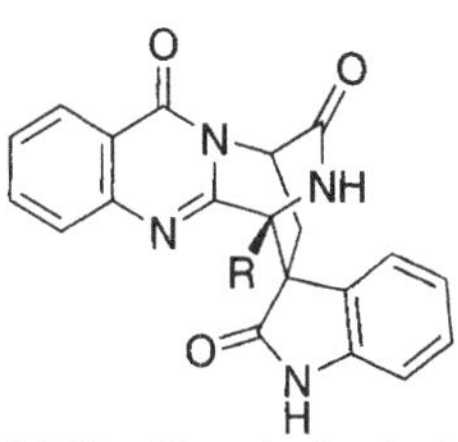

24: R = Me alantrypinoine
25: R = CH_2OH serantrypinone

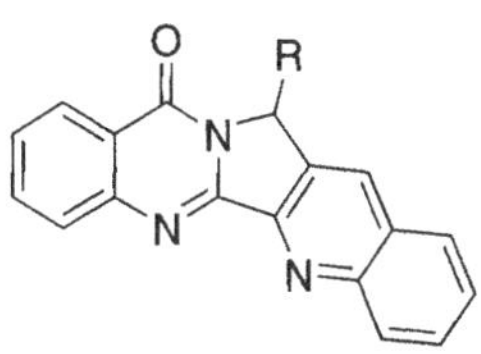

26: R = H luotonin A
27: R = OH luotonin B
28: R = OMe luotonin E

Fig. 2 Examples of quinazoline natural products

Chinese medicine Wu-Chu-Yu and interesting pharmacological properties of **16** have been found [66]. The indolopyridoquinazoline skeleton has been found also in various sources as the congeners such as 1,2-dihydroxy [67], 2-methoxy [68], 3-hydroxy [69], 7-hydroxy [70], and 7,8-dihydroxy rutaecarpines, etc. [71]. Indolo[2,1-*b*]quinazoline-6,12-dione known as the antibiotic tryptanthrin (**17**) has also a long history [72], and it was isolated from various sources [73] including *Strobilanthes Crusia O. Kuntze*, used as a traditional remedy for dermatophytic infection such as athlete's foot in Okinawa [74], and yet new interesting biological activities such as inhibition of HGF (hepatocyte growth factor) production have been found [75].

The structural diversity of fungal quinazolines has been broadened with the discovery of asperlicin (**20**) along with asperlicins B, C, D, and E, produced by *Aspergillus alliaceus*, which is a potent cholecystokinin (CCK) antagonists [76–79]. A series of new quinazoline alkaloids fused with benzodiazepinone were also isolated from a fungus culture of *Penicillium* sp., wherein benzomalvin A (**18**) is prototypical member [80, 81]. The simplest sclerotigenin (**19**) was recently isolated from organic extracts of sclerotia of *Penecillium sclerotigenum* (NRRIL,3461) [82], and many other benzodiazepinoquinazoline alkaloids such as circumdatin A–G were also isolated from the fungus *Aspergillus ochraceus* [83–85]. The quinazoline alkaloid hinckdentine A isolated from the marine bryozoan *Hincksinoflustra denticulate* collected from the eastern coast of Tasmania contains a novel hexahydroazepino[4′,5′:2,3]indolo[1,2-*c*]quinazoline ring system providing a challenging synthetic target [86]. A group of quinazoline alkaloids involving the pyrazino[2,1-*b*]quinazoline-3,6-dione substructure was found also in fungal metabolites. For example, the simplest member glyantrypine (**21**) was isolated from *Aspergillus clavatus* [87]. Fumiquinazolines A–E, F (**22**) and G (**23**) were isolated from a strain of *Aspergillus fumigatus* found in *Pseudolabrus* marine fish [88, 89]. Fumiquinazolines H and I were isolated from a fungus (*Acremonium* sp.) [90]. Fiscalines A–C were also isolated from *Neosartorya fischeri* [91, 92]. Novel spiro-type pyrazinoquinazolines such as spiroquinazoline from *Aspergillus flavipes* [93], alantrypinone (**24**) [94], and serantrypinone (**25**) [95], both from *Penicillium thymicola* were isolated. *N*-Acetylardeemin [96] having the pyrazinoquinazoline substructure isolated from *Aspergillus fischeri* is known as one of the most potent MDR (multidrug resistance) inhibitor [97, 98]. These alkaloids often exhibit very interesting biological properties and have drawn considerable interest of synthetic chemists as discussed in Sect. 3. Several new pyrroloquinazolinoquinoline alkaloids luotonin A (**26**), B (**27**), and E (**28**), etc. featuring a unique structure with a quinoline and a quinazoline fused together have been isolated from the aerial parts of *Peganum nigellastrum* (Zygophyllaceae) [99, 100]. Among them **26** has triggered considerable synthetic works due to its biological activity such as topoisomerase I inhibitor [101] like antitumor camptothecin [102] and has been reviewed recently [29]. Details are given in Sect. 3.5.

Studies on quinazoline compounds and quinazoline natural products have a long history as above, however, remarkable progress of synthetic methodology applicable to synthesis of quinazoline alkaloids and related molecules has been attained during the last decade. In this review article, the topical synthetic methodologies such as, for examples, iminophosphorane mediated synthesis (aza-Wittig methodology), use of organometallic reagents, microwave assisted synthesis, and solid phase synthesis, etc. will be discussed retrospectively in Sect. 2. Some selected examples of quinazoline alkaloids synthesis by these methodologies will be discussed in Sect. 3. On the other hand, quinazoline skeleton has been utilized very elaborately for the designing of unique aziridination reagent [103, 104], quinazolinap ligands to use as the asymmetric catalyst [105], heterocalixarenes [106] and electroluminescent devices [107], etc., however, these topics are not discussed in this article due to limitations of space.

2
Progress in Synthetic Methodology of Quinazolines

The topical synthetic methodologies such as iminophosphorane mediated synthesis (aza-Wittig methodology), microwave-assisted synthesis, solid-phase synthesis, and application of organometallic reagents, etc., will be discussed retrospectively focusing on the pathways to quinazoline, quinazoline-4-one and their derivatives.

2.1
Aza-Wittig Methodology in Quinazoline Synthesis (Iminophosphorane Mediated Synthesis)

The utilization of the aza-Wittig method for the synthesis of heterocyclic natural products is spreading out to a variety of nitrogen heterocycles from simple alkaloids to complex functionalized natural products [108–111]. In this review, emphasis will be placed on our results in this area, and in addition on related areas. Topics to be covered include recent progress of quinazoline syntheses by the aza-Wittig method (i.e., iminophosphorane mediated synthesis) focusing on the tandem Staudinger/intramolecular aza-Wittig reaction including quinazolinone annelation, chemoselectivity in the cyclization. Application to synthesis of quinazoline natural products such as benzomalvins, and some related examples based on the so-called Eguchi aza-Wittig protocol will be discussed in Sect. 3.

2.1.1 The Aza–Wittig Reaction

The aza–Wittig reaction [112–117] is the nitrogen analog of the Wittig olefination process and involves the reaction of an iminophosphorane [118–121] with a carbonyl group. The reaction provides an excellent method for the construction of carbon-nitrogen double bonds via inter and intramolecular aza–Wittig reaction.

The reaction is useful in the synthesis of acyclic imines [122–124] and heterocumulenes [112–117] and in the intramolecular formation of carbon-nitrogen double bonds in heterocyclic synthesis [112–117]. On the other hand aza–Wittig type reactions of iminophosphoranes with carbon dioxides, carbon disulphides, isocyanates, isothiocyanates and ketenes render access to functionalized heterocumulenes as highly reactive intermediates able to undergo a plethora of heterocyclization reactions [112–117].

Stability, basicity, and nucleophilicity of iminophosphoranes are mainly determined by the substituents at the nitrogen atom. General reactivity trends of *N*- and *P*-substituted series (R′ and R″, respectively) are summarized as follows [108–110, 112, 118–121]:

$$R'N = PPh_3 \gg ArN = PPh_3 \gg R'CON = PPh_3 \sim R'SO_2N = PPh_3$$
$$R'N = P(R'')_3 > R'N = P(Ar)_3 > R'N = P(OR'')_3 > R'N = P(OPh)_3$$

Carbonyl groups of aldehydes, ketones, acid halides, and heterocumulenes are generally reactive [108–110, 112–117]. In the intramolecular version, amide, imide, and ester carbonyl groups are also reactive giving rise to iminocyclization (see Sect. 2.1.3) [112–117].

2.1.2 Synthesis of Quinazoline Skeleton by the Aza–Wittig Reaction Followed by Various Cyclizations

The intermolecular version has been utilized for various heterocyclic syntheses including quinazoline ring via initial imine formation followed by electrocyclization, cycloaddition, and nucleophilic cyclization, etc. as developed by Molina (tandem aza–Wittig/cyclization strategy) [115], Saito (aza–Wittig/electrocyclization method) [125, 126] and Wamhoff (three-component reaction) [112]. These general principles include pyrimido annulation as summarized in Scheme 1. Following these general strategies, a number of quinazoline derivatives such as dihydroquinazolines (**29a,b**) [126], (**30a–c**) [127], quinazolin-4-one derivatives (**31a,b**) [128], (**32a,b,c**) [128, 129], (**32a**) [130, 131], and **33** [132–135] are synthesized starting from anthranilic acid derivatives or the corresponding iminophosphorane or the pyrimidine derivative (I) (Fig. 3).

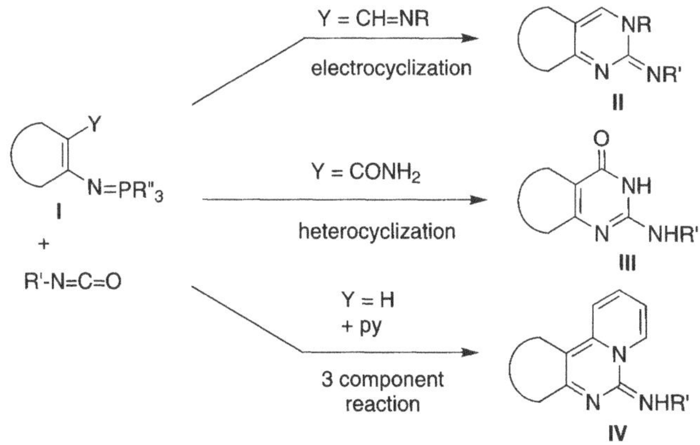

Scheme 1 Principles of tandem aza-Wittig/heterocyclizations

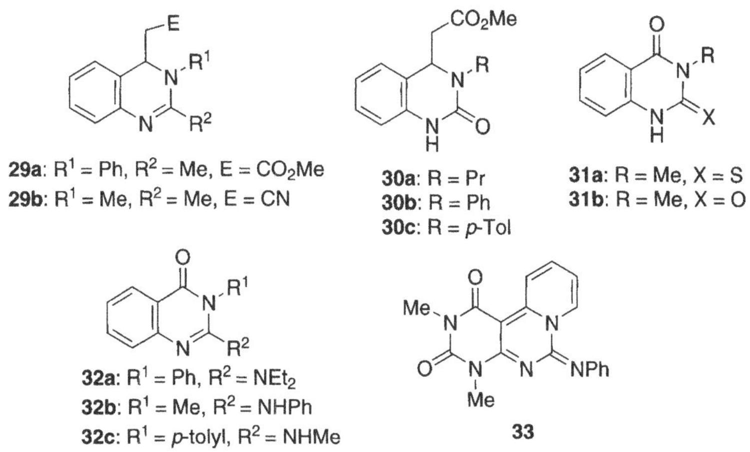

Fig. 3 Examples of quinazolines prepared by tandem aza-Wittig/electrocyclization, heterocyclization, three-component reaction

Rossi et al. have utilized the tandem aza-Wittig/electrocyclization principle for synthesis of quinazoline ring (**35**) and (**36**) starting from *N*-imidoyl iminophosphorane (**34**) [136, 137] (Scheme 2). Other unique synthetic strategies with *N*-vinyliminophosphoranes by Nitta [138], Palacios [139], and benzotriazolyl derivatives by Katrizky [140] have also been developed demonstrating the maturity and excellent prospects of iminophosphorane-mediated syntheses.

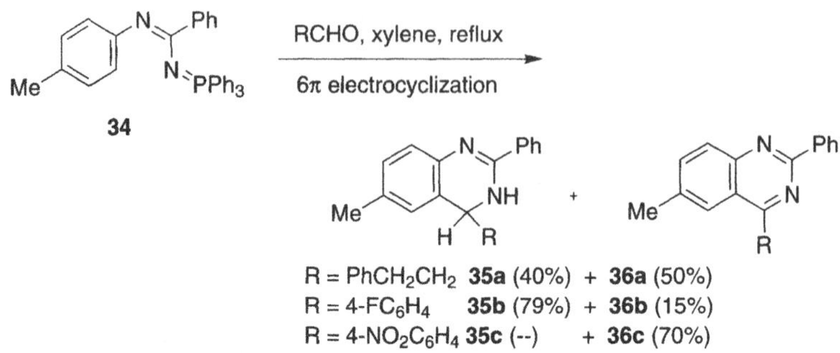

Scheme 2 Example of tandem aza-Wittig/electrocyclization [136, 137]

2.1.3 The Intramolecular Aza-Wittig Reaction

The intramolecular version of the aza-Wittig reaction provides a direct route to heterocyclic ring (Scheme 3). The reactivity depends on substituents Y and R′ at the carbonyl group, and on Z and R″ at the iminophosphorane; thus, a useful application has to consider the followings:

1. the ring size (formation of 5–7-membered rings ≫ 4-membered ring),
2. the carbonyl reactivity (COR, COAr, RCOOR, RCONRCOR ≫ RNCOR),
3. the substituents on P ($PR''_3 > P(Ar)_3 > P(OR'')_3$),
4. the substituent on N (CH_2, Ar, C = C, CO),
5. the ring strain: OS value ≤ 20 kcal/mol (the difference of heat of formation between unsaturated and saturated analogs) [113, 114].

The ring closure to a non-cumulated sulfoxide via an intramolecular aza-Wittig type reaction to construct S = N linkage has been reported recently by Hemming et al. [141].

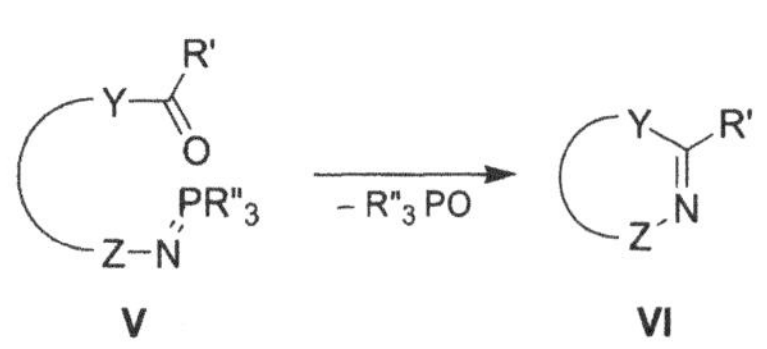

Scheme 3 Principle of intramolecular aza-Wittig method

2.1.4 Quinazolinone Annelation

We had interest in utilizing 2-phosphoranylideneamino-benzoyl derivatives as building blocks, particularly in view of anthranilic acids as important biological precursors of various alkaloids such as glomerine, vascine, and microbial products like tryptanthrin and anthramycine [54].

Thus, acylation of *N*-methylamides **38** with 2-azidobenzoyl chloride **37** (readily available from 2-azidobenzoic acid [142]) forms imides **39a–c**, which upon treatment with triphenylphosphine (TPP) in the course of consecutive Staudinger reaction/intramolecular aza–Wittig reaction yield exclusively 3-methylquinazolin-4(3*H*)-ones **40a–c** quantitatively (Scheme 4) [143, 144]. This procedure provides simple and efficient quinazolinone annelation of amides and lactams.

The intramolecular iminocyclization of *N*-(2-azidomethylphenyl)succinimide **41a** and -glutarimide **41b** proceeds cleanly with TPP to afford fused quinazolines **42a** and **42b**, respectively, in high yields (Scheme 5) [145]. On the other hand, the iminocyclization of **43b** to **44b** requires the use of TBP and heating at 140 °C. Reduction of the nitro group of **44b** affords antitumor agent batracylin (NSC-320846) **44c** (Scheme 5) [146].

A facile route to benzoannelated macrocyclic 1,4-diamines has been developed via quinazolinone annelation of lactams, followed by reductive ring

Scheme 4 Quinazolinone annelation method [143–145]

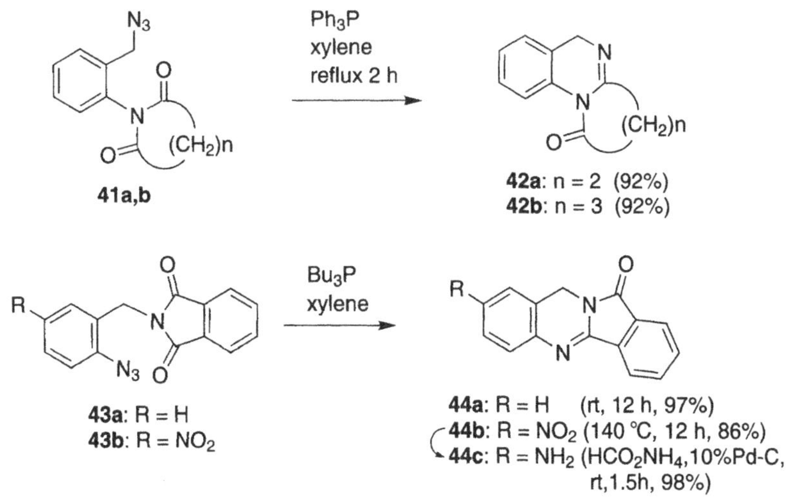

Scheme 5 Imination of imidecarbonyl, synthesis of batracylin [146]

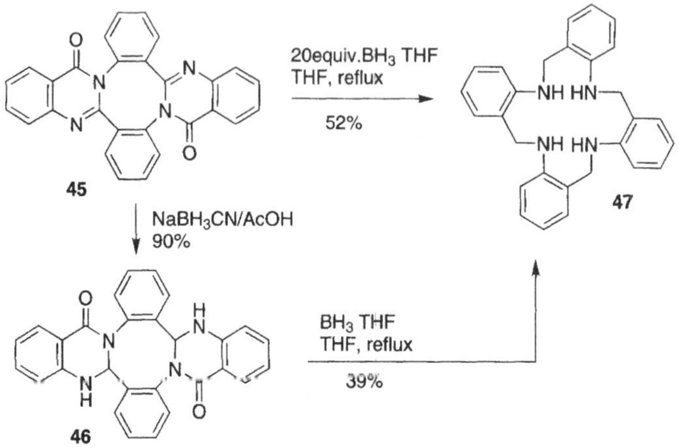

Scheme 6 Bisquinazolinone annelation and reductive ring enlargement [147, 148]

enlargement [147, 148]. Synthesis of 16-membered tetrabenzotetraazamacrocycle **47** is an example of molecular design based on quinazoline quinazolinone annelation. Bisquinazolinone annelation of dianthranilide gave bisquinazolinone **45** (82%), which was converted to **47** via reductive ring enlargement with $BH_3 \cdot THF$, and to dihydroquinazoline derivative **46** with $NaBH_4CN$ reduction (Scheme 6).

Examples of quinazoline alkaloids synthesis by the quinazolinone annelation method are given in Sect. 3.

2.2 Microwave-Assisted Synthesis of Quinazoline Compounds

Microwave-assisted organic synthesis is becoming popular with organic chemists and comprehensive reviews are available in recent year [149, 150]. Microwave heating is very convenient to use in organic synthesis. The heating is instantaneous, very specific, and there is no contact required between the energy source and the reaction vessel. Recent interest has been focused on "dry media" synthesis and particularly on solvent-free procedure using various mineral oxides [151, 152] and solvent-less reactions with neat reactants in the absence of a catalyst or solid support [153]. Furthermore, the diversity generating potential of multi-component reactions (MCRs) has been recognized and their utility in preparing libraries to screen for functional molecules is well appreciated. Consequently, the design of novel MCRs is an important field of research [154, 155]. In this section some selected literature examples of quinazoline synthesis by these methodologies is discussed.

2.2.1
Microwave-Assisted Niementowski Quinazoline Synthesis

The striking improvement in the Niementowski quinazoline synthesis [156, 157] has been fulfilled using microwave irradiation by Besson et al. [158, 159] (Scheme 7). Using microwave irradiation and/or Appel's salt, new efficient routes to various substituted and fused quinazolines have been developed by Besson et al. [158, 159] and also by others [160, 161].

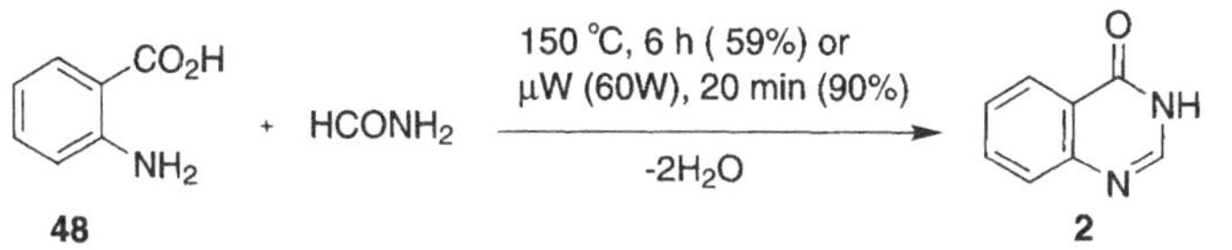

Scheme 7 Microwave-assisted Niementowski quinazoline synthesis [158, 159]

2.2.2
Microwave-Assisted Multi-Component One-Pot Synthesis of Quinazolines

One-pot synthesis of 4(3*H*)-quinazolinones from **48**, amines, and formic acid (or orthoesters) was developed by Rad-Moghadam et al. in 1998 [162], and recently more detailed procedures using inorganic solid support and neat one-pot procedure under microwave irradiation have been developed by Dandia et al. [163] (Scheme 8), and Liu et al. [164]. Also facile one-pot synthesis of 2,4(1*H*,3*H*)-quinazolinediones has been developed recently as a green chemical procedure by Nikpour et al. [165].

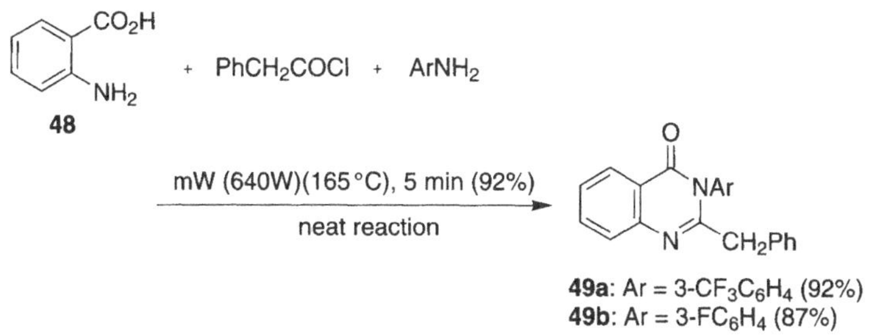

Scheme 8 One-pot synthesis of 4(3*H*)-quinazolinones [162–165]

2.3
Solid-Phase Synthesis of Quinazoline Compounds

The use of polymer-supported reagents in organic synthesis effectively reduces workup and purification to a simple filtration. Polymer-supported reagents can also be used in conjunction with the microwave heating, that has become an increasingly popular tool for combinatorial synthesis as well as green chemical synthesis in recent years [153, 166–168]. Solid-phase heterocyclic chemistry has been reviewed recently by Krchňák and Holladay [168]. Application of this methodology for synthesis of quinazoline compounds is increasing rapidly, however, only a few examples are discussed in this section (see also Sect. 3.4.2).

2.3.1
Polymer-Supported Aza–Wittig Reaction

A perfluoroalkyl-tagged triphenylphosphine and aminomethyl polystyrene based solid-phase bound phosphine were applied for the parallel synthesis of 4(3*H*)-quinazolinones (**2**) via the intramolecular aza–Wittig reaction by Bannwarth et al. [169]. Both reagents provide a library of **2** by a simple work-up procedure in good yields. A diverse library of benzodiazepine-quinazolinone alkaloids (the circumdatin family, cf 3.3) has been prepared by Thomas with a polymer supported phosphine-mediated intramolecular aza–Wittig reaction as the key step, a novel modified Eguchi aza–Wittig protocol (Scheme 9) [170]. The multi-arrayed library generation strategy starts out from readily accessible benzodiazepinediones **50** and anthranilic acids, and all library members were purified by preparative reversed-phase HPLC, yielding 283 isolated pure products **51a,b** and **52a,b** from 384 individual reactions.

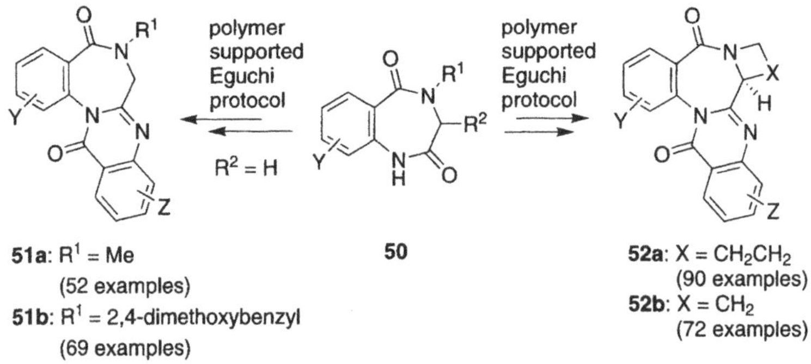

Scheme 9 Benzodiazepine-quinazoline alkaloids library by polymer supported phosphine-mediated aza–Wittig reaction [170]

2.3.2
Solid-Phase Synthesis of Quinazoline Derivatives

To date, a number of solid-phase syntheses of 4(3*H*)-quinazolinones [171], quinazoline-2,4-diones including monothioxo analogs [172, 173], quinazolines [174, 175], and dihydroquinazolines [176–178] have been developed. As a recent example, synthesis of 2-arylamino-3,4-dihydroquinazolines (**56**) by

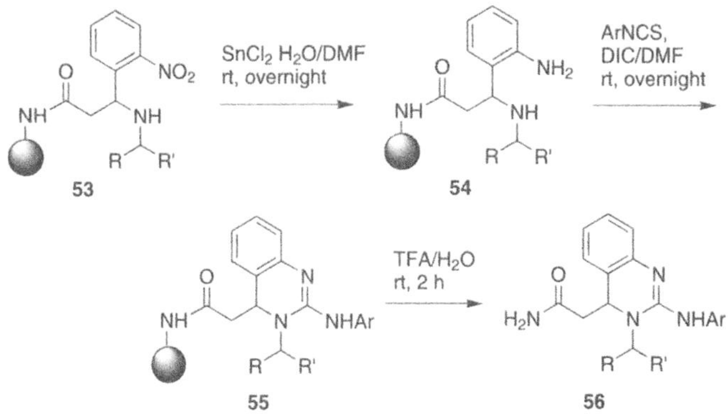

Scheme 10 Solid-phase synthesis of 2-arylamino-3,4-dihydroquinazolines [177]

Lam et al. is outlined partly in Scheme 10 [177]. The nitro group of resin-bound intermediate **53** was reduced to 1,3-deiamine **54** which gave **55** by carbodiimide-mediated cyclization. Final products **56** (19 examples) were obtained after TFA cleavage followed by HPLC in 46–83% yields with 72–94% purity.

2.4 Other New Synthesis of Quinazoline Derivatives

Among a number of synthetic reactions of quinazoline derivatives [2–4, 19–32], some selected recent examples are discussed below briefly.

2.4.1 Quinazoline Synthesis by Use of Organometallic Reagents

Useful routes to various quinazolines from 2-aminobenzonitrile **57** using organometallic reagents have been developed by Bergman et al. [179, 180] (Scheme 11). Synthesis of quinazoline derivatives by direct lithiation method

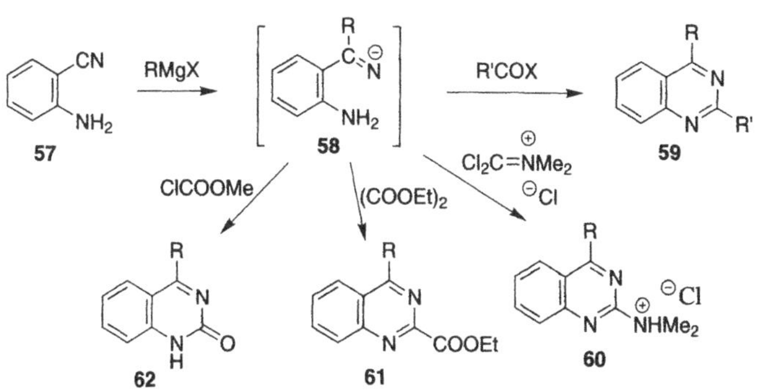

Scheme 11 Quinazoline derivatives from 2-aminobenzonitrile using Grignard reagents [179, 180]

has been reviewed [3], and was utilized for synthesis of luotonin A (see Sect. 3.5).

2.4.2
Metal Catalyzed Synthesis of Quinazoline Derivatives

Since Watanabe's synthesis of 4(3*H*)-quinazolinones in 1993 via transition-metal catalyzed reductive *N*-heterocyclization [181], several catalytic methods for quinazoline synthesis have been developed [182–186]. For example, palladium-catalyzed cyclocarbonylations of halides with appropriate reactants provided regioselective synthesis of 4(3*H*)-quinazolinone derivatives [182] and indoloquinazolines [184]. Also selenium-catalyzed reductive *N*-heterocyclization to quinazolinones has been developed by Sonoda et al. [183]. Copper-catalyzed heteroannulation with alkynes has been developed as highly region- and stereoselective route to 2-(2-arylvinyl)-1,2,3,4-tetrahydroquinazolin-4-ones **64** by Kundu et al. [185] (Scheme 12). Recently, condensation of anthranylamide with various aldehydes to 4-quinazolinones has been found to give excellent yields in the presence of cupric chloride [186].

For synthesis of quinazoline derivatives, various coupling reactions have been utilized after synthesis of quinazoline-2,4(1*H*,3*H*)-diones via palladium-catalyzed oxidative coupling by Hirota et al. [187]. For example, synthesis of diarylquinazolines by iron-catalyzed cross-coupling reaction [188], and diaminoquinazolinones by palladium-catalyzed amination [171] have been developed. Synthetic applications to quinazoline alkaloids are given in Sect. 3.

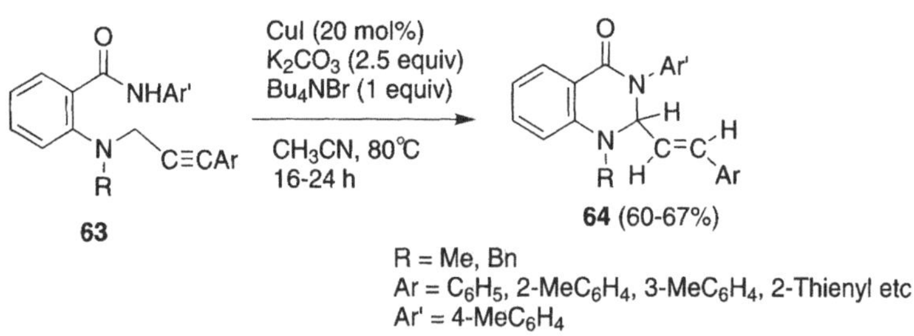

Scheme 12 Synthesis of 2-(2-arylvinyl)-1,2,3,4-tetrahydroquinazolin-4-ones [185]

2.4.3
Quinazoline Synthesis by Intramolecular Nucleophilic Heterocyclization Reactions

A new efficient synthesis of 3-arylquinazolin-4-ones **68** from anthranylamides **65** has been developed by Natsugari et al. (Scheme 13) [189]. The cyclization of **66** by method A based on Mazurkiewicz–Ganesan cyclization [190–194]

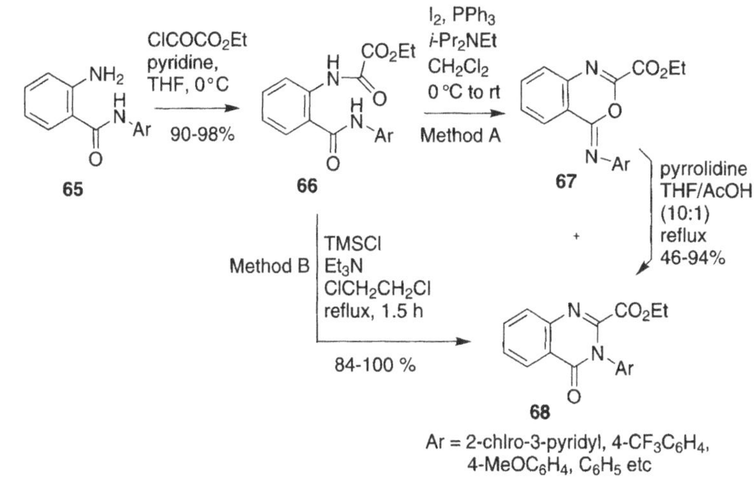

Scheme 13 Intramolecular heterocyclization route to 3-aryl-4(3*H*)-quinazolinones [189]

gave **68** in 46–94% yields after two-step process as shown but a tedious chromatographic separation was required. Method B based on the protocol in a patent literature [195] using modified conditions provided more efficient and convenient route to **68** for improved yields and easier work-up. The ester group of **68** was converted to the corresponding carboxamides. Other unique quinazoline syntheses by defluorinative nucleophilic cyclizations have been reported by Hynes et al. [196], Kotsuki et al. [197], and by defluorinative rearrangement by Uneyama et al. [198]. Synthesis of aminoquinazolines by heterocyclization was developed also by several groups [199–201].

2.4.4
Quinazoline Synthesis by Pericyclic Reactions

After Rossi's synthesis of quinazolines via 6π-electrocyclization (see Scheme 2), a few but unique quinazoline synthesis by electrocyclization were reported [202, 203]. On the other hand, rapid analogue synthetic methods of substituted 1,2,3-triazolo[1,5-*a*]quinazolines based on cycloaddition/condensation of 2-azidobenzoic acid or related compounds were developed by

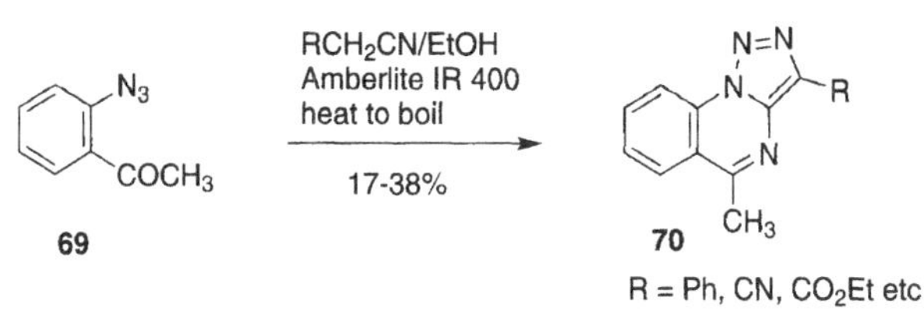

Scheme 14 Cycloaddition route to 1,2,3-triazolo[1,5-*a*]quinazoline [205]

Tennant [204], Smalley et al. [205], and Jones et al. [206]. For example, 1,2,3-triazolo[1,5-*a*]quinazolines **70** were obtained in moderate yields by base-catalyzed cycloaddition/condensation reaction (Scheme 14) [205]. Synthesis of 4(3*H*)-quinazolinones and -imines via 1,3-dipolar cycloaddition has been reported also [207, 208]. Novel formation of quinazolines via thermal ring contraction of 3*H*-1,4-benzodiazepines is found by Sashida et al. [209]. Synthesis of quinazoline alkaloids by cycloadditions is discussed in Sect. 3.

2.4.5 Other Notable Examples of Quinazoline Synthesis

Quinazoline synthesis using carbon dioxide under mild conditions has been developed by Mizuno et al. [210, 211]. The simple solvent free synthesis of quinazoline-2,4-diones under supercritical carbon dioxide and a catalytic amount of base provides an industrially benign approach and meets the challenges of green chemistry as previously discussed the microwave-assisted and/or solid-phase syntheses (see Sects. 2.2 and 2.3).

3 Synthesis of Quinazoline Alkaloids

As stated in Sect. 1, there are a number of reviews [19–29] and monographs [30–32] on the quinazolines and quinazoline alkaloids. Also the synthesis of heterocyclic natural products by the aza-Wittig method has been reviewed recently by Eguchi [108–110], and Fresneda and Molina have reviewed timely the application of iminophosphorane-based methodologies for the synthesis of natural products [111]. In this section, selected topics on synthesis of quinazoline alkaloids will be discussed retrospectively focusing on these synthetic methodologies.

3.1 Pyrroloquinazolinone Alkaloids

3.1.1 Vasicinone and Related Alkaloids

Application of quinazolinone annelation method (see Sect. 2.1.4) to pyrrolidinone **71** provides a facile synthesis of deoxyvacisinone **11** [143–145] (Scheme 15). Cyclization of 2-azidobenzoyl derivative **72** via iminophosphorane **73** to 2,3-dihydropyrrolo[2,1-*b*]quinazolin-9(1*H*)-one (**11**) proceeds more rapidly with tributylphosphine (TBP) than with TPP in accordance with the general reactivity trend; however, steric effects should also be considered as an important factor in these aza-Wittig reactions.

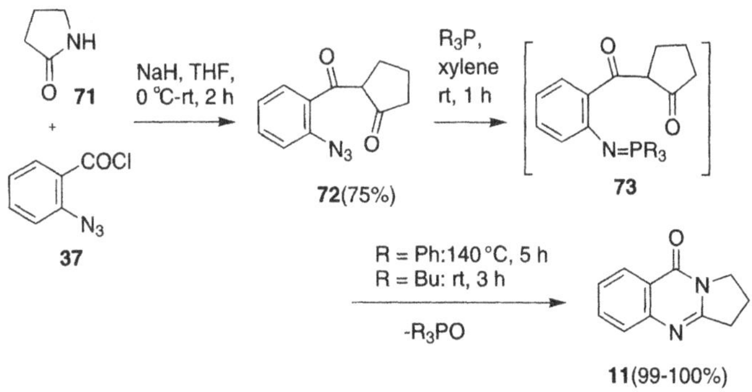

Scheme 15 Simple synthesis of deoxvasicinone [143, 144]

Quinazolinone annelation of the *O*-protected chiral pyrolidinone **74** (derived from L-aspartic acid) forms pyrrolo[2,1-*b*]quinazolin-9(1*H*)-one **75**; subsequent desilylation affords (*S*)-(−)-vasicinone **10**, which is identical with the natural *l*-product (Scheme 16) [212, 213]. Asymmetric oxidation of deoxyvasicinone **11** (via the imine enolate) with either (*R*)- or (*S*)-Davis' oxaziridine reagent (10-camphorsulfonyloxaziridine) [214, 215] provides a convenient route to both enantiomers, thus confirming the recently revised stereochemistry of natural vasicinone (Scheme 16) [212, 213]. Recently another approaches to optically active pyrrolo[2,1-*b*]quinazolinones **10** have been reported by Kamal et al. (lipase-catalyzed resolution) [56], and Argade et al. (asymmetric synthesis from (*S*)-acetoxysuccinic anhydride) [216]. One-pot synthesis of **11**, and related alkaloids has been also developed by utilizing microwave irradiation by Liu et al. [217]. Biogenetically patterned short-step synthesis of pyrroloquinazolinone alkaloids is well established by Onaka [218], and for many other synthesis, see the references cited in these papers.

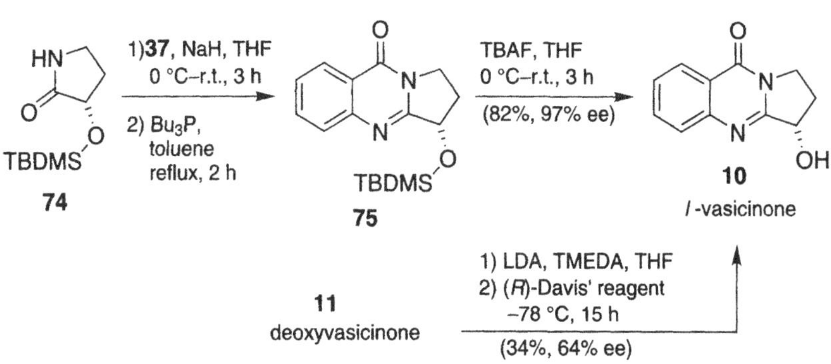

Scheme 16 Synthesis of (*S*)-(−)-vasicinone [212]

3.1.2 Chemoselectivity of Intramolecular Aza–Wittig Cyclizations

The chemoselectivity of bifunctional systems is important for the selective ring construction by intramolecular aza–Wittig reactions. The selective cyclization of methyl 1-(2-azidobenzoyl)-5-oxo-L-prolinate **76a** with TBP and TPP involves the ester carbonyl group rather than the imidoyl carbonyl group affording **77a** selectively; but both **77a** and **78a** are formed with triethylphosphite based on the product ratios **77**/**78** as determined by ^{1}H NMR spectra of the reaction mixture (Scheme 17, Table 1) [219, 220]. The isolated **77a** is sensitive to moisture and is converted quantitatively into **78a** by treatment with a catalytic amount of conc. HCl in THF at rt for 3 h. The imidocarbonyl group of amide **76b** is more reactive yielding **78b** exclusively. The product ratio of 7-membered rings **77a,b** versus 6-membered rings

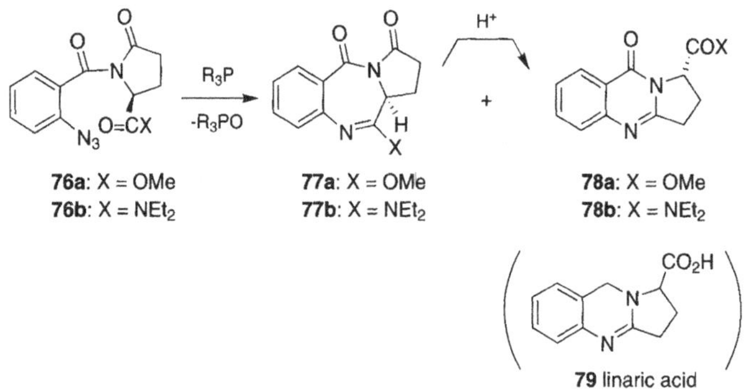

Scheme 17 Chemoselectivity of intramolecular aza–Wittig cyclization [219, 220]

Table 1 Chemoselectivity of the Staudinger/aza–Wittig tandem reaction of **76a,b** [219, 220]

Entry	**76**: X	R	Reaction conditions[a]	Yield [%][b]	**77**:**78**[c]
1	OMe	Bu	rt, 3 h	79	88:12
2	OMe	Ph	rt, 4 h	63	97:3
3	OMe	OEt	rt, 6 h	45	36:64
4	OMe	OEt	rt, 6 h; 80 °C, 12 h	69	36:64
5	NEt_2	Bu	rt, 3 h; 80 °C, 4 h	91	trace:> 99
6	NEt_2	Ph	rt, 4 h; 140 °C, 6.5 h	98	trace:> 99

[a] In benzene or xylene, 1.1 equivalent of R_3P
[b] Yields were determined after conversion into **78**
[c] Determined by ^{1}H NMR

78a,b depends on the carbonyl function and on the steric effect exerted by the substituents of the phosphorane reagents [144]: The bulkier TBP and TPP form predominantly **77a**; the 6-membered **78a** is formed preferentially with the smaller triethylphosphite. These results may be useful for the synthetic design of alkaloids and related heterocycles. It should be mentioned also that thus readily obtainable **78a,b** are the deoxyvasicinone derivatives, a quinazolinone analogue of recently isolated natural product linaric acid **79** [221].

3.2
Indolopyridoquinazoline- and Indoloquinazoline Alkaloids

3.2.1
Rutaecarpine

A successful simple application of the quinazolinone annelation is the short step synthesis of rutecarpine **16**, an indolopyridoquinazoline alkaloid of Evodia rutaecarpa, from 2,3,4,9-tetrahydro-1*H*-*β*-carbolin-1-one **80** via 2-azidobenzoyl derivative **81** (Scheme 18) [222]. Even after Kametani's practical iminoketene/amides condensation synthesis [227, 228], many other synthetic routes to **16** and related systems utilizing Fischer indole synthesis [223–225] or cyclocondensation [65, 226, 229, 230], etc. have been developed; among these syntheses, above quinazolinone annelation route is one of the most facile and concise one. Recently very concise Pd-assisted biaryl coupling route to **16** as well as to luotonin A (**26**) and B (**27**) has been developed by Harayama et al. [231, 232] (see Sect. 3.5).

3.2.2
Tryptanthrin

The simple synthesis of antimycotic alkaloid tryptanthrin **17** via the quinazolinone annelation is summarized also in Scheme 18. Quinazolinone annelation of commercially available isatin **84** under the shown conditions gave **17** in 43% overall yield (Scheme 18) [222]. As the second route, oxindole **82** was annelated to indolo[2,1-*b*]quinazolin-12(6*H*)-one **83**, which could be converted to **17** by air-oxidation but only in the modest yield [222]. Recently an efficient procedure for synthesis **17**, **16** and luotonin A (**26**), etc. has been reported using cyclocondensation of iminochloride and methylanthranylate, followed by further manipulations by Jahng et al. [229]. Several known attractive routes to **17** involves cyclocondensation of isatin with isatoic anhydride by Mitscher et al. [230], $(Ru)_3(CO)_{12}$-catalyzed reductive *N*-heterocyclization by Watanabe et al. [182, 223].

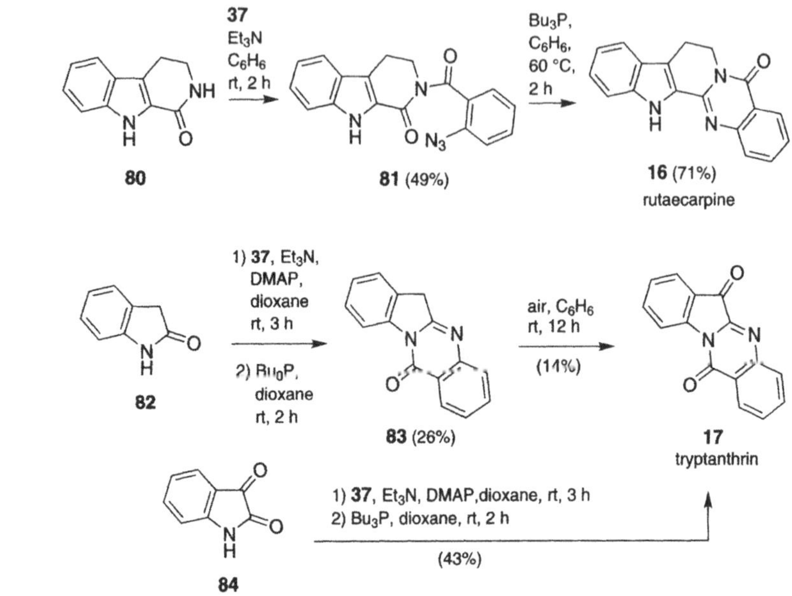

Scheme 18 Synthesis of rutaecarpine and tryptanthrin [222]

3.3 Quinazolinobenzodiazepine Alkaloids

A series of quinazoline alkaloids fused with 1,4-benzodiazepinone are isolated from different fungal species or their cultures. These alkaloids including benzolmalvine A (**18**) [80, 81], sclerotigenin (**19**) [82], circumdatin F [83–85], asperlicin (**20**) [76–79] and their analogues consist of two anthranilic acid and an amino acid forming a novel quinazolino [1,7]benzodiazepine ring systems. They show interesting biological effects [76–85] and their total syntheses have been studied actively by several groups. Selected examples will be discussed below.

3.3.1 (-)-Benzomalvin A

Eguchi et al. have successfully applied the intramolecular aza-Wittig method to the synthesis of the neurokinin receptor antagonists (–)-benzomavin A **18** and B **90**, which contain a 1,4-benzodiazepine ring and a 4-quinazolinone ring, and can be regarded as L-phenylalanine derivatives composed of two anthranilic acid moieties [233, 234]. This synthesis utilized the 1,4-benzodiazepine synthesis [235, 236] combined with the quinazolinone annelation reaction as explained in Scheme 19. Amino acid ester **85** provided (–)-benzomalvin A **18** with 99.7% ee (based on HPLC analysis) [237–239]. Bromi-

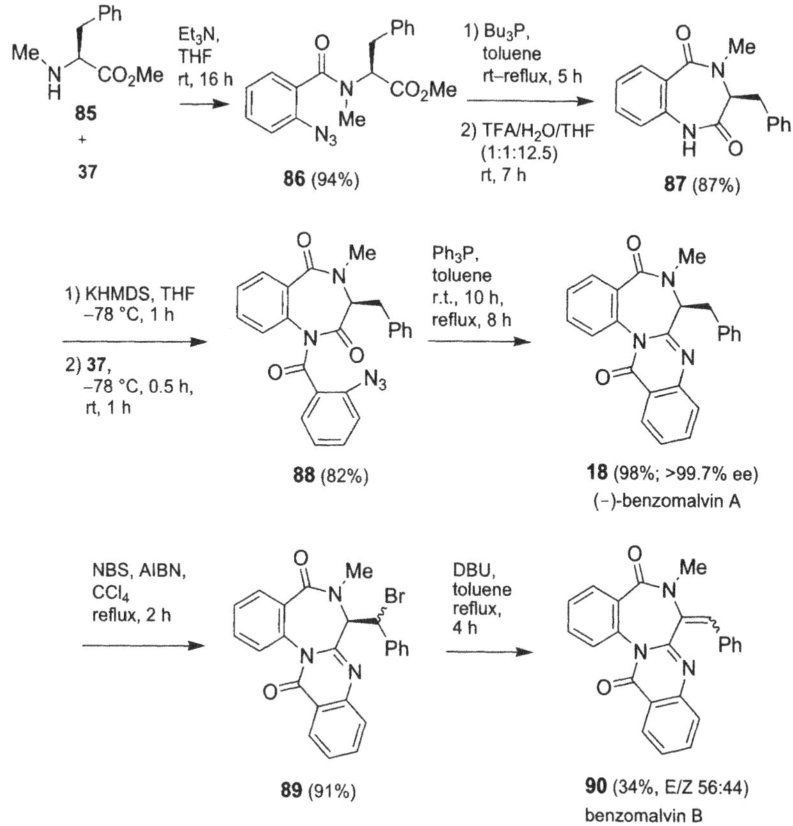

Scheme 19 Synthesis of benzomalvins A and B [233, 234]

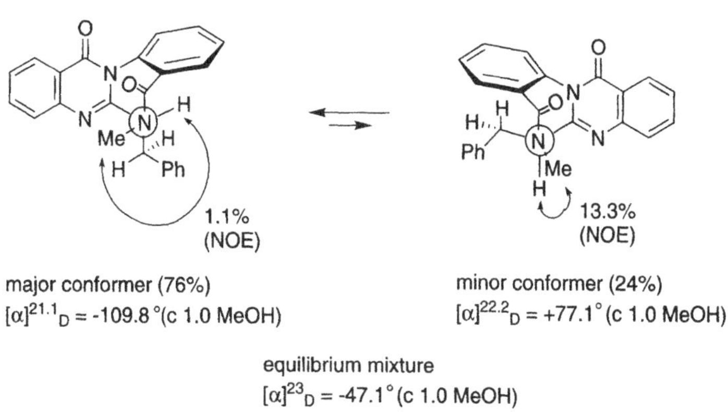

Fig. 4 Isomerization of (–)-benzomalvin A [234]

nation of **18**, followed by dehydrobromination gave benzomalvin B as an E/Z mixture.

(–)-Benzomalvin A was found to be very unstable in solution affording a 76 : 24 equilibrium mixture of major and minor conformers at 23 °C for 280 h in $CDCl_3$ as determined by ^{1}H NMR spectra (Fig. 4). The NOE measurements, and X-ray crystallographic analysis of the separable minor conformer, supported the assigned spatial proximity between *N*-Me and H-7 as shown. The minor conformer was shown to be (+)-invertmer based on the optical rotation [233, 234].

3.3.2 Asperlicins, Circumdatin F, and Sclerotigenin

The quinazolinone annelation procedure described above by acylation of an amide group with 2-azidobenzoylchloride (**37**) followed by the Staudinger/intramolecular aza-Wittig tandem reaction together with intramolecular aza-Wittig methods (see Sect. 2.2.2) has become known as the Eguchi aza-Wittig protocol [26, 111, 240] after its successful applications by Snider et al. to the synthesis of quinazolinobenzodiazepine alkaloids (Scheme 20) and the pyrazino[2,1-*b*]quinazoline scaffold in the fumiquinazoline alkaloids (see Sect. 3.4.1) and to the preparation of ardeemin by Danishefsky et al. (see Sect. 3.4.3). As shown in Scheme 20, this method has been utilized for the synthesis of asperlicin C (**93a**), circumdatin F (**93b**) and sclerotigenin (**19**) by selective acylation of the more acidic anilide nitrogen of benzodiazepinediones **91** without the need for protecting groups, followed by TBP-induced cyclization of the 1-(2-azidobenzoyl)-1,4-benzodiazepine-2,5-dione derivative **92** [241].

Utilizing the Eguchi aza-Wittig protocol Snider et al. achieved the total synthesis of the more complex antibiotic (–)-asperlicin **20**, known also for its potent cholecystokinin antagonist activity [76–79] (Scheme 21) [242]. L-Tryptophan-derived benzodiazepinedione **94** is converted into the fused quinazolinone **95**.

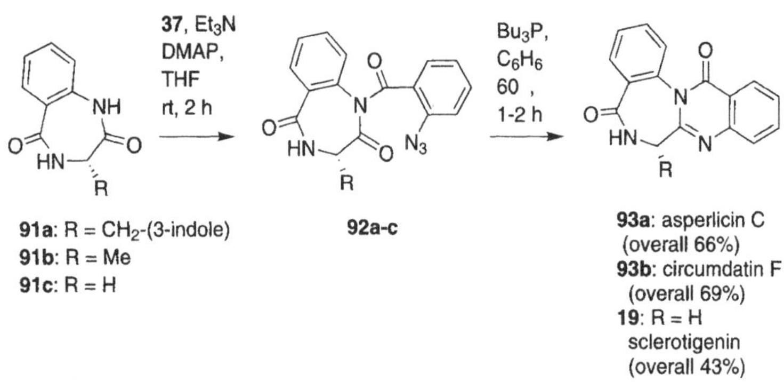

Scheme 20 Synthesis of asperlicin C, circumdatin F, and sclerotigenin [241]

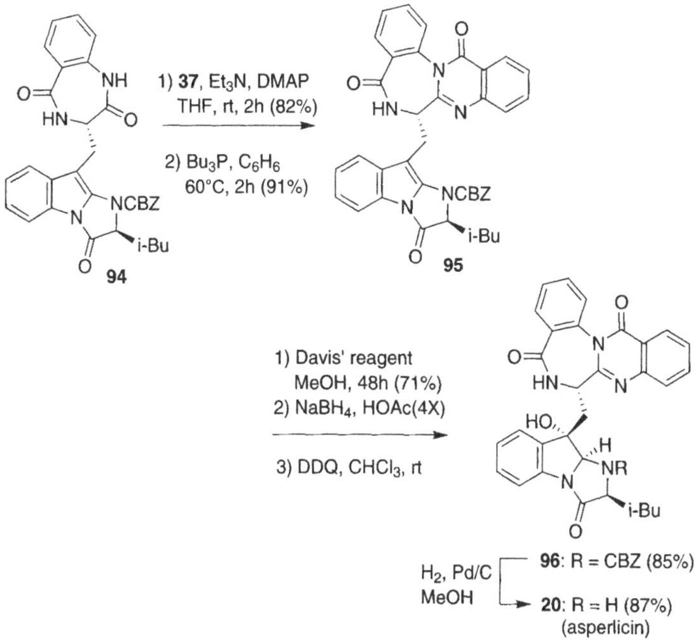

Scheme 21 Synthesis of asperlicin [242]

Epoxidation of **95** with Davis' oxaziridine reagent [214, 215], followed by $NaBH_4$ reduction and DDQ oxidation affords **96**, which upon deprotection by hydrogenolysis gives asperlicin **20** (15 steps from L-tryptophan with 8% overall yield).

Total synthesis of asperlicin C and E via condensation route by Bock et al. is well known [76–79, 243]. The first total synthesis of asperlicin D via both cyclodehydration and intramolecular aza-Wittig routes has been fulfilled very recently by Al-Said et al. [244], and the iminophosphorane intermediates having secondary amide protons have been shown to provide a one-step entry to quinazolino[1,4]benzodiazepine ring system like asperlicin D (a positional isomer of asperlicin C). A facile route to *N*-benzylsclerotigenin has been reported also by this group [245]. On the other hand, total synthesis of circumdatin F (**19**) (and circumdatin C) via dehydrative cyclization (or benzoxazine) route has been also fulfilled by Bergman et al. [246].

3.4 Pyrazinoquinazoline Alkaloids

Several groups of quinazoline alkaloids involving pyrazino[2,1-*b*]quinazoline-3,6-dione substructure are found in fungal metabolites. These include glyantrypine (**21**) [87], fumiquinazolines A–E, F (**22**), G (**23**) [88, 89], H and I [90], and fiscalines A–C [91, 92]. Novel spiro-type pyrazinoquinazolines such as spiroquinazoline [93], alantrypinone (**24**) [94], and serantrypinone (**25**) [95] are also included. *N*-Acetylardeemin [96] having the pyrazinoquinazoline substructure is known as one of the most potent MDR inhibitor [97, 98]. Like this example, these alkaloids often exhibit very interesting biological properties (see Sect. 1) and have drawn considerable interest of synthetic chemists. Recently comprehensive review on the chemistry of pyrazino[2,1-*b*]quinazoline-3,6-dione by Avendaño and Menéndez has appeared [20], and in this section, some of selected examples are discussed.

3.4.1
Fumiquinazolines, Glyantrypine, and Fiscaline B

Snider et al. has utilized the quinazolinone annelation method (see Sect. 2.2) for the first synthesis of (+)-fumiquinazoline G 23 [248] (the natural product is the (–)-enantiomer; however, a 3 : 2 equilibrium mixture is obtained with base from either fumiquinazoline F or G [89]). (+)-Fumiquinazoline G 23 is the simplest member of a quinazolinone fused to a piperazine-2,5-dione ring. Snider et al. has developed a general procedure for photochemically deprotectable *N*-(2-nitrobenzyl)piperazine-2,5-dione **97b**, which provides a short and efficient synthesis of fumiqunazoline G (**23**) (Scheme 22) [241]. Subsequently, more efficient synthesis of (–)-fumiquinazoline F (**22**) and (–)-glyantrypine (**21**) has been developed by Avendano, Menendenz, Söllhuber et al. (Scheme 22) [252–255]. The regioselective acylation of unprotected 3-arylmethylpiperazine-2,5-diones **97a** and **97c** (derived from methyl tryptophanate) with TMSCl, Et_3N, and 2-azidobenzoylchloride (**37**) is fol-

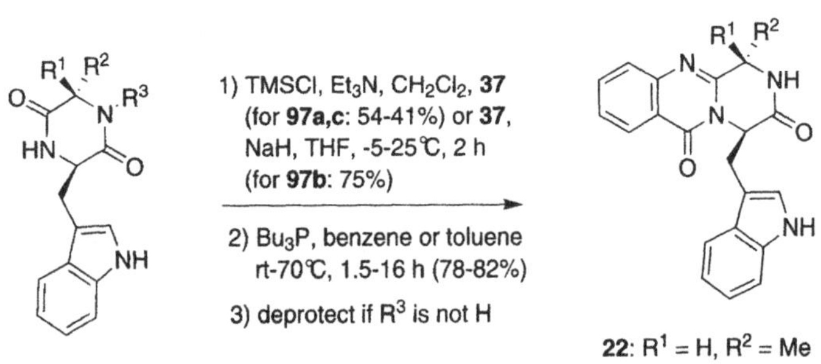

97a
R^1 = H, R^2 = Me, R^3 = H
97b
R^1 = Me, R^2 = H, R^3 = 2-nitrobenzyl
97c
R^1 = H, R^2 = H, R^3 = H

22: R^1 = H, R^2 = Me
(fumiquinazoline F)
(34% overall yield from **97a**)
23: R^1 = Me, R^2 = H
(fumiquinazoline G)
(49% overall yield from **97b**)
21: R^1 = H, R^2 = H
(glyantrypine)
(44% overall yield from **97c**)

Scheme 22 Synthesis of fumiquinazolines F, G, and glyantrypine [247–255]

lowed by the intramolecular Staudinger/aza-Wittig protocol (Scheme 22). The Staudinger/intramolecular aza-Wittig reaction proceeds cleanly affording **21–23** in good yields [252–255]. Fiscalin B [91, 92] was synthesized via direct alkylation of *N*-Boc-3-indolylmethyl bromide of 1-isopropyl-2,4-dihydro-1*H*-pyrazino[2,1-*b*]quinazoline-3,6-dione by this group [255].

On the other hand, new dehydrative cyclization route of tripeptides based on the Mazurkiewicz protocol [190–192] leading to fumiquinazolines F **22**, G **23**, and fiscalin B has been developed by Ganesan et al. [191, 192]. This Mazurkiewicz-Ganesan cyclization of starting tripeptides has been proven to proceed via the iminooxazine intermediate followed by rearrangements to the pyrazinoquinazolinones by Snider [250] and Hart [256, 257], and were used for synthesis of fumiquinazolines (A, B, I [249], or C, E, and H [250]), alantrypinone [256, 257] (see Sect. 3.4.4), and quinazolinobenzodiazepine alkaloids [244–246], etc. Wang, Ganesan, and Sim has applied this method to solid-phase synthesis of fumiquinazolines [258] and related alkalolids [259] (see Sect. 3.4.2).

3.4.2
Verrucines A, B, and Anacine

The first total solid-phase syntheses of verrucines A, B, and anacine [260–262] have been fulfilled utilizing the Mazurkiewicz–Ganesan cyclization by Wang and Sim (Scheme 23) [259]. Starting with Fmoc-L-Gly(Trt)-Sasrin-

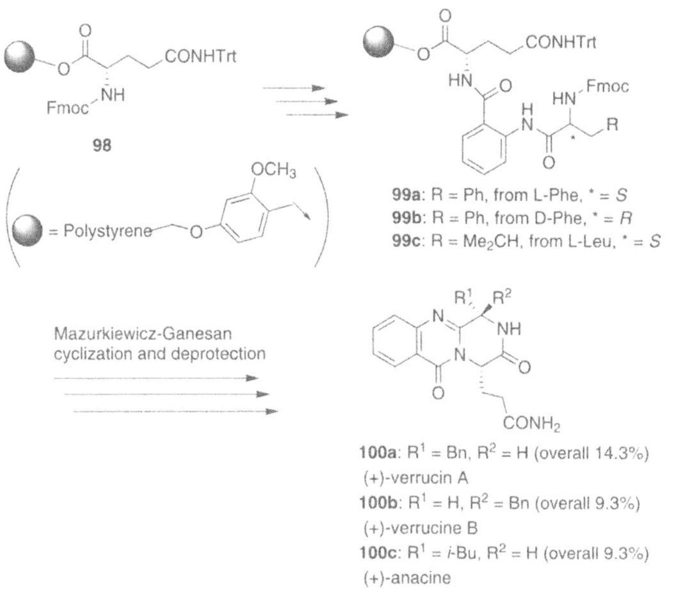

Scheme 23 Solid-phase total synthesis of verrucins A, B, and anacine [259]

resin (**98**), (+)-verrucines A (**100a**), B (**100b**), and (+)-anacine (**100c**) were synthesized in seven steps. These syntheses proved unambiguously that **100a** and **100b** are the derivatives of D-phenylalanine and L-glutamine, and **100c** and its natural diastereomer as 1,4-*syn*- and *anti*-quinazolines but not originally proposed benzodiazepines [261].

3.4.3 Ardeemin and *N*-Acetylardeemin

The quinazolinone annelation method has been applied to piperazinedione **101** derived from L-tryptophan methyl ester by Danishefsky et al. for the synthesis of ardeemin **102a** and *N*-acetylardeemin **102b** [96], one of the most potent known agents for reversal of multiple drug resistance (MDR) [97, 98] featuring a hexahydropyrrolo[2,3-*b*]indole scaffold substituted at the benzylic ring junction with the 1,1-dimethylallyl ("reverse-prenyl") group [263, 264] (Scheme 24). Acylation of **101** with KHMDS and **37** works well to give the corresponding imide in variable yields (50–80%); for technical reasons, on larger scale preparations the use of 2-azidobenzoic anhydride is preferred over the acid chloride **37** [264]. The Staudinger/intramolecular aza-Wittig reaction of the imide product with TBP gave aredeemin **102a**, which was acetylated to 5-*N*-acetylardeemin **102b** (9 steps with 12.5% overall yield from bis(Boc)tryptophan methyl ester). A seco analogue of ardeemin has been synthesized by Menéndez et al. also using the quinazolinone annelation method [265]. De-"prenyl" ardemeen has been synthesized also by Söllhuber et al. by stereoselective acid-promoted cyclization of 1,4-anti-4-alkyl-

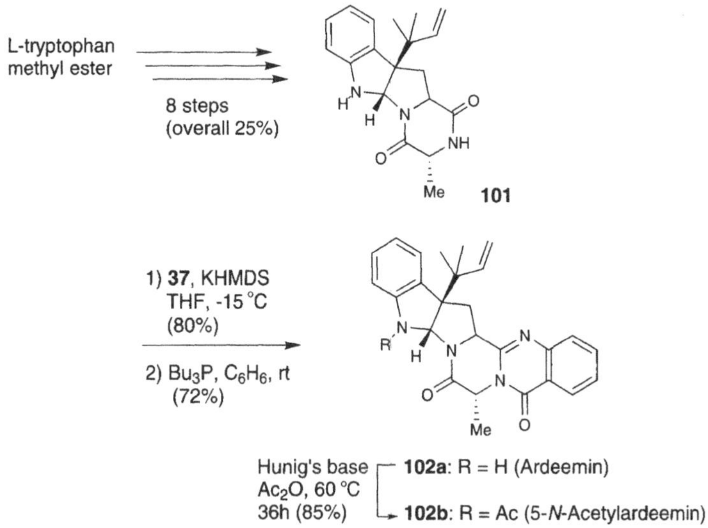

Scheme 24 Synthesis of ardeemine and *N*-acetylardeemine [264]

1-(3-indolylmethyl)pyrazino[2,1-*b*]quinazoline-3,6-diones in four steps 45% overall yield starting from *N*-2-aminobenzoyl-D-Ala methyl ester [266]. For straightforward synthesis of didehydro analogues of *N*-acetylardeemin from tryptamine is reported by Avendaño et al. [267].

3.4.4 Alantrypinone, etc.

Novel spiro-type pyrazinoquinazoline alkaloids such as spiroquinazoline from *Aspergillus flavipes* [93], (+)-alantrypinone (**24**) [94], and (–)-serantrypinone (**25**) [95], both from *Penicillium thymicola* possess a tricyclic pyrazinoquinazolinedione base bridged by a 3-methyleneoxindole substructure. The unusual structural feature and biological property [93] lead to efforts of two groups toward their total synthesis. Hart et al. fulfilled total synthesis of *ent*-**24** via *N*-acyliminium ion cyclization that converts enamide **103** [255] to bridged indole **104** (Scheme 25) [256, 257]. Subsequent NBS-mediated oxidative rearrangement of the bridged indole **104** to the oxindole led to the spirocyclic structure of *ent*-**24** in 12% overall yield from isatoic anhydride by 10 steps. However, *ent*-17-epi-alantrypinone **105** was formed in 50 : 50 to 40 : 60 ratio to *ent*-**24** [256]. On the other hand, Kende et al. has fulfilled a concise total synthesis of (±)-alantrypinone **24** by unique hetero Diels–Alder route (Scheme 26) [268, 269].

The Diels–Alder reaction of azadiene **106** prepared from 1-methyl-2,4-dihydro-1*H*-pyrazino[2,1-*b*]quinazoline-3,6-dione [255] with 3-methylene-oxindole **107** proceeds smoothly in chloroform at rt to afford *exo*- (**108**) and *endo*-isomer (**109**) in 55 and 18% yields, respectively. Mild hydrolysis of

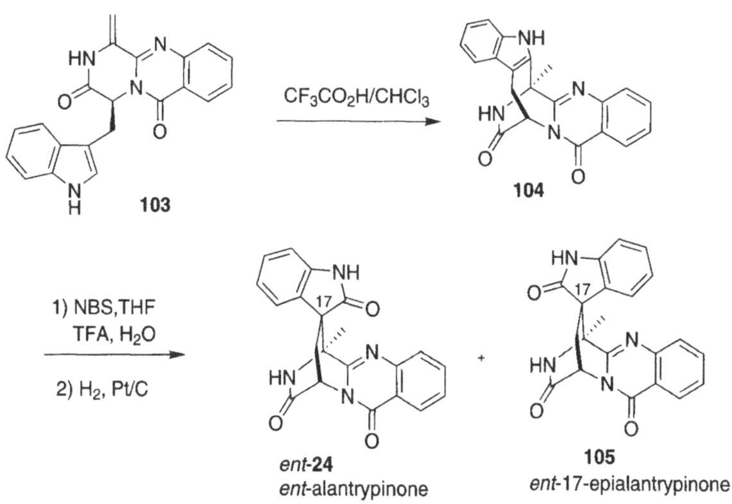

Scheme 25 Synthesis of ent-alantrypinone [256, 257]

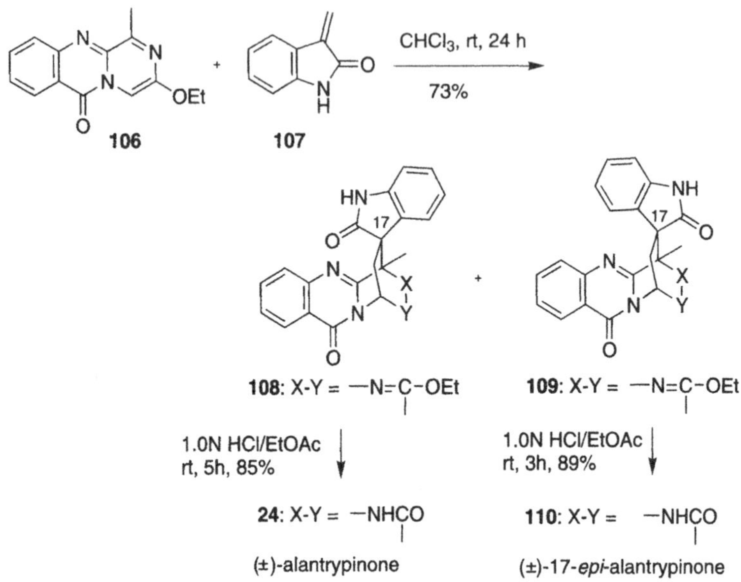

Scheme 26 Hetero Diels–Alder route to (±)-alantrypinone [268, 269]

these adducts gave (±)-alantrypinone **24** and (±)-17-*epi*-alantrypinone **110**, respectively (Scheme 26). Based on thermal equibration results in various conditions, the interesting epimerizations between **108** and **109**, and **24** and **110** are rationalized by an intramolecular rearrangement involving an anionic retro-Mannich reaction [268, 269]. (For synthesis of oxygen-briged spiro-type fumiquinazolines C, and H see [248, 250])

3.5 Pyrroloquinazolinoquinoline Alkaloids

3.5.1 Luotonins A, B, and E

Luotonin A (**26**), a novel pyrroloquinazolinoquinoline alkaloid isolated from the aerial parts of *Peganum nigellastrum Bunge* by Nomura et al. [99, 100], showed a cytotoxic activity against mouse leukemia cells (P-388) and an inhibitory activity against topoisomerase I and II [101]. The biological activity of luotonin A coupled with the structural similarity with the topoisomerase inhibitor camptothecin (**111** in Scheme 27) [102] have triggered a number of synthetic approaches to this alkaloid and of course also analogues thereof. Very comprehensive review on luotonin A by Ma et al. has appeared recently [29], and here, only selected examples of total synthesis will be discussed focusing on synthetic routes. A number of total syntheses of luotonin A reported heretofore are classified into three main routes by ring-construction strategies. Route I constructs B (or B, C) ring by Friedländer quinoline synthesis [270, 271], and or by intramolecular Povarov reaction [272] (Scheme 27). Route II constructs ring C by Heck coupling [273–275], and or by Mitsunobu cyclization [276] (Scheme 28), and route III constructs C, D (or D) rings by cycloadditions including Kametani quinazolinone synthesis by amide condensation [227] (Scheme 29).

Friedländer condensation of pyrrolo[2,1-*b*]quinazoline-3,9-dione (**112**) with anthranylaldehyde **114** gave luotonin A **26** in 36% yield (Scheme 27) [277–279].

Dione **112** was obtained from **10** (Jones oxidation, 56%), **11** (SeO_2 oxidation, 42%), and **113** (ozonolysis, 63%). One-pot synthesis of **26** from **10** is devised also [277]. Intramolecular hetero Diels–Alder (Povarov) approach has been developed by Twin and Batey [280]. Intramolecular Povarov reaction between aldehyde precursor **116** prepared from acetate **115** with aniline under the shown conditions afforded **26** in 51% yield presumably after in situ oxidation of the initially formed 1,2-dihydroquinoline (Scheme 27).

The route II approaches are summarized in Scheme 28. The very concise synthesis of luotonin A **26** as well as rutaecarpine **16** (see Sect. 3.2.1)

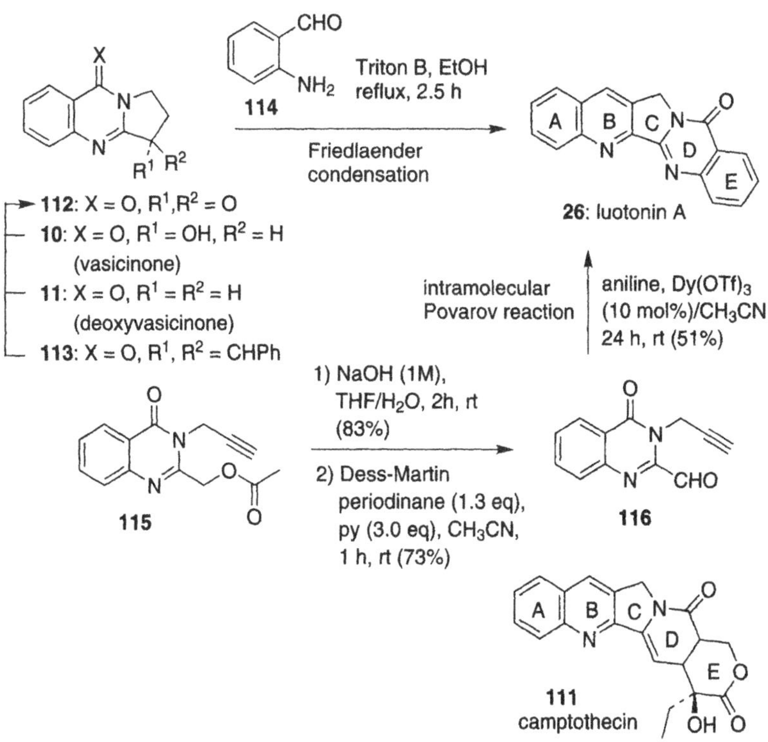

Scheme 27 Route I synthesis of luotonin A [277–280]

via a biaryl coupling using Heck reaction [273–275] has been developed by Harayama et al. [231, 232]. The required 3-[2-bromoquinolin-3-yl]-4(3*H*)-quinazolinone **118** was prepared by *N*-alkylation of **2** with the quinolyl bromide **117**, and the intramolecular biaryl coupling of **118** with the Pd reagent afforded **26** in good yields (Scheme 28). Synthesis of luotonin B **27** was fulfilled also by the NBS-mediated regioselective hydroxylation of **26** [231, 232]. On the other hand, Mhaske and Argade fulfilled total synthesis of luotonins A (**26**), B (**27**), and E (**29**) by utilizing regioselective quinazolinone-directed *ortho* lithiation of quinazolinoylquinoline **119** as the key step (Scheme 28) [281]. The reaction of dilithiated intermediate **120** with formaldehyde yielded the desired **121** in high yield. The Mitsunobu cyclization [276] of **121** furnished **26** in 95% yield. The reaction of **120** with DMF furnished **27** via intermediate **122**. The PCC oxidation of **121** fur-

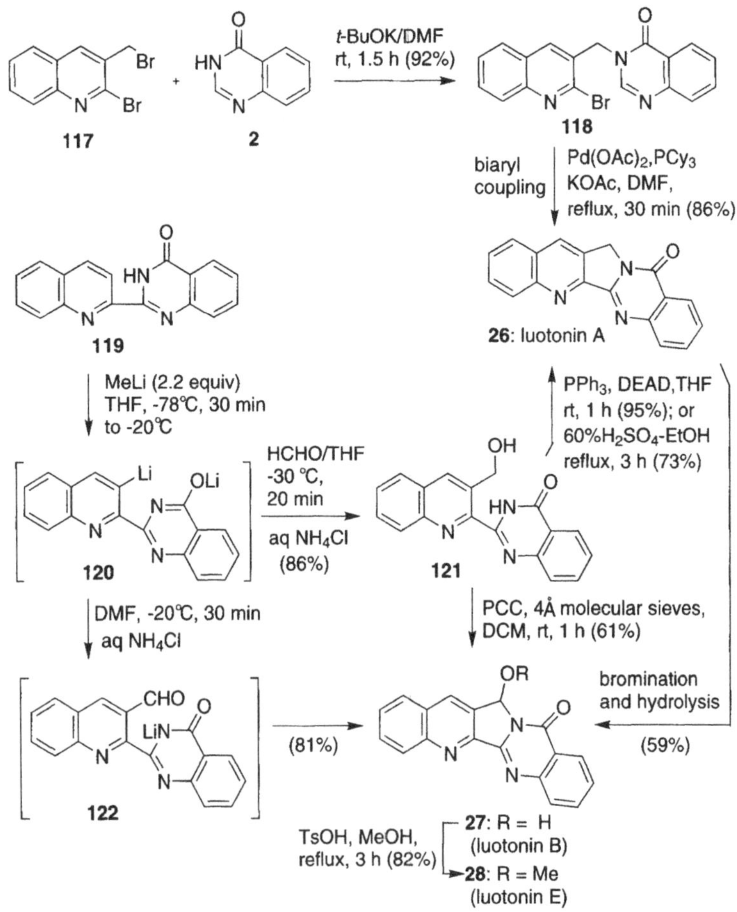

Scheme 28 Route II synthesis of luotonin A, B, and E [231, 232, 281, 282]

nished also **27**. On treatment with TsOH/MeOH, **27** was converted to luotonin E **28**. Similar synthesis of luotonins A, B, and E without use of lithiation has been reported also by Chavan and Sivappa [282]. They succeeded acid-catalyzed cyclodehydration of **121** to afford **26** without use of the Mitsunobu conditions that require laborious column chromatography to obtain pure luotonin A.

The route III involves an efficient total synthesis of luotonin A via intramolecular hetero Diels–Alder reactions by Ihara et al. [283, 284], and several syntheses of luotonin A utilizing well-known Kametani's quinazolinone synthesis via iminoketene cycloadditions or analogues as the final key-step by several groups [229, 285–288] (Scheme 29). In the hetero Diels—Alder route, a key precursor, *N*-(2-cyanoquinolin-3-ylmethyl)-2-methoxybenzamide (**122**) prepared from the corresponding 2-bromo derivative by palladium-catalyzed coupling with copper (I) cyanide underwent the intramolecular hetero Diels–Alder reaction via a putative intermediate **123** in the presence of chlorotrimethylsilane and zinc chloride to furnish luotonin A (Scheme 29). On the other hand, first total synthesis of luotonin A via

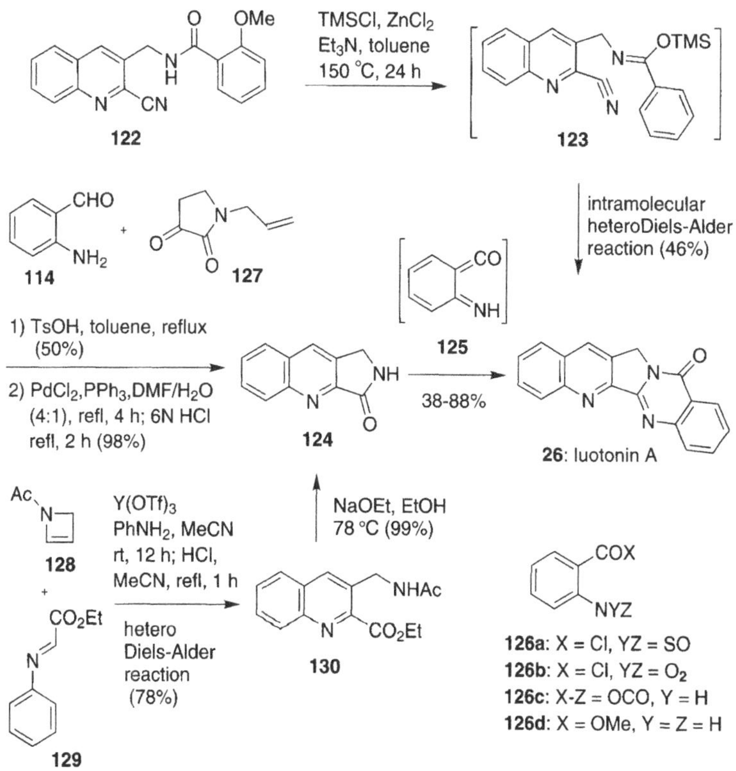

Scheme 29 Route III synthesis of luotonin A [283–288]

quinazolinone annelation of 3-oxopyrrolo[3,4-*b*]quinoline (**124**) has been reported by Ganesan et al. in 1998 [285]. They prepared required amide **124** by Danishefsky procedure [289] but the reaction with 2-sulfinylaminobenzoyl chloride (**126a**) under the Kametani's conditions [227, 228] was not successful due to the insolubility of **124**. Use of lithium bis(trimethylsilyl)amide for deprotonation of **124** allowed successful quinazolinone annelation with **126a** in THF at rt for 2 h to furnish **26** in 85% yield (Scheme 29) [285]. Dallavalle et al. developed one-pot, three-step synthesis of **26** from **124** [286]. They prepared **124** by Friedländer condensation of *N*-allylpyrrolidine-2,3-dione **127** with anthranylaldehyde **114**, followed by deprotection of allylic group after palladium-catalyzed isomerization to enamide function (Scheme 29). A one-pot sequence of acylation of **124** with 2-nitrobenzoylchloride (**126b**) (NaH, THF, 60 °C, 1 h), reduction of the nitro group (Fe, AcOH/EtOH, reflux, 2 h) and subsequent ring-closure led to **26** (38%) [286]. Mw-assisted rapid synthesis of **26** (85% yield) via the cyclocondensation of the amide **124** with isatoic anhydride **126c** has been fullfilled by Yadav et al. under solvent-free conditions [287]. Jahng et al. developed facile synthesis of **26** as well as rutaecarpine and tryptanthrin via cyclocondensation of iminochloride derived from lactams with methyl anthranylate **126d.** Application of this quinazolinone annelation for **124** furnished **26** in 88% yield [229]. A unique hetero Diels—Alder synthesis of the key intermediate **124** was developed by Stevenson et al. [288]. A Lewis acid catalyzed [4 + 2] cycloaddition reaction between azetin **128** and imine **129** yielded quinoline derivative **130**, which underwent base-catalyzed cyclization to give the amide **124** in a high yield (Scheme 29). For synthesis of luotonin A analogues and SAR studies, see the review of Ma et al. [29] and references cited therein.

3.6 Antimalarial Quinazoline Alkaloids Having a Piperidine Ring

3.6.1 Febrifugine and Isofebrifugine

Febrifugine (**14**) is the antimalarial alkaloid that is isolated from *Dichroa febrifuga* (Chinese name: Cháng Shan) or *Hydrangea umbellate* with isofebrifugine (**15**) [60–63]. Recently, Kobayashi et al. corrected the error in the absolute stereostructure of **14** and **15** as shown in Scheme 30. It is well known that **15** is transformed into **14** by heating [60]. Due to their attractive biological activity, a number of syntheses of these alkaloids have been reported to date [290–299]. In this section, some of selected topics will be discussed briefly.

Kobayashi et al. fulfilled first asymmetric synthesis of febrifugine utilizing tin(II)-catalyzed asymmetric aldol-type reaction and lanthanide-catalyzed

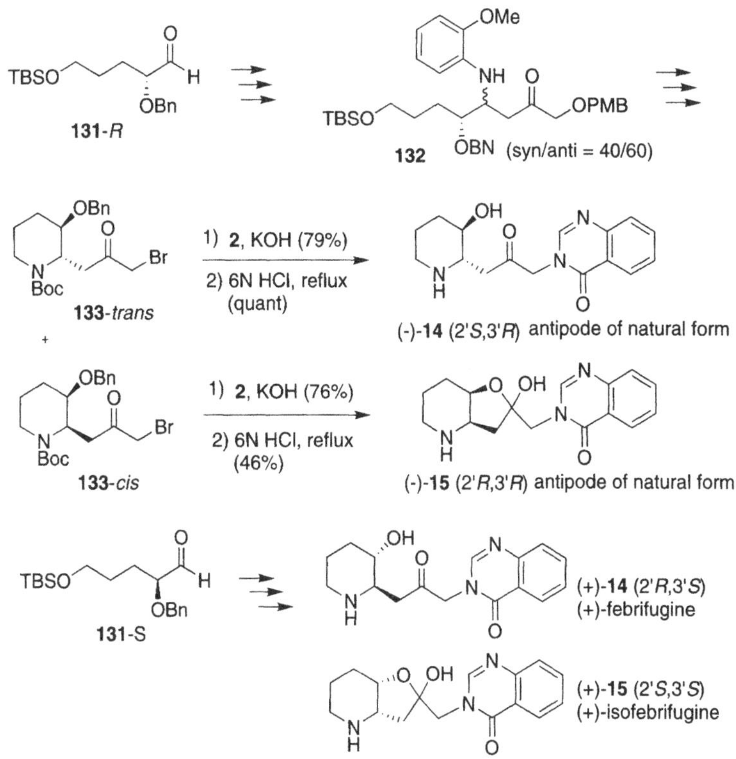

Scheme 30 Synthesis of stereoisomers of febrifugine and isofebrifugine [290, 291]

Mannich-type three-component reaction as the key steps (Scheme 30) [290, 291].

The key intermediate **132** was prepared by a novel aqueous three-component reaction involving **131**-*R*, 2-methoxyaniline, 2-methoxy-3-(*p*-methoxybenzyloxy)-1-propene using ytterbium dodecyl sulfate, $Yb(DS)_3$ (Lewis acid-surfactant-combined catalyst). Piperidine derivatives **133**-*trans* and **133**-*cis* were prepared from **132**, and coupled with 4-hydroxyquinazoline (**2**) using potassium hydroxide, respectively [300], followed by deprotection afforded antipode of natural febrifugine and isofebrifugine, respectively. Similarly another pair of **14** and **15** were prepared from **131**-*S*. These unambiguous total syntheses of stereoisomers revised the absolute configurations of natural febrifugine and isofebrifugine from (2′S,3′R) and (2′R,3′R) to (2′R,3′S) and (2′S,3′S) (Scheme 30). Thereafter, many unique synthetic routes to **14** and **15** have been developed by several groups. For example, Kobayashi et al. has developed more flexible new routes to febrifugine and isofebrifugine using the $Sc(OTf)_3$ (or other Lewis acid)-catalyzed coupling reaction between tin(II) enolate (or silyl enol ether) of 3-acetonyl-4(3*H*)-quinazolinone and 2,3-diacetoxy-*N*-benzyloxycarbonylpiperidine as the key-step [292, 296–298].

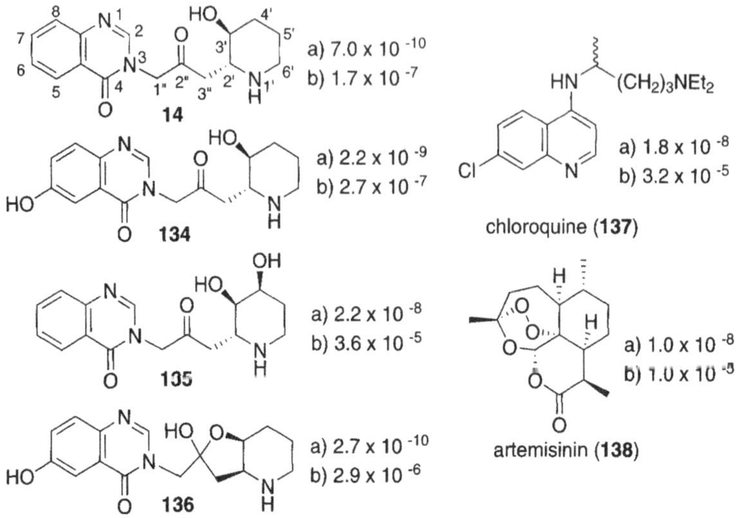

a) antimalarial activity against *P. falciparum* FCR-3 in vitro: EC_{50} (M)
b) cytotoxicity against FM3A mouse mammary cells: EC_{50} (M)

Fig. 5 Examples of febrifugine metabolites and their antimalarial activities against *P. falciparum* in vitro [62]

Other new syntheses of **14** and/or **15** have been developed using dioxabicyclo[3.2.1]octenone chiral block by Ogasawara et al. [293], yeast reduction by Takeuchi et al. [294], 1,3-dipolar cycloaddition by Hatakeyama et al. [295], and the reductive deamination/recyclization reaction of the proline derivative with SmI_2 by Honda et al. [299], etc.

Although febrifugine **14** demonstrated outstanding antimalarial activity, both in vitro and in vivo with no resistant parasite to **14** reported, its use as an antimalarial drug has been precluded due to its strong emetic properties and other undesirable side effects [301]. Many analogues of **14** and **15** are searched with the goal of preserving the strong antimalarial activity, while dramatically reducing side effects. Oshima et al. analyzed the metabolites of **14** and **15**, etc. by mouse liver S9 and have synthesized their analogues for the biological tests [62]. Some of these examples are shown with their antimalarial activity (a) and cytotoxity (b) in Fig. 5. Compounds **134** and **136** having 6-hydroxyl group exhibited powerful antimalarial activities. Compounds **134** and **135** exhibited comparable or stronger antimalarial activity than the clinically used medicines, chloroquine (**137**) and artemisinin (**138**) against *P. berghei* in vivo (mice), and furthermore, **134** was less toxic than the parent alkaloid **14**. From these findings, **134** has been suggested to be a good candidate as a new antimalarial drug by Oshima et al. [62].

4 Concluding Remarks

Recent progress in quinazoline alkaloids and related chemistry has been reviewed retrospectively. Emphasis is placed on recent developments of synthetic methodologies of quinazoline compounds, including aza–Wittig methodology, microwave-assisted synthesis, solid-phase synthesis, and other new synthetic reactions of quinazoline compounds using organometallic reagents, metal-catalyzed reactions, heterocyclizations, and pericyclic reactions. Syntheses of quinazoline alkaloids by the aza–Wittig methodology focusing on the application of the intramolecular Staudinger/aza–Wittig protocol, and some other methodologies are discussed. The efficient and regioselective construction of quinazoline alkaloids and related ring-structures has been shown to be successful even for synthesis of highly polyfunctional quinazoline alkaloids.

Plants and other natural resources used as traditional medicines have been widely explored in drug discovery. Bioassay-directed isolation followed by identification and characterization of bioactive compounds leads a development for new medicinal drugs. Quinazoline alkaloids are one of attractive natural products leading to many drug developments. During the improvement stage of such lead compounds, rational drug design-based modification affords synthetic analogues with increased activity, decreased toxicity, or improved pharmacological availability as exemplified by suggestion of a good candidate as a new antimalarial drug from antimalarial quinazoline alkaloid febrifugine (Sect. 3.6). From this point of view, simple and convenient method for selective synthesis of quinazoline compounds such as quinazolinone annelation combined with microwave-assisted synthesis and use of polymer-reagents would be in demand continuously as one of the indispensable strategies for development of new drugs and related fine materials.

Acknowledgements The author express his sincere gratitude to the dedicated, hardworking, skillful efforts of his coworkers whose names appear in references cited. The author would like to express his appreciation to the Ministry of Education, Science, Sports and Culture, Japan for the generous support by Grants-in-Aid for Scientific Research.

References

1. Merck Index (2001) 13th edn, No 8141. Merck Publishing Group, Rahway, NJ
2. Reddy PS, Reddy PP, Vasantha T (2003) Heterocycles 60:183
3. El-Hiti GA (2000) Heterocycles 53:1839
4. El-Hiti GA, Abdel-Megeed MF (2005) Heterocycles 65:3007
5. Jantova S, Stankovsky S, Spirkova K (2004) Biol Bratislava 59:741
6. Tran TP, Ellsworth EL, Stier MA, Domagala JM, Showalter HDH, Gracheck SJ, Shapiro MA, Joannides TE, Singh R (2004) Bioorg Med Chem Lett 14:4405

7. Kuneš J, Bažant J, Pour M, Waisser K, Šlosárek M, Janota J (2000) Il Farmaco 55:725
8. Dandia A, Singh R, Sarawgi P (2005) J Fluorine Chem 126:307
9. Ram VJ, Farhanullah, Tripathi BK, Srivastava AK (2003) Bioorg Med Chem 11:2439
10. Gineinah MM, El-Sherbeny MA, Nasr MN, Maarou AR (2002) Arch Pharm Pharm Med Chem 11:556
11. Jindal DP, Bhatti RS, Ahlawat S, Gupta R (2002) Eur J Med Chem 37:419
12. Decker M (2005) Eur J Med Chem 40:305
13. Rosowsky A, Wright JE, Vaidya CM, Forsch RA (2000) Pharmacol Therap 85:191
14. Gangjee A, Kothare M, Kisliuk RL (2000) J Heterocycl Chem 37:1097
15. Foster BA, Coffrey HA, Morin MJ, Rastinejad F (1999) Science 286:250
16. Forsch RA, Wright JE, Rosowsky A (2002) Bioorg Med Chem 10:2067
17. Abdel-Jalil RJ, Aldoqum HM, Ayoub MT, Voelter W (2005) Heterocycles 65:2061
18. Levitzky A (2003) Acc Chem Res 36:462
19. Witt A, Bergman J (2003) Curr Org Chem 7:659
20. Avendaño CA, Menéndez JC (2003) Curr Org Chem 7:1
21. Michael JP (2004) Nat Prod Rep 21:650
22. Michael JP (2003) Nat Prod Rep 20:476
23. Michael JP (2002) Nat Prod Rep 19:742
24. Michael JP (2001) Nat Prod Rep 18:543
25. Michael JP (2000) Nat Prod Rep 17:603
26. Michael JP (1999) Nat Prod Rep 16:697
27. Michael JP (1998) Nat Prod Rep 15:595
28. Michael JP (1997) Nat Prod Rep 14:11
29. Ma Z, Hano Y, Nomura T (2005) Heterocycles 65:2203
30. Armarego WLF (1979) In: Katrizky AR, Boulton AJ (eds) Adv Heterocycl Chem, vol. 24. Academic, New York, p. 1
31. Brown DJ (1984) In: Boulton AJ, McKillop A (eds) Comprehensive Heterocyclic Chemistry, vol. 3(2B). Pergamon Press, Oxford, p. 57
32. Brown DJ (1996) Quinazolines. In: The Chemistry of Heterocyclic Compounds, supplement 1, vol. 55. Wiley, New York
33. Merck Index (2001) 13th edn, No. 5987. Merck Publishing Group, Rahway, NJ
34. Kacker IK, Zaheer SH (1951) J Indian Chem Soc 28:344
35. van Zyl EF (2001) Forensic Sci Int 122:142
36. Merck Index (2001) 13th edn, No. 3749. Merck Publishing Group, Rahway, NJ
37. Merck Index (2001) 13th edn, No. 5987. Merck Publishing Group, Rahway, NJ
38. Jackman GB, Petrow V, Stephenson O (1960) J Pharm Pharmacol 12:529
39. Merck Index (2001) 13th edn, No. 7803. Merck Publishing Group, Rahway, NJ
40. Merck Index (2001) 13th edn, No. 1475. Merck Publishing Group, Rahway, NJ
41. Merck Index (2001) 13th edn, No. 3466. Merck Publishing Group, Rahway, NJ
42. Mohri S (2001) J Synth Org Chem Jpn 59:514
43. Stinson SC (2000) Chem Eng News 78:58
44. Goto T, Kishi Y, Takahadhi S, Hirata Y (1965) Tetrahedron 21:2059
45. Tsuda K, Ikuma S, Kawamura M, Tachikawa R, Sakai K, Tamura C, Amakasu O (1964) Chem Pharm Bull 12:1357
46. Woodward RB (1964) Pure Appl Chem 9:49
47. Merck Index (2001) 13th edn, No. 9318. Merck Publishing Group, Rahway, NJ
48. Mosher HS, Fuhrman FA, Buchwald HD, Fisher HG (1964) Science 144:1100
49. Matsui T, Ohtsuka Y, Sakai K (2000) Yakugaku Zasshi 120:825
50. Simidu U, Noguchi T, Hwang DF, Hashimoto K (1987) Appl Environ Microbiol 53:1714

51. Kishi Y, Fukuyama T, Aratani M, Nakatubo F, Goto T, Kakoi H (1972) J Am Chem Soc 94:9219
52. Ohyabu N, Nishikawa T, Isobe M (2003) J Am Chem Soc 125:8798
53. Hinman A, Du Bois J (2003) J Am Chem Soc 125:11510
54. Groeger D (1984) Biosynthesis of Anthranilic Acid Derived Alkaloids. In: van der Plas HC, Ötvös L, Simonyi M (eds) Bio-Organic Heterocycles. Synthetic, Physical and Pharmacological Aspects. Elsevier, Amsterdam, p 165
55. Merck Index (2001) 13th edn, No. 10000. Merck Publishing Group, Rahway, NJ
56. Kamal A, Ramana KV, Rao MV (2001) J Org Chem 66:997
57. Yoshida S, Aoyagi T, Harada S, Matsuda N, Ikeda T, Naganawa H, Hamada M, Takeuchi T (1991) J Antibiot (Tokyo) 44:111
58. Bergman J, Brynoff A (1990) Tetrahedron 46:1295
59. Bergman J (1997) J Chem Res (S):224
60. Koepfli JA, Mead JF, Brockman JA Jr (1947) J Am Chem Soc 69:1837
61. Koepfli JB, Mead JF, Brockman JA Jr (1949) J Am Chem Soc 71:1048
62. Hirai S, Kikuchi H, KIim H-S, Begum K, Wataya Y, Tasaka H, Miyazawa Y, Yamamoto K, Oshima Y (2003) J Med Chem 46:4351
63. Merck Index (2001) 13th edn, No. 3975. Merck Publishing Group, Rahway, NJ
64. Merck Index (2001) 13th edn, No. 8377. Merck Publishing Group, Rahway, NJ
65. Bergman J, Bergman S (1985) J Org Chem 50:1246
66. Sheu J-R (1999) Cardiovasc Drug Rev 17:237
67. Wattanapiromsakul C, Forster PI, Water PG (2003) Phytochemistry 64:609
68. Christopher E, Bedir E, Dunbar C, Khan IA, Okunji CO, Schuster BM, Iwu MM (2003) Helv Chim Acta 86:2914
69. Li X-C, Dunnbar DC, ElSohly HN, Walker LA, Clark AM (2001) Phytochemistry 58:627
70. Wu T-S, Yeh J-H, Wu R-L, Chen K-T, Lin L-C, Chen C-F (1995) Heterocycles 41:1071
71. Ikuta A, Urabe H, Nakamura T (1998) J Nat Prod 61:1012
72. Bergman J, Lindstrom J-O, Tilstam U (1985) Tetrahedron 41:2879
73. Bergman J (1999) Phytochemistry 28:3547
74. Honda G, Tabata M (1979) Planta Medica 36:85
75. Motoki T, Takami Y, Yagi Y, Tai A, Yamamoto I, Gohda E (2005) Biol Pharm Bull 28:260
76. Merck Index (2001) 13th edn, No. 848. Merck Publishing Group, Rahway, NJ
77. Goetz MA, Lopez M, Monaghan RL, Chang RSL, Lotti VJ, Chen TB (1985) J Antibiot 38:1633
78. Liesch JM, Hensens OD, Springer JP, Chang RSL, Lotti VJ (1985) J Antibiot 38:1638
79. Sun HH, Byard SJ, Cooper R (1994) J Antibiot 47:599
80. Sun HH, Barrow CJ, Sedlock DM, Gillum AM, Cooper RJ (1994) J Antibiot 47:515
81. Sun HH, Barrow CJ, Cooper R (1995) J Nat Prod 58:1575
82. Joshi BK, Gloer JB, Wicklow DT, Dowd PF (1999) J Nat Prod 62:650
83. Rahbaek L, Breinholt J, Frisvad C, Christphersen C (1999) J Org Chem 64:1689
84. Rahabaek L, Breinholt J (1999) J Nat Prod 62:904
85. Dai J-R, Carte BK, Sidebottom PJ, Yew ALS, Ng S-B, Huang Y, Butler MS (2001) J Nat Prod 64:125
86. Blackman AJ, Hambly TW, Pricker K, Taylor WC, Thirasasna N (1987) Tetrahedron Lett 28:5561
87. Penn J, Mantle OC, Bilton JN, Sheppard RN (1992) J Chem Soc Perkin Trans 1:1495
88. Numata A, Takahashi C, Matsushita T, Miyamoto T, Kawai K, Usami Y, Matsumura E, Inoue M, Oishi H, Shingu T (1992) Tetrahedron Lett 33:1621

89. Takahashi C, Matsushita T, Doi M, Minoura K, Shingu T, Kumeda Y, Numata A (1995) J Chem Soc Perkin Trans 1:2345
90. Belofsky GN, Anguera M, Jensen PP, Fenical W, Kock M (2000) Chem Eur J 6:1355
91. Wong S-M, Musza LL, Kydd GC, Kullnig R, Gillum AM, Cooper R (1993) J Antibiot 46:545
92. Fujimoto H, Negishi E, Yamaguchi K, Nishi N, Yamazaki M (1996) Chem Pharm Bull 44:1843
93. Barrow CJ, Sun HH (1994) J Nat Prod 57:471
94. Larsen TO, Frydenvang K, Frisvad JC, Christophersen C (1998) J Nat Prod 61:1154
95. Ariza MR, Larsen TO, Petersen BO, Duus JØ, Christopersen C, Barrero AF (2001) J Nat Prod 64:1590
96. Hochlowski JE, Mullally MM, Spanton SG, Whittern DN, Hill P, McAlpine JB (1993) J Antibiot 46:380
97. Karwowski JP, Jackson M, Rasmussen RR, Humphrey PE, Poddig JB, Kohl WL, Scherr MH, Kadam S, McAlpine JB (1993) J Antibiot 46:374
98. Teodori E, Dei S, Scapecchi S, Gualtieri F (2002) Il Farmaco 57:385
99. Ma ZZ, Hano Y, Nomura T, Chen YJ (1997) Heterocycles 46:541
100. Ma ZZ, Nomura T, Chen YJ (1999) Heterocycles 51:1883
101. Ma ZZ, Hano Y, Nomura T, Chen YJ (2004) Bioorg Med Chem Lett 14:1193
102. Cagir A, Jones SH, Gao R, Eisenhauer BM, Hecht SM (2003) J Am Chem Soc 125:13628
103. Atkinson RA (1999) Tetrahedron 55:1519
104. Atkinson RS, Meades CK (2001) J Chem Soc Perkin Trans 1:1518
105. Connolly DJ, Lacey PM, McCarthy M, Saunders CP, Carroll A-M, Richard Goddard R, Guiry PJ (2004) J Org Chem 69:6572
106. Kumar S, Hundal G, Paul D, Hundal MS, Singh H (2000) J Chem Soc Perkin Trans 1:2295
107. Wang S, Lou H, Liu Y, Yu G, Lu P, Zhu D (2001) J Mater Chem 11:2971
108. Eguchi S, Okano T, Okawa T (1997) Recent Res Dev Org Chem 1:337
109. Eguchi S, Okano T, Okawa T (1999) Chem Abstr 130:252510
110. Eguchi S (2005) ARKIVOV (ii) 98
111. Fresneda PM, Molina P (2004) Synlett 1
112. Wamhoff H, Richard G, Stoelben S (1995) In: Katrizky AR, Boulton AJ (eds) Adv Heterocycl Chem, vol. 64. Academic, New York, p 159
113. Eguchi S, Matsushita Y, Yamashita K (1992) Org Prep Proced Int 24:211
114. Eguchi S, Okano T (1993) J Synth Org Chem Jpn 51:203
115. Molina P, Vilaplana MJ (1994) Synthesis 1197
116. Gusar NI (1991) Russ Chem Rev 60:146
117. Ming-Wu D, Zhao-Jie L (2001) Chin J Org Chem 21:1
118. Barluenga J, Palacios F (1991) Org Prep Proced Int 23:1
119. Gololobov YG, Kasukhin LF (1992) Tetrahedron 48:1353
120. Gololobov YG, Zhmurova IN, Kasukhin LF (1981) Tetrahedron 37:437
121. Shalev DE, Chiacchiera SM, Radkowsky AE, Kosower EM (1996) J Org Chem 61:1689
122. Hemming K, Bevan MJ, Loukou C, Patel SD, Renaudeau D (2000) Synlett 1565
123. Charette AB, Boezio AA, Janes MK (2000) Org Lett 2:3777
124. Hanquet G, Colobert F, Lanners S, Solladié G (2003) ARKIVOC vii:328
125. Saito T, Nakane M, Endo M, Yamashita H, Oyamada Y, Motoki S (1986) Chem Lett 135
126. Saito T, Tsuda K, Saito Y (1996) Tetrahedron Lett 37:209
127. Molina P, Aller E, Lorenzo A (1998) Synthesis 283

128. Molina P, Aljarin M, Vidal A (1988) Tetrahedron Lett 29:3849
129. Molina P, Alajarin M, Vidal A (1989) Tetrahedron 45:4263
130. Ding M-W, Zeng G-P, Wu T-J (2000) Synth Commun 30:1599
131. Liang Y, Ding M-W, Liu Z-J, Liu X-P (2003) Synth Commun 33:2843
132. Wamhoff H, Schmidt A, Nieger M (1991) Tetrahedron Lett 32:4473
133. Wamhoff H, Schmidt A (1993) J Org Chem 58:6976
134. Wamhoff H, Schmidt A (1993) Heterocycles 35:1055
135. Wamhoff H, Wintersohl H, Stoelben S, Paasch J, Nai-jue Z, Fang G (1990) Libigs Ann Chem 901
136. Rossi E, Celentano G, Stradi A (1990) Tetrahedron Lett 31:903
137. Rossi E, Calabrese D, Farma F (1991) Tetrahedron 47:5819
138. Miyabara H, Takayasu T, Nitta M (1999) J Chem Soc Perkin Trans 1:3199
139. Palacios F, Alonso C, Rubiales G (1997) J Org Chem 62:1146
140. Katritzky AR, Nair SK, Khokhlova T, Akhmedov NGJ (2003) J Org Chem 68:5724
141. Hemming K, Loukou C, Elkatip S, Smalley RK (2004) Synlett 101
142. Ardakan MN, Smalley RK, Smith RH (1983) J Chem Soc Perkin Trans 1:2501
143. Takeuchi H, Eguchi S (1989) Tetrahedron Lett 30:3313
144. Takeuchi H, Hagiwara S, Eguchi S (1989) Tetrahedron 45:6375
145. Takeuchi H, Eguchi S (1989) Chem Commun 602
146. Eguchi S, Goto S (1994) Heterocyclic Commun 1:51
147. Takeuchi H, Matsushita Y, Eguchi S (1991) J Org Chem 56:1535
148. Eguchi S, Matsushita Y, Takeuchi H (1992) J Org Chem 57:6975
149. Lidström P, Tierney J, Wathey B, Westman J (2001) Tetrahedron 57:9225
150. Kappe CO (2004) Angew Chem Int Ed 43:6250
151. Varma R (1999) Green Chem 1:43
152. Lidström P, Tierney J, Wathey B, Westman J (2001) Tetrahedron 57:9199
153. Jeelnik M, Varma RS, Polanc S, Koèevar M (2002) Green Chem 4:35
154. Domling A, Ugi I (2000) Angew Chem Int Ed Engl 39:3168
155. Nair V, Rajesh C, Vinod AU, Bindu S, Sreekanth AR, Mathen JS, Balagopal L (2003) Acc Chem Res 36:899
156. v Niementowski S (1895) J Prakt Chem 51:564
157. Merck Index 13th edn, ONR-279. Merck Publishing Group, Rahway, NJ
158. Alexandre F-R, Berecibar A, Wrigglesworth R, Besson T (2003) Tetrahedron 59:1413
159. Pereira MdF, Picot L, Guillon J, Léger J-M, Jarry C, Thiéry V, Besson T (2005) Tetrahedron Lett 46:3445
160. Desai AR, Desai KR (2005) ARKIVOC xiii:98
161. Seijas JA, V-Tato MP, Martinez MM (2000) Tetrahedron Lett 41:2215
162. Rad-Moghadam K, Khajavi MS (1998) J Chem Res (S):702
163. Dandia A, Singh R, Sarawgi P (2005) J Fluorine Chem 126:307
164. Liu J-F, Lee J, Dalton AM, Bi G, Baldino CM, McElory E, Brown M (2005) Tetrahedron Lett 46:1241
165. Nikpou F, Paibast T (2005) Chem Lett 34:1438
166. Bräse S, Gil C, Knepper K (2002) Bioorg Med Chem 10:2415
167. Franzén RG (2000) J Comb Chem 2:195
168. Kirchňák V, Holladay MW (2002) Chem Rev 102:61
169. Barthelemy S, Schneider S, Bannwarth W (2002) Tetrahedron Lett 43:807
170. Grieder A, Thomas AW (2003) Synthesis 1707
171. Wéber C, Demeter Á, Szendrei G, Greiner I (2003) Tetrahedron Lett 44:7533
172. Makino S, Nakanishi E, Tsuji T (2001) Tetrahedron Lett 42:1749
173. Schwinn D, Bannwarth W (2003) Helv Chim Acta 86:188

174. Cobb JM, Fiorini MT, Goddard CR, Theoclitou ME, Abell C (1999) Tetrahedron Lett 40:1045
175. Makino S, Suzuki N, Nakanishi E, Tsuji T (2000) Synlett 1670
176. Zhang J, Barker J, Lou B, Saneil H (2001) Tetrahedron Lett 42:8405
177. Song A, Mařík J, Lam KS (2004) Tetrahedron Lett 45:2727
178. Wang F, Hauske JR (1997) Tetrahedron Lett 38:8651
179. Bergman J, Brynolf A, Elman B, Vuolinen A (1986) Tetrahedron 42:3697
180. Wiklund P, Bergman J (2003) Org Biomol Chem 1:367
181. Akazome M, Kondo T, Watanabe Y (1993) J Org Chem 58:310
182. Larksarp C, Alper H (2000) J Org Chem 65:2775
183. Nishiyama Y, Hirose M, Kitagaito W, Sonoda N (2002) Tetrahedron Lett 43:1855
184. Battistuzzi G, Cacchi S, Fabrizi G, Marinelli F, Parisi LM (2002) Org Lett 4:1355
185. Kundu NG, Chaudhuri G (2001) Tetrahedron 57:6833
186. Abdel-Jalil RJ, Voelter W, Saeed M (2004) Tetrahedron Lett 45:3475
187. Hirota K, Kuki H, Maki Y (1994) Heterocycles 37:563
188. Furstner A, Leitner A, Mendez M, Krause H (2002) J Am Chem Soc 124:13856
189. Fuwa H, Kobayashi T, Tokitoh T, Torii Y, Natsugari H (2005) Tetrahedron 61:4297
190. Mazurkiewicz R (1989) Monatsh Chem 120:973
191. Wang H, Ganesan A (1998) J Org Chem 63:2432
192. Wang H, Ganesan A (2000) J Org Chem 65:1022
193. Snider BB, He F (1999) J Org Chem 64:1397
194. Snider BB, Zeng H (2003) Heterocycles 61:173
195. Miyata K, Sakai Y, Tsuda Y, PCT Int Appl WO9708153
196. Hynes JB, Tomažič A, Parrish CA (1991) J Heterocyclic Chem 28:1357
197. Kotsuki H, Sakai H, Morimoto H, Suenaga H (1993) Synlett 1993
198. Hao J, Ohkura H, Amii H, Uneyama K (2000) J Chem Soc Chem Commun 1883
199. Gruner M, Rehwald M, Eckert K, Gewald K (2000) Heterocycles 53:2363
200. Zieliński W, Kudelko A (2000) Monatsh Chem 131:895
201. Szczepankiewicz W, Suwiński J, Bujok R (2000) Tetrahedron 56:9343
202. Bobošíková M, Clegg W, Coles SJ, Dandárová M, Hursthouse MB, Kiss T, Krutošiková A, Liptaj T, Prónayová N, Ramsden CA (2001) J Chem Soc Perkin Trans 1:680
203. Janowski WK, Prager RH, Smith JA (2000) J Chem Soc Perkin Trans 1:3212
204. Tennant G (1996) J Chem Soc (C):2290
205. Porter TC, Smalley RK, Teguiche M, Purwono B (1997) Synthesis 773
206. Jones P, Chambers M (2002) Tetrahedron 58:9973
207. Erba E, Pocar D, Trimarco P (2005) Tetrahedron 61:5778
208. Al-Talib M, Jochims JC, Hamed A, Wang Q, Ismail AE-H (1992) Synthesis 697
209. Kaname M, Tsuchiya T, Sashida H (1999) Heterocycles 51:2407
210. Mizuno T, Iwai T, Ishino Y (2004) Tetrahedron Lett 45:7073
211. Mizuno T, Okamoto N, Ito T, Miyata T (2000) Heteroatom Chem 11:428
212. Eguchi S, Suzuki T, Okawa T, Matsushita Y (1996) J Org Chem 61:7316
213. Joshi BS, Newton MG, Lee DW, Barber AD, Pelletier SW (1996) Tetrahedron Asymm 7:25
214. Davis FA, Towson JC, Vashi DB, Reddy RT, McCauley JP, Harakal ME, Gosciniak DJ (1990) J Org Chem 55:1254
215. Davis FA, Reddy RE, Kasu PVN, Portonovo PS, Carrol PJ (1997) J Org Chem 62:3625
216. Mhaske SB, Argade NP (2001) J Org Chem 66:9038
217. Liu J-F, Ye P, Sprague K, Sargent K, Yohannes D, Baldino CM, Wilson CJ, Ng S-C (2005) Org Lett 7:3363
218. Onaka T (1971) Tetrahedron Lett 12:4387

219. Okawa T, Sugimori T, Eguchi S (1996) Chem Lett 25:843
220. Okawa T, Sugimori T, Eguchi S, Kakehi A (1998) Heterocycles 47:375
221. Hua H, Cheng M, Li X, Pei Y (2002) Chem Pharm Bull 50:1393
222. Eguchi S, Takeuchi H, Matsushita Y (1992) Heterocycles 33:153
223. Mhaske SB, Argade NP (2004) Tetrahedron 60:3417
224. Chavan SP, Sivappa R (2004) Tetrahedron Lett 45:997
225. Chang HW, Kim SI, Jung H, Jahng Y (2003) Heterocycles 60:1359
226. Mohanta PK, Kim K (2002) Tetrahedron Lett 43:3993
227. Kametani T, Chu VL, Higa T, Koizumi M, Ihara M, Fukumoto K (1977) J Am Chem Soc 99:2306
228. Kametani T, Figa T, Loc CV, Ihara M, Koizumi M, Fukumoto K (1976) J Am Chem Soc 98:6186
229. Lee ES, Park J-G, Jahng Y (2003) Tetrahedron Lett 44:1883
230. Mitscher LA, Wong W-C, DeMeulenaere T, Suiko J, Drake S (1981) Heterocycles 15:1017
231. Harayama T, Hori A, Serban G, Morikami Y, Matsumoto T, Abe H, Takeuchi Y (2004) Tetrahedron Lett 60:10645
232. Harayama T, Morikami Y, Shigeta Y, Abe H, Takeuchi Y (2003) Synlett 847
233. Sugimori T, Okawa T, Eguchi S, Yashima E, Okamoto Y (1997) Chem Lett 869
234. Sugimori T, Okawa T, Eguchi S, Kakehi A, Yashima E, Okamoto Y (1998) Tetrahedron 54:7997
235. Eguchi S, Yamashita K, Matsushita Y (1992) Synlett 295
236. Eguchi S, Yamashita K, Matsushita Y, Kakehi A (1995) J Org Chem 60:4006
237. Chankvetadze B, Yashima E, Okamoto Y (1994) J Chromatgr A 670:39
238. Yashima E, Yamamoto C, Okamoto Y (1995) Polym J 27:856
239. Yamamoto C, Okamoto Y (2004) Bull Chem Soc Jpn 77:227
240. The intramolecular version of the tandem Staudinger/aza-Wittig reaction is often referred to as Eguchi aza-Wittig protocol (cf. references 14f, 60, 119–120, etc.)
241. Snider BB, Busuyek MV (2001) Tetrahedron 57:3301
242. Foxman HF, Snider BB (1998) J Amer Chem Soc 120:6417
243. Bock MG, DiPardo RM, Pitzenberger SM, Homnik CF, Springer JP, Freidinger RM (1987) J Org Chem 52:1644
244. Al-Said NH, Al-Qaisi LS (2006) Tetrahedron Lett 47:693
245. Al-Said NH, Ishtaiwi ZN (2005) Acta Chim Slov 52:328
246. Witt A, Bergman J (2001) J Org Chem 66:2784
247. He F, Snider BB (1997) Synlett 483
248. Snider BB, Zeng H (2003) J Org Chem 68:545
249. Snider BB, Zeng H (2000) Org Lett 2:4103
250. Snider BB, Zeng H (2002) Org Lett 4:1087
251. He F, Snider BB (1999) J Org Chem 64:1397
252. Cledera P, Avendaño C, Menéndez JC (1998) Tetrahedron 54:12349
253. Cledera P, Avendaño C, Menéndez JC (2000) J Org Chem 65:1743
254. Hernándes F, Lumetzberger A, Avendaño C, Söllhuber M (2001) Synlett :387
255. Hernández F, Buenadicha FL, Avendaño C, Söllhuber M (2001) Tetrahedron Asymmetry 12:3387
256. Hart DJ, Mongomedov N (1999) Tetrahedron Lett 5429
257. Hart DJ, Magomedov NA (2001) J Am Chem Soc 123:5892
258. Wang H, Ganesan A (2000) J Comb Chem 2:186
259. Wang H, Sim MM (2001) J Nat Prod 64:1497
260. Larsen TO, Franzyk H, Jensen SR (1999) J Nat Prod 62:1578

261. Boyes-Korkis J, Gurney KA, Penn J, Mantle PG, Bilton JN, Sheppard RN (1993) J Nat Prod 56:1707
262. Larsen TO, Svendsen A, Smedsgaard J (2001) Appl Environ Microbiol 67:3630
263. Marsden SP, Depew KM, Danishefsky SJ (1994) J Am Chem Soc 116:11143
264. Depew KM, Marsden SP, Zatorska D, Zatorski A, Bornmann WG, Danishefsy SJ (1999) J Am Chem Soc 121:11953
265. Caballero E, Avendaño C, Menéndez JC (2000) Heterocycles 53:1765
266. Hernándes F, Avendaño C, Söllhuber M (2003) Tetrahedron Lett 44:3367
267. Sánchez JD, Ramos MT, Avendaño C (2000) Tetrahedron Lett 41:245
268. Chen Z, Fan J, Kende AS (2004) J Org Chem 69:79
269. Kende AS, Fan J, Chen Z (2003) Org Lett 5:3205
270. Friedländer P (1882) Chem Ber 15:2572
271. Eckert K (1981) Angew Chem Int Ed 20:208
272. Povarov LS (1967) Russ Chem Rev 36:656
273. Heck RF, Nolley JP Jr (1972) J Org Chem 37:2320
274. Heck RF (1982) Org React NY 27:345
275. Cabri W, Candiani I (1995) Acc Chem Res 28:2
276. Mitsunobu O (1981) Synthesis 1
277. Ma Z-Z, Hano Y, Nomura T, Chen Y-J (1999) Heterocycles 51:1593
278. Kelly TR, Chamberland S, Silva RA (1999) Tetrahedron Lett 40:2723
279. Molina P, Tárraga A, G-Tejero A (2000), Synthesis :1523
280. Twin H, Batey RA (2004) Org Lett 6:4913
281. Mhaske SB, Argade NP (2004) J Org Chem 69:4563
282. Chavan SP, Sivappa R (2004) Tetrahedron 60:9931
283. Ihara M, Komori C, Toyota M (2002) Heterocycles 56:101
284. Toyota M, Komori C, Ihara M (2003) ARKIVOC viii:15
285. Wang H, Ganesan A (1998) Tetrahedron Lett 39:9097
286. Dallavalle S, Merlini L (2002) Tetrahedron Lett 43:1835
287. Yadav JS, Reddy BVS (2002) Tetrahedron Lett 43:1905
288. Osborne D, Stevenson PJ (2002) Tetrahedron Lett 43:5469
289. Danishefsky S, Bryson TA, Puthenpurayl J (1975) J Org Chem 40:796
290. Kobayashi S, Ueno M, Suzuki R, Ishitani H (1999) Tetrahedron Lett 40:2175
291. Kobayashi S, Ueno M, Suzuki R, Ishitani H, Kim H-S, Wataya Y (1999) J Org Chem 64:6833
292. Okitsu O, Suzuki R, Kobayashi S (2000) Synlett :989
293. Taniguchi T, Ogasawara K (2000) Org Lett 2:3193
294. Takeuchi Y, Azuma K, Takakura K, Abe H, Kim H-S, Wataya Y, Harayama T (2001) Tetrahedron 57:1213
295. Ooi H, Urushibara A, Esumi T, Iwabuchi Y, Hatakeyama S (2001) Org Lett 3:953
296. Sugiura M, Kobayashi S (2001) Org Lett 3:477
297. Sugiura M, Hagio H, Hirabayashi R, Kobayashi S (2001) J Org Chem 66:809
298. Sugiura M, Hagio H, Hirabayashi R, Kobayashi S (2001) J Am Chem Soc 123:12510
299. Katoh M, Matsune R, Nagase H, Honda T (2004) Tetrahedron Lett 45:5221
300. Ranganathan D, Farrooqui F (1984) Tetrahedron Lett 25:5701
301. Wiesner J, Ortmann R, Jomaa H, Schlitzer M (2003) Angew Chem Int Ed 42:5274

Top Heterocycl Chem (2006) 6: 157–179
DOI 10.1007/7081_039

Published online: 8 June 2006

Bioactive Heterocyclic Alkaloids of Marine Origin

Masaki Kita[1] · Daisuke Uemura[2] (✉)

[1]Research Center for Materials Science, Nagoya University, Furo-cho, Chikusa, 464-8602 Nagoya, Japan

[2]Department of Chemistry, Graduate School of Science and Institute for Advanced Research, Nagoya University, Furo-cho, Chikusa, 464-8602 Nagoya, Japan
uemura@chem3.chem.nagoya-u.ac.jp

Abstract Many kinds of alkaloids with extraordinary structures and significant biological activities have been isolated from marine organisms. This work features the structures, biological activities, and biogenesis of novel heterocyclic marine alkaloids, which control biologically and physiologically intriguing phenomena. Pinnatoxins and pteriatoxins, potent shellfish poisons, were isolated from the Okinawan bivalve *Pinna* sp. and *Pteria* sp. Norzoanthamine hydrochloride, isolated from the colonial zoanthid *Zoanthus* sp., suppresses decreases in bone weight and strength in ovariectomized mice. Symbioimine, an amphoteric iminium metabolite from the dinoflagellate *Symbiodinium* sp., inhibits osteoclast differentiation. Other novel alkaloids, such as pinnamine, pinnaic acids, halichlorine, and zamamistatin, are also described.

Keywords Heterocyclic marine alkaloids · Shellfish poison · Anti-osteoporosis · Super-carbon-chain compound

1 Introduction

Alkaloids are nitrogen-containing compounds that occur naturally not only in plants but also in microorganisms, marine organisms, and animals. Many kinds of alkaloids with extraordinary structures and significant biological activities have been isolated from marine organisms [1,2]. They continue to provide lead structures in the search for new drugs or biological probes for physiological studies. As new and more complicated diseases are encountered worldwide, the importance of novel bioactive alkaloids has increased due to their potential application in chemotherapy.

Many kinds of bioactive nitrogenous compounds, such as peptides, indols, oxazoles, and thiazoles, have been identified from marine invertebrates [3,4]. The true origins or progenitors of these metabolites have been suggested to be microorganisms, i.e., microalgae, bacteria, and fungi. These microorganisms are carried through symbiosis, association, a food chain, and other forms of nutrient-dependency with host animals [5–7]. Consequently, the isolation of bioactive metabolites from cultured marine microorganisms, such as symbiotic dinoflagellates and bacteria, as well as from their host animals, has been well investigated. Several alkaloidal metabolites isolated from cyanobacteria have been suggested to help to inhibit predation by marine herbivores, such as fish and sea urchins. However, the real role of most marine bioactive alkaloids in the ecosystem has not been well clarified.

In our ongoing search for bioactive metabolites from marine organisms, several novel heterocyclic alkaloids, such as pinnatoxins, norzoanthamine, pinnaic acids, zamamistatin, and symbioimine, have been isolated. This work features the structures, biological activities, and biogenesis of these bioactive heterocyclic marine alkaloids, along with up-to-date topics.

2 Pinnatoxins and Pteriatoxins, Potent Macrocyclic Iminium Toxins from an Okinawan Bivalve

2.1 Isolation and Structure of Pinnatoxins

Shellfish of the genus *Pinna* live mainly in shallow waters of the temperate and tropical zones of the Indian and Pacific Oceans [8]. The adductor muscle of this bivalve is eaten in Japan and China, and food poisoning resulting from its ingestion occurs frequently. Although it has been suggested that this poisoning is caused by bacterial infection or neurotoxins, the true causative agent was ambiguous. Chinese investigators have reported that a toxic extract from *P. attenuata*, referred to as pinnatoxin, is a Ca^{2+} channel activator [9]. We have successfully isolated pinnatoxin A (**1**), a mixture of pinnatoxins B and C (**2**, **3**), and pinnatoxin D (**4**) from *P. muricata* as a major cause of food poisoning [10–13].

The structures and stereochemistries of pinnatoxins have been clarified by extensive analysis using NMR experiments and positive ion ESI MS/MS. Pin-

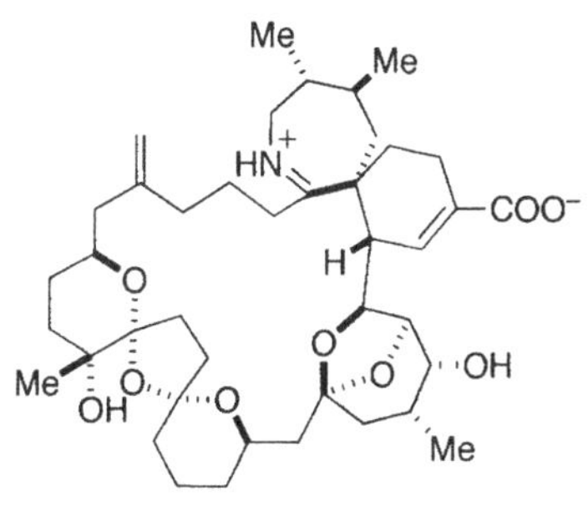

Pinnatoxin A (**1**)

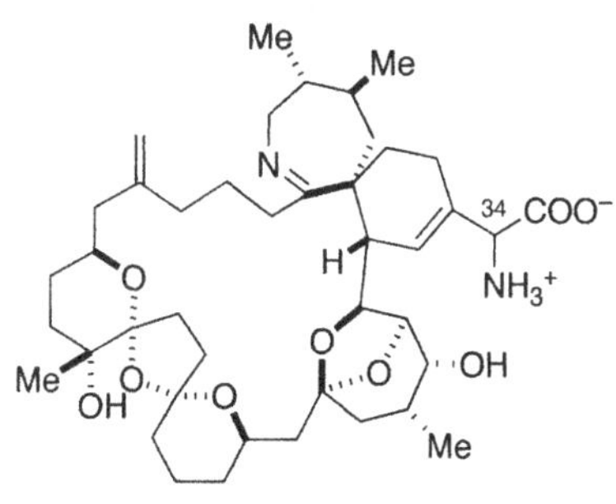

Pinnatoxin B (**2**) : 34*S* isomer
Pinnatoxin C (**3**) : 34*R* isomer

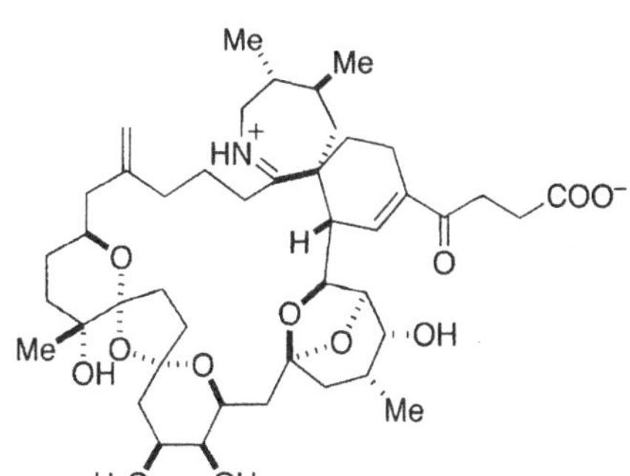

Pinnatoxin D (**4**)

natoxins consist of a 20-membered ring, i.e., with 5,6-bicyclo, 6,7-azaspiro, and 6,5,6-triketal moieties in their structure. In particular, they contain a carboxylate anion and an iminium or ammonium cation. Recently, Kishi's group achieved the total synthesis of **1** and *ent*-**1** [14], and the absolute stereochemistry of pinnatoxin A (**1**) has been confirmed. Interestingly, while natural **1** showed significant acute toxicity, its antipode *ent*-**1** was not toxic [15].

Pinnatoxins B (**2**) and C (**3**), the most toxic constituents in the pinnatoxin series, were isolated as a 1 : 1 mixture [13]. The molecular formula of both pinnatoxins B (**2**) and C (**3**) was determined to be $C_{42}H_{64}N_2O_9$ by ESIMS, which reflects a 29 MS unit (CH_3N) increase compared to that of pinnatoxin A (**1**). In positive ion ESI MS/MS analysis, a series of prominent fragment ions

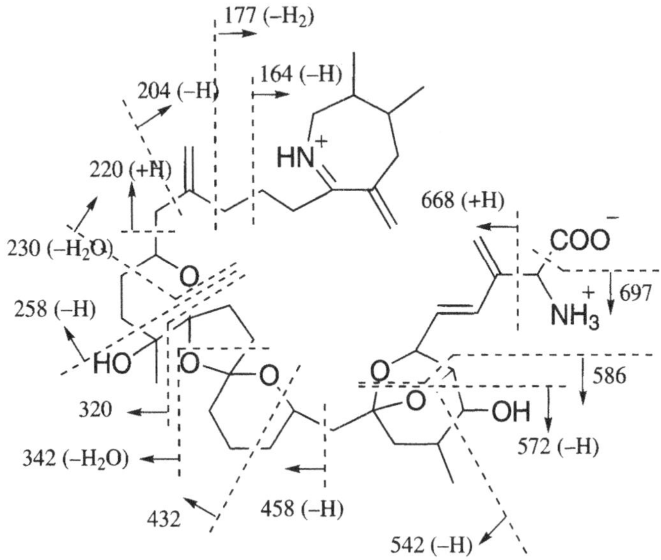

Fig. 1 ESI-MS/MS fragmentation pattern of pinnatoxins B (**2**) and C (**3**)

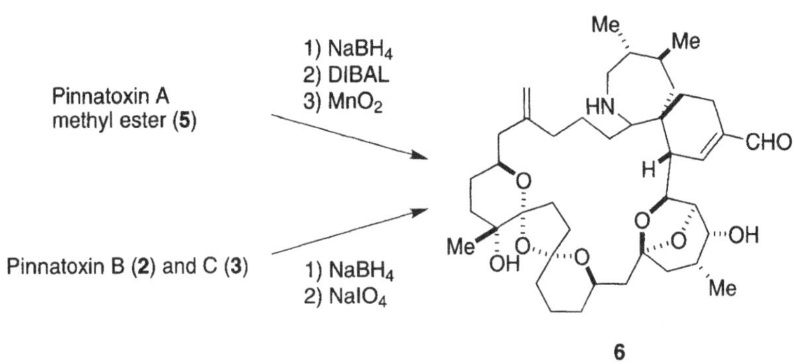

Scheme 1

were generated by a cyclohexane ring-opening reaction (retro-Diels–Alder reaction), followed by bond cleavage of carbocycles (Fig. 1).

The stereochemistry of the macrocyclic moiety in pinnatoxins B (**2**) and C (**3**) was determined as follows. Reduction of the imino group in **2** and **3** with $NaBH_4$ followed by oxidative cleavage with $NaIO_4$ provided aldehyde **6** (Scheme 1). Aldehyde **6** was also obtained by the reduction of iminium and a carboxylic acid moiety in pinnatoxin A methyl ester (**5**) followed by oxidation of the resulting alcohol. Since the spectroscopic data of **6** derived from **2** and **3** and that from **5** were identical, the relative stereochemistry of the macrocyclic core in **2** and **3** was confirmed to be the same as that in pinnatoxin A (**1**).

2.2 Biological Activity and Biogenesis of Pinnatoxins

Pinnatoxin A (**1**) showed potent acute toxicity against mice (LD_{99} 180 μg/kg (*i.p.*)) with characteristic neurotoxic symptoms. Pinnatoxin A activated Ca^{2+} channels. Pinnatoxins B (**2**) and C (**3**) were the most toxic constituents in the pinnatoxin series, which makes them as potent as tetrodotoxin (LD_{99} 22 μg/kg). Although pinnatoxin D (**4**) showed weaker acute toxicity than the other pinnatoxins (LD_{50} > 10 μg/MU), **4** showed the strongest cytotoxicity against the murine leukemia cell line P388 (IC_{50} 2.5 μg/ml).

The backbone of pinnatoxins and their analogues could be configured from C1 to C34 in a single carbon chain, in a polyketide biogenetic pathway (Fig. 2) [10]. This biosynthetic proposal entails an intramolecular Diels–Alder reaction to construct a cyclohexene ring (G-ring) followed by imine formation to establish a 6,7-spiro-ring system. Because of the structural similarity of the imine moiety adjacent to the spirocyclic core, other macrocyclic imines, i.e. pteriatoxins, spirolides, gymnodimine, may also be biosynthesized via the same intramolecular Diels–Alder reaction. Indeed, Kishi and his coworkers achieved the total synthesis of **1** using a biomimetic intramolecular Diels–Alder reaction as shown in Scheme 2 [14].

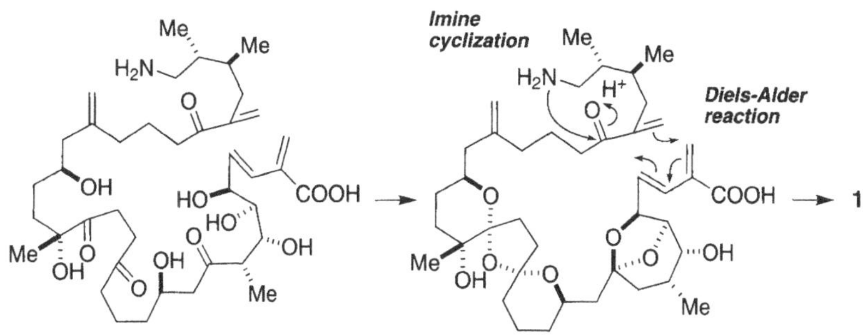

Fig. 2 Proposed biogenesis of pinnatoxin A (**1**)

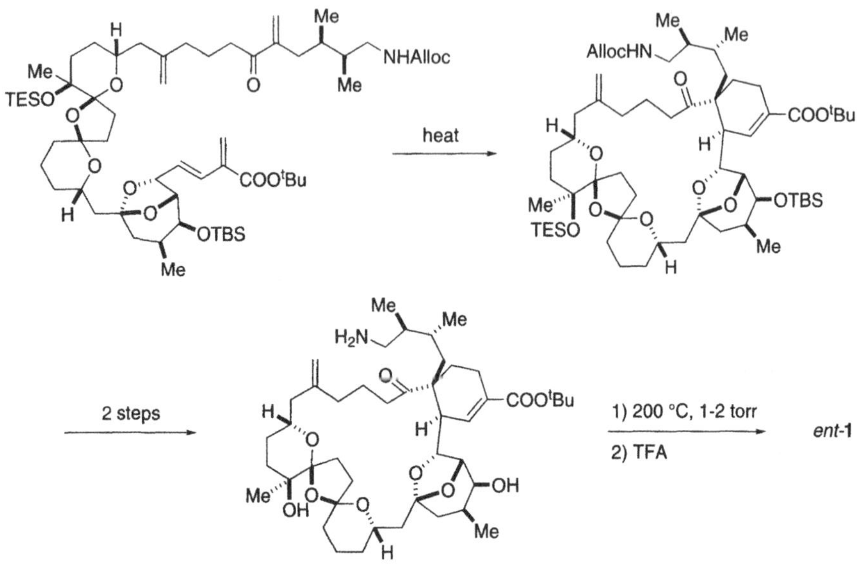

Scheme 2

2.3 Pteriatoxins, Pinnatoxin Analogs from the Okinawan Bivalve *Pteria penguin*

In our study of shellfish poisons, we observed that a moray eel vomits the viscera of the Okinawan bivalve *Pteria penguin*. We found that the aqueous EtOH extract of viscera of *P. penguin* showed potent acute toxicity against mice along with severe convulsion. Guided by this toxicity, pteriatoxins A (**7**), B, and C (**8**, **9**: a 1 : 1 mixture) were isolated as extremely toxic and minor components [16]. Although the isolated yields of pteriatoxins were too low (less than 20 μg) to deduce their structures by usual NMR analysis, a nano-mole-order structure determination of pteriatoxins was achieved by a detailed analysis of ESI MS/MS. As a consequence, pteriatoxins were determined to

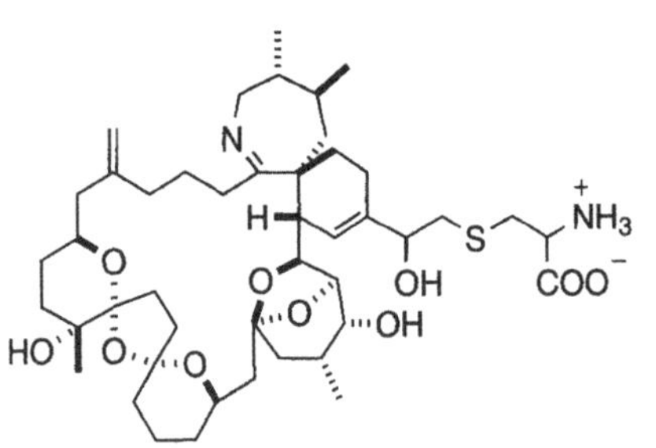
Pteriatoxin A (**7**)

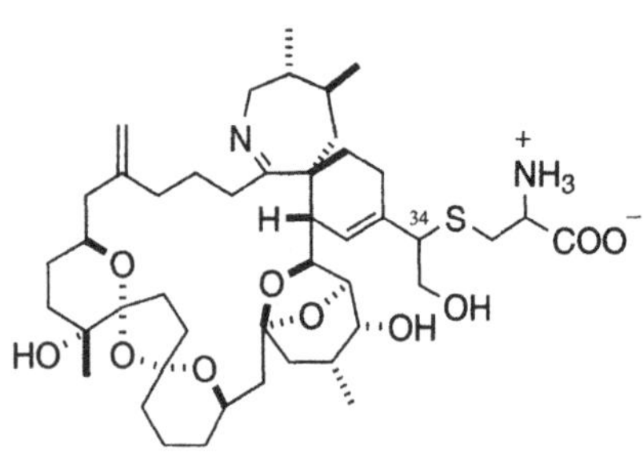
Pteriatoxin B (**8**)
Pteriatoxin C (**9**)

be pinnatoxin analogs containing a cysteine moiety. The position of duplicate signals in the ^{1}H NMR spectrum suggested that pteriatoxins B (**8**) and C (**9**) are C-34 epimers, like **2** and **3**.

Pteriatoxins A (**7**), B, and C (**8**, **9**) showed significant acute toxicity against mice with LD_{99} values of 100 and 8 μg/kg, respectively. The toxic symptoms of pteriatoxins were also similar to those of pinnatoxins. Extracts from the digestive glands of several species that are closely related to *Pinna* sp., including *P. muricata*, *P. attenuata*, *P. atropupurea*, and the commonly eaten shellfish *Atrina pectinata*, all produced the same symptoms of poisoning in mice. Thus, these shellfish may become toxic as a result of feeding on common toxic organisms such as dinoflagellates.

2.4 Pinnamine, an Alkaloidal Marine Toxin from *Pinna muricata*

In a continuation of our work on pinnatoxins, a novel marine alkaloid, pinnamine (**10**), was isolated from the Okinawan bivalve *P. muricata*. Pinnamine exhibited acute toxicity against mice, with characteristic toxic symptoms, such as scurrying around and convulsion (LD_{99} 0.5 mg/kg) [17]. The structure of pinnamine (**10**) was determined to be a unique alkaloid containing a 9-azabicyclo[4.2.1]nonane moiety and a dihydro-γ-pyrone ring. The absolute stereostructure was determined by an analysis of the circular dichroism spectrum [18].

The structure and toxic symptoms of pinnamine resemble those of anatoxin-a (**11**) [19, 20], a potent postsynaptic depolarizing neurotoxin known as very fast death factor (VFDF), and atropine [21], a representative suppressor of the

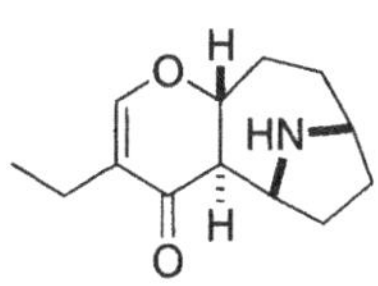

Pinnamine (**10**)

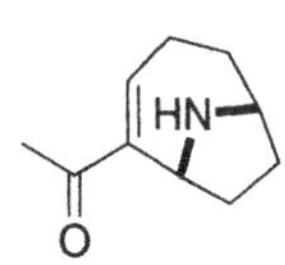

Anatoxin-a (**11**)

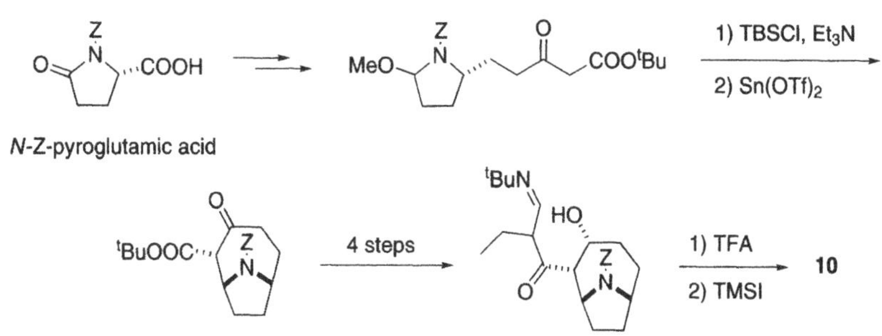

Scheme 3

parasympathetic nervous system. Thus, the toxic expression of pinnamine, similar to that of atropine, may result from excitability of the cerebrum.

Recently, an enantioselective synthesis of pinnamine (**10**) has been achieved (Scheme 3) [22]. The 9-aza-bicyclo[4.2.1]nonane moiety in **10** was constructed by convergence of the β-keto ester into the silyl enol ether followed by Lewis acid treatment. Synthetic pinnamine was found to correspond uniquely to the natural compound based on a comparison of their spectral data including their CD spectra.

3 Norzoanthamine, A Significant Inhibitor of Osteoclast

3.1 Structure of Zoanthamines

Norzoanthamine (**12**), zoanthamine (**13**), and its homologues **14–21** were isolated from the colonial zoanthid *Zoanthus* sp. [23–26]. The relative stereochemistry of norzoanthamines was determined by X-ray analysis. The absolute stereochemistry of norzoanthamine was established by a modified Mosher's method [27].

Interestingly, equilibration was observed between the lactone structure and iminium structure on norzoanthamine (**12**). The NMR spectrum of norzoanthamine hydrochloride in CD_3OD implied the presence of an iminium structure (**22**, $\delta_{C\text{-}10} = 193.3$) but not a lactone structure **12** in norzoanthamine (Scheme 4). The zwitter iminium structure was also demonstrated by transformation into methyl ester **23** by the treatment of **12** with $CH_3I - Ag_2O$. On the other hand, hydrolysis of **23** with aqueous HCl led to the recovery of **12**. Recently, zooxathellamine (**21**) was isolated from the cultured symbiotic dinoflagellate *Symbiodinium* sp. [28]. As with norzoanthamine (**12**), **21** also adopted a zwitter ion structure with carboxylate and iminium moieties in D_2O, but had a lactone structure in either $CDCl_3$ or CD_3OD.

3.2 Biological Activities of Zoanthamines

IL-6 (Interleukin-6) is known to stimulate osteoclast formation, and the suppression of secretion can be effective in the prevention of osteoporosis. Norzoanthamine (**12**) and norzoanthamine hydrochloride (**22**) inhibit IL-6 induction at values of 13 and 4.7 μg/ml, respectively [29]. Meanwhile, the effect of norzoanthamine hydrochloride (**22**) on bone weight and strength was tested in ovariectomized mice, an animal model of postmenopausal osteoporosis (Fig. 3). Norzoanthamine hydrochloride (**22**) (0.08 mg/kg/day, p.o.) significantly suppressed the decrease in femoral weight caused by ovariec-

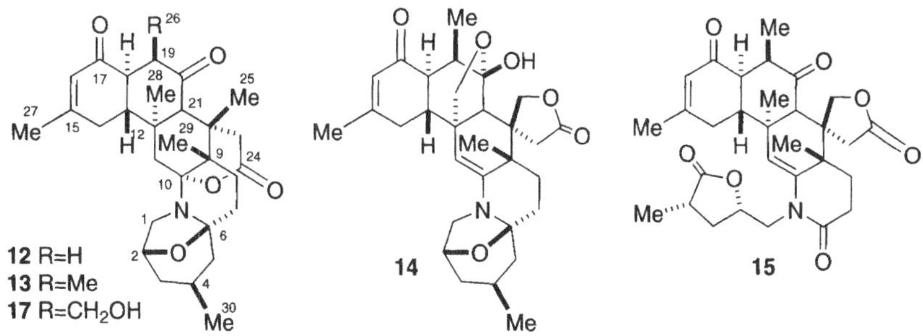

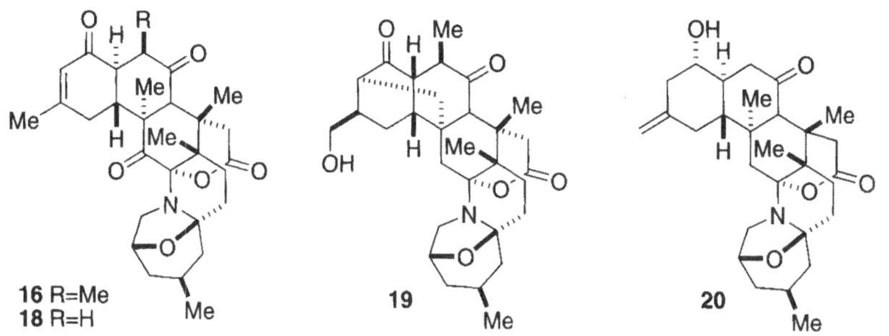

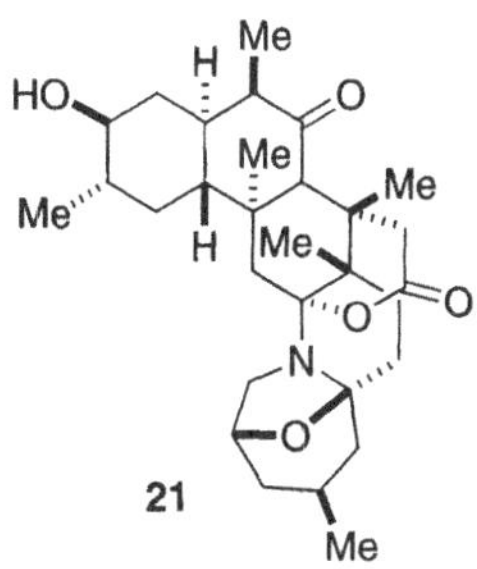

Norzoanthamine (**12**)
Zoanthamine (**13**)
Zoanthenamine (**14**)
Zoanthamide (**15**)
Zoanthaminone (**16**)
Oxyzoanthamine (**17**)
Norzoanthaminone (**18**)
Cyclozoanthamine (**19**)
Epinorzoanthamine (**20**)
Zooxathellamine (**21**)

tomy [30, 31]. The primary spongiosa did not significantly increase, and the morphology of the metaphysis remained nearly normal. It is known that uterine hypertrophy is a serious side effect of 17β-estradiol [32]. In contrast, norzoanthamine hydrochloride (**22**) did not have an estrogen-like side effect on reproductive organs. Thus, the action mechanism of norzoanthamine hydrochloride is suggested to differ from that of estrogen.

Osteoporosis is caused by an imbalance between bone resorption and bone formation, which results in bone loss and fractures after mineral flux occurs [33]. The frequency of fracture is significantly increased in patients with osteoporosis, and hip fracture in elderly patients with osteoporosis is a very serious problem because it often limits their quality of life [34]. In addition to preventing the loss of bone mass, maintenance of the mechanical strength of bone tissue is a very important point to consider in the development of anti-

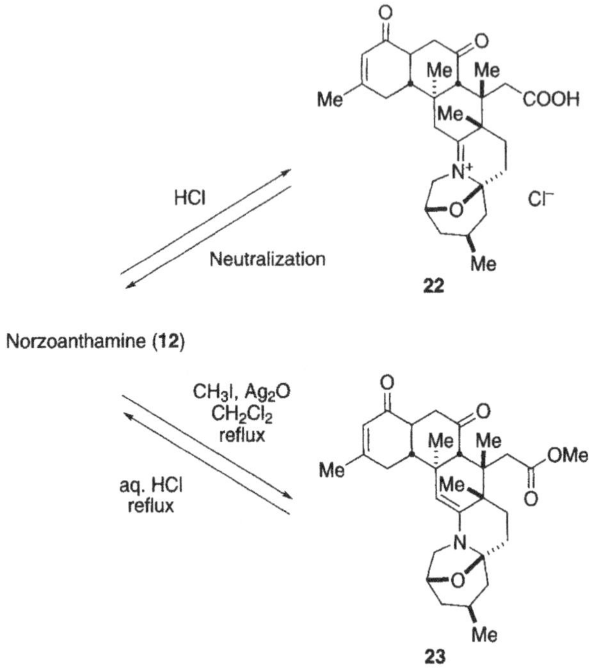

Scheme 4

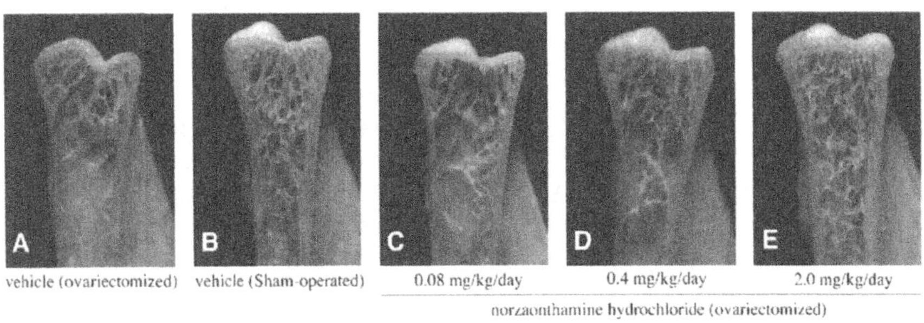

Fig. 3 Effect of norzoanthamine hydrochloride (22) on humeralis morphology in ovariectomized mice

osteoporotic drugs. From this point of view, norzoanthamine hydrochloride (22) is considered to be a potent candidate.

3.3
Biogenesis of Zoanthamines

On the basis of their molecular formulas, zoanthamines have been regarded as terpenoids, but the biogenetic pathway of zoanthamines remains unclear. We have proposed a polyketide biogenetic pathway for zoanthamines, as

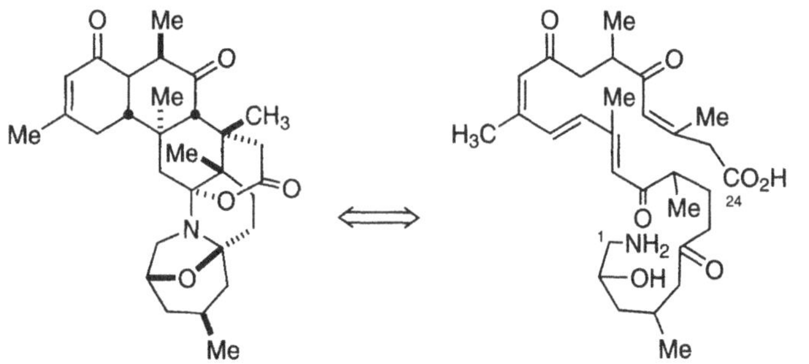

Fig. 4 Proposed biogenesis of zoanthamine

shown in Fig. 4 [23]. A single carbon chain precursor (C1 to C24) could be cyclized to give a complex polycyclic structure, like **13**. Meanwhile, feeding experiments with a labeled compound for zooxathellamine (**21**) have been examined, and a polyketide biosynthetic pathway has been supported [28].

Recently, a total synthesis of norzoanthamine (**12**) has been achieved using an intramolecular Diels–Alder reaction as a key step for constructing the requisite chiral triene [35]. This synthesis may be a powerful tool for advancing the study of norzoanthamine as a therapeutic drug.

4 Pinnaic Acids and Halichlorine, Novel Marine Azaspirocycles

4.1 Pinnaic Acids, Potent $cPLA_2$ Inhibitors

Specific inhibitors of phospholipase A_2 (PLA_2) have been considered as potential drugs for the treatment of inflammation and other disease states, since PLA_2 is linked to the initial step in the cascade of enzymatic reactions that lead to the generation of inflammatory mediators [36, 37]. Marine natural products such as manoalide [38] and luffariellolide [39] have been reported to be potent PLA_2 inhibitors [40, 41]. A cytosolic 85-kDa phospholipase ($cPLA_2$) exhibits specificity for the release of arachidonic acid from membrane phospholipids [42–44]. Therefore, compounds that inhibit $cPLA_2$ activity have been targeted as anti-inflammatory agents.

Pinnaic acid (**24**) and tauropinnaic acid (**25**) were isolated from the viscera of *P. muricata* [45]. Both **24** and **25** have a unique 6-azaspiro[4.5]decane moiety. The gross structure of **24** was also confirmed by a comparison of the EI-MS fragment peaks with the corresponding peaks of **25**. The relative stereochemistry of **25** was deduced from phase-sensitive NOE correlations.

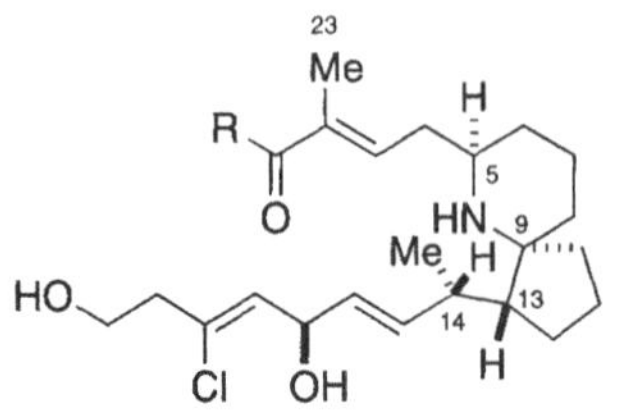

Pinnaic acid (**24**) : R = OH
Tauropinnaic acid (**25**) : R = $NHCH_2CH_2SO_3H$

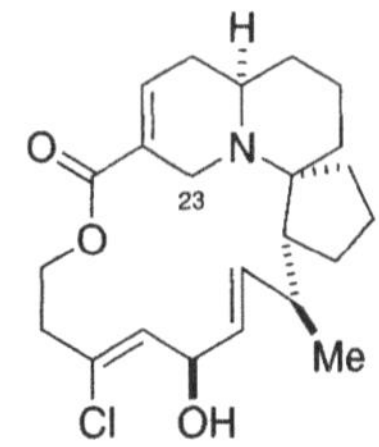

Halichlorine (**26**)

Pinnaic acid (**24**) and tauropinnaic acid (**25**) inhibited $cPLA_2$ activity in vitro with IC_{50} values of 0.2 mM and 0.09 mM, respectively. The activities of pinnaic acids were not so strong, but were still interesting, since $cPLA_2$ inhibitors are rare.

4.2
Halichlorine, An Inhibitor of VCAM-1 Induction

Adhesion molecules are involved in the process of adhesion between cells and the extracellular matrix in the formation of multicellular bodies. The activity of adhesive molecules is very important for the maintenance of function and performance [46]. The clinical application of adhesion molecules as anti-inflammatory agents and immunosuppressive agents are possible, provided that the function of the adhesive molecules can be controlled.

A simple model of multistage adhesion between leukocyte and vascular cells is proposed. This process can be classified into four stages, i.e., (1) rolling, (2) triggering, (3) strong adhesion, and (4) transmigration. VCAM-1 (vascular cell adhesion molecule-1) is affected during the phase of strong adhesion [47]. Thus, drugs that block the induced expression of VCAM-1 may be useful for treating atherosclerosis, coronary artery diseases, angina, and noncardiovascular inflammatory diseases.

Halichlorine (**26**) was isolated from the marine sponge *Halichondria okadai* Kadota [48]. Halichlorine consists of a sterically hindered 15-membered lactone, an azabicyclo[4.4.0]ring, and a 5,6-spiro ring moiety. The relative stereochemistry of **26** was confirmed mainly by the coupling constants and NOESY spectral data. Furthermore, an oxidative degradation product of **26** was synthesized from D-(−)-tartaric acid to determine the absolute stereochemistry of halichlorine [49]. Halichlorine inhibits the induction of VCAM-1 at IC_{50} 7 μg/ml. Although VCAM-1 and ICAM (intercellular adhesion molecule 1) belong to the same immunoglobulin superfamily, halichlorine does not affect ICAM (IC_{50} > 100 μg/ml).

4.3
Biogenesis and Synthesis of Pinnaic Acids and Halichlorine

The structure of pinnaic acid (**24**) from the bivalve *P. muricata* has been shown to be closely similar to that of halichlorine (**26**) from the marine sponge *H. okadai*. Each carbon atom has been tentatively numbered according to the supposed biogenetic formation of the N-C23 bond. Thus, these two bioactive metabolites may each be biosynthesized by symbiotic marine microorganisms.

Both pinnaic acids and halichlorine have attracted the attention of synthetic chemists. To date, 15 research groups have published synthetic studies [50]. The Danishefsky group has achieved the total synthesis of pinnaic acid [51, 52] and halichlorine [53, 54] in an asymmetric manner. Since pinnaic acid is a zwitterionic molecule, the NMR spectrum is quite sensitive to the measurement conditions. Recently, a racemic total synthesis of pinnaic acid (**24**) has been achieved (Scheme 5) [55]. A detailed comparison of the ^{1}H-NMR spectra of both synthetic and natural samples supported Danishefsky's revision of the configuration at C14.

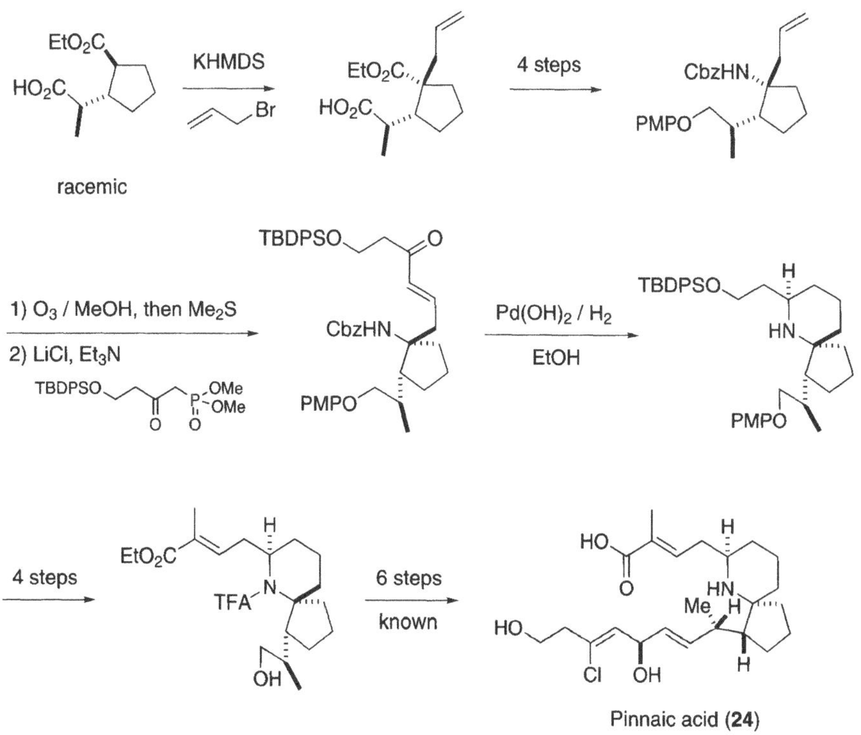

Scheme 5

5
Zamamistatin, a Significant Antibacterial Bromotyrosine Derivative from a Marine Sponge

Bacteria and diatoms are present soon after immersion in seawater, resulting in a biofilm that covers the surface. The establishment of this microfouling biofilm layer is soon followed by macrofouling by barnacles, mussels, and algae. To prevent such macrofouling, several metallic compounds, such as bis(tributyltin)oxide (TBTO), have been used in antifouling paints. However, since their use is restricted to prevent environmental pollution the development of environmentally acceptable antifouling agents has been expected. From marine algae, which are known to be one of the largest producers of biomass in the marine environment, several substances with potent antifouling activity have been isolated, such as fatty acids, lipopeptides, amides, alkaloids, terpenoids, lactones, pyrroles, and steroids [56].

We have especially searched for compounds to prevent microfouling, which would consequently prevent such macrofouling [57]. Antibacterial activity against the marine bacteria *Rhodospirillum salexigens* SCRC 113 strain with adhering properties was selected as a bioassay to identify such compounds. In our continuing search for such compounds, zamamistatin, a novel bromotyrosine derivative, was isolated from the Okinawan sponge *Pseudoceratina purpurea* [58]. Zamamistatin exhibited significant antibacterial activity against *R. salexigens* (21 mm, 1.6 μg/disk).

The molecular formula of zamamistatin was determined to be $C_{18}H_{18}Br_4N_2O_6$. Observation of only nine carbon signals by ^{13}C NMR and its optical rotation value suggested that zamamistatin was an optically active dimer with a C_2 symmetrical structure. On the basis of the analysis of 2D-NMR spectra, the structure of zamamistatin was elucidated to be **28**, possess-

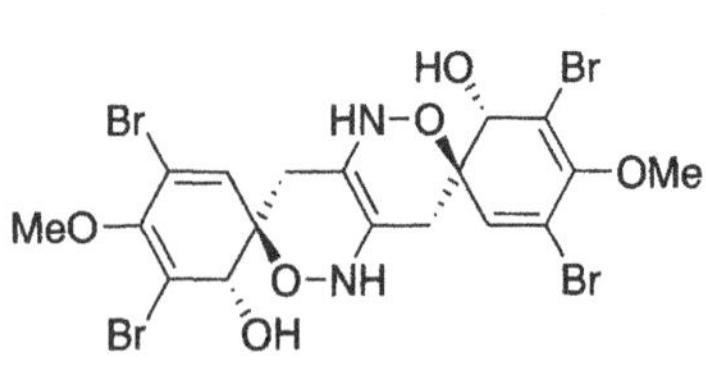

Zamamistatin (**27**)

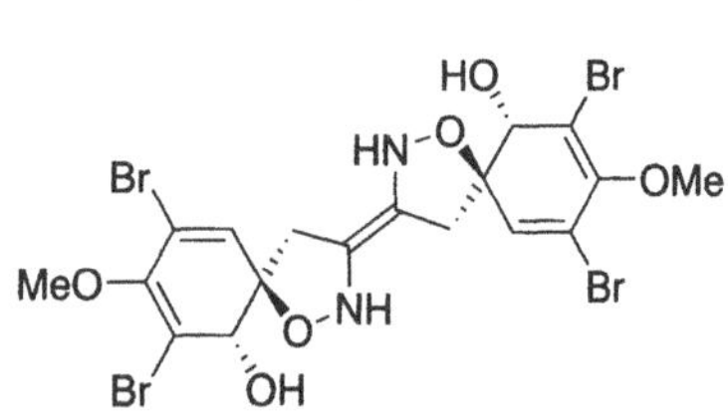

28

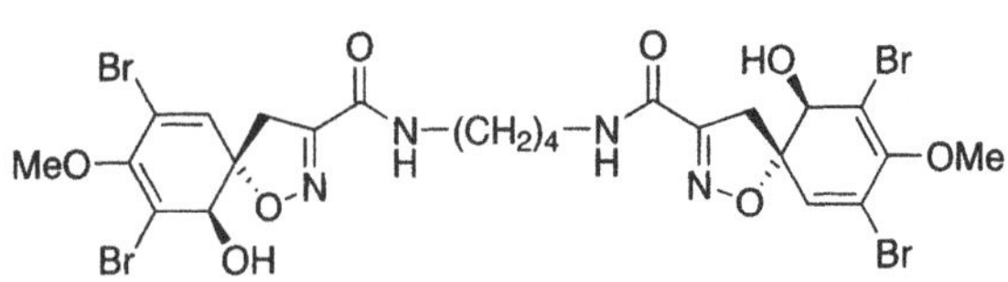

Aerothionin (**29**)

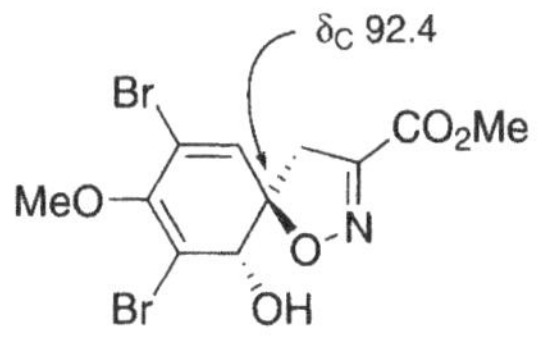

30

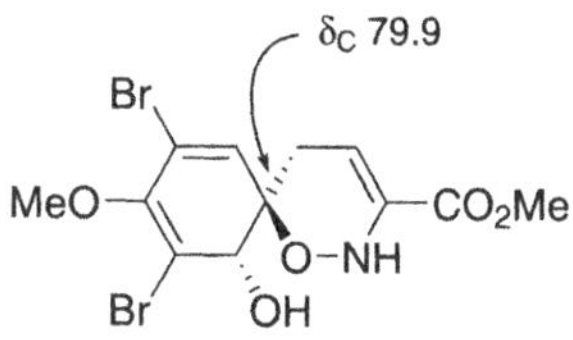

31

ing an *exo*-type dimer of an isooxazoline ring like aerothionin **29** [59, 60]. Recently, however, zamamistatin has been re-isolated and its structure has been revised [61]. A dihydro-1, 2-oxazine methyl ester **30** and a isoxazoline methyl ester **31** [62] were synthesized as model compounds. The spirocarbon signal (C-6) for **30** and **31** appeared at δ_C 92.4 and 79.9, respectively, while the carbon signal in the natural compound appeared at δ_C 74.3. Thus, the structure of natural zamamistatin has been revised to be **27**, an *endo*-type dimer of an aza-oxa-spiro[6.6] unit possessing a dihydro-1,2-oxazine ring moiety.

6
Symbioimines, Potential Antiresorptive Drugs

Large polyol and polyether compounds, such as palytoxin, halichondrin, ciguatoxin, and maitotoxin, are characteristic marine secondary metabolites. These compounds are composed of a long carbon backbone functionalized by oxygen, and have been called "super-carbon-chain compounds" [63]. Interestingly, several super-carbon-chain compounds consist of a single chain starting from a carboxylic acid (C-terminus) and an amine moiety (N-terminus, which are sometimes acylated), thus they can be considered to be huge amino acids, i.e., palytoxin (**32**) (115 straight carbon chain) [64], zooxanthellatoxin-A (**33**) (106 carbons) [65], and azaspiracid 1 (**34**) (40 carbons) [66, 67] (Fig. 5). This concept may also be applicable to pinnatoxins and zoanthamines, which could be biosynthesized from a single carbon chain precursor possessing a terminal amino group, as shown in Figs. 2 and 4.

The symbiotic marine dinoflagellate *Symbiodinium* sp., which is a type of zooxanthellae, is found in a wide range of marine invertebrates, and produces several large bioactive polyol compounds, such as zooxanthellatoxins and zooxanthellamides [65, 68–70]. In our continuing search for biologically active compounds, two unique amphoteric iminium compounds were isolated from this dinoflagellate: symbioimine (**35**) and neosymbioimine (**36**) [71–73].

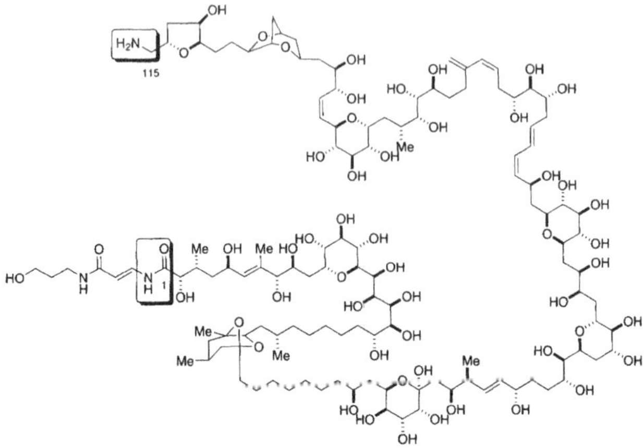

Palytoxin (**32**)

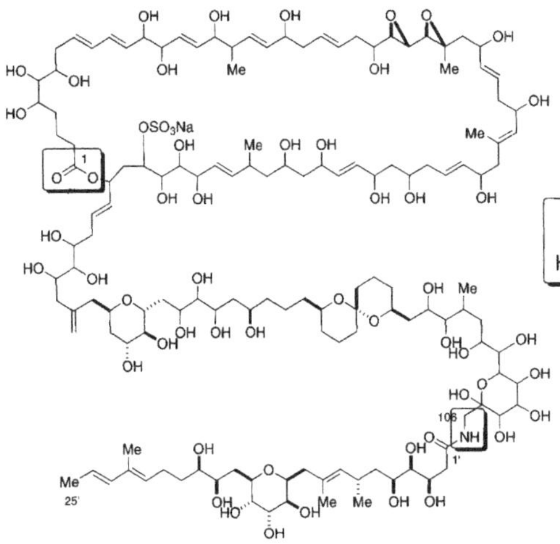

Zooxanthellatoxin A (**33**)

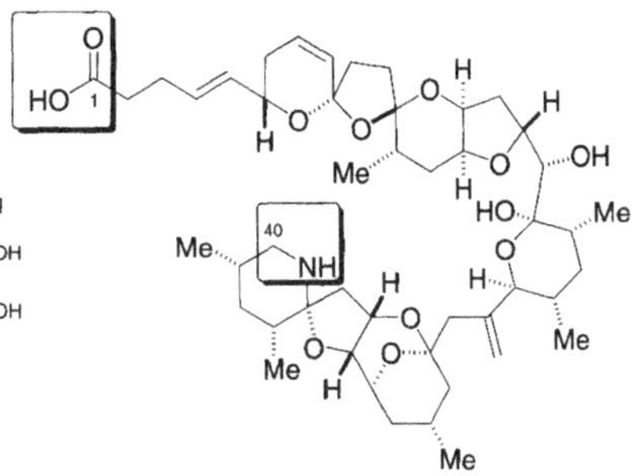

Azaspiroacid 1 (**34**)

Fig. 5 Super-carbon-chain compounds with a terminal amino group

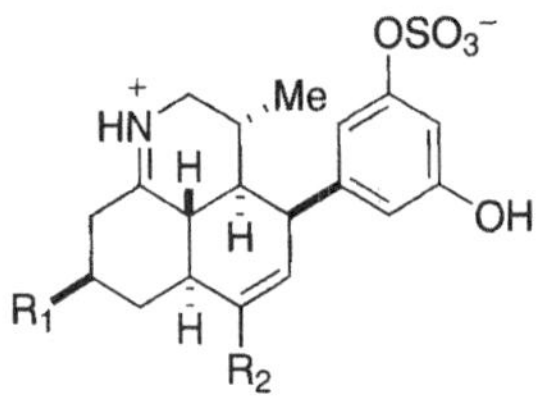

35 : $R_1 = R_2 = H$ (Symbioimine)
36 : $R_1 = R_2 = Me$ (Neosymbioimine)

6.1 Structure of Symbioimines

IR spectrum of symbioimine (**35**) showed absorption bands for hydroxyl (3450 cm^{-1}), iminium (1690 cm^{-1}), and sulfate (1240, 1140, 1050 cm^{-1}) groups. The ^{13}C NMR signal at 188.0 (C-5) implied the presence of an iminium functionality in this water-soluble amphoteric compound. Its structure, which consists of a characteristic 6,6,6-tricyclic iminium ring possessing an aryl sulfate moiety, was deduced by 2D-NMR analysis (Fig. 6).

The relative stereochemistry of **35** was deduced as follows. The large magnitudes of $J_{1a,2}$ = 12.0 Hz, $J_{2,3}$ = 11.1 Hz, $J_{3,4}$ = 11.1 Hz, $J_{4,9}$ = 11.9 Hz, $J_{7a,8a}$ = 12.9 Hz, and $J_{8a,9}$ = 12.9 Hz suggested that all seven of these protons, H-1a, H-2, H-3, H-4, H-7a, H-8a, and H-9a, were oriented in *anti* arrangements with respect to the tricyclic ring. Thus, three six-membered rings may show *trans* ring fusion with each other and that the methyl group (C-19) may be oriented in a pseudo-equatorial conformation with respect to the six-membered iminium ring with a twist-boat conformation. NOEs were observed for H-4/H-14 and H-4/H-18, suggesting that the aryl moiety may be oriented in a pseudo-axial conformation with respect to the cyclohexene ring with a twist-boat conformation. Finally, the stereostructure of **35** was confirmed by X-ray crystallographic analysis. The absolute stereochemistry of **35** was confirmed to be 2*R*, 3*R*, 4*S*, 9*R*, 12*S*, based on the value of the Flack parameter 0.03(13).

Neosymbioimine (**36**) was found to be a congener of **35** possessing a 6,6,6-tricyclic iminium ring, an aryl sulfate moiety, and three methyl groups.

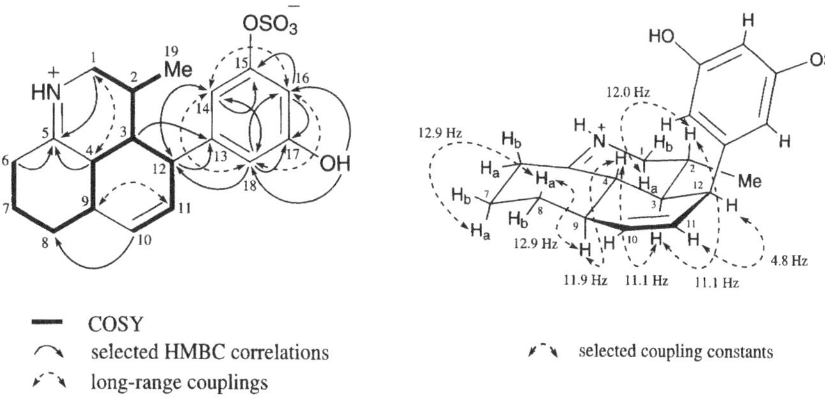

Fig. 6 Planar and stereo structure of symbioimine (**35**) based on 2D-NMR analysis

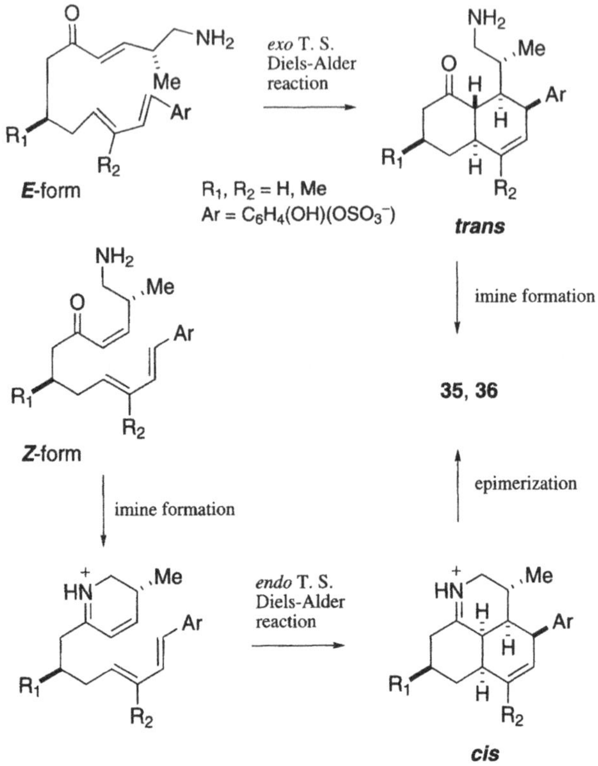

Scheme 6

6.2 Biogenesis of Symbioimines

The unique structures of symbioimines, including a 6,6,6-tricyclic iminium ring, can be explained by the plausible biogenetic pathway shown in Scheme 6, [72, 74]. An intramolecular *exo* transition-state Diels–Alder reaction followed by imine cyclization could form the carbon framework of **35** and **36** stereospecifically, as in the case of pinnatoxins (Fig. 2). Since the 6,6,6-tricyclic iminium ring moiety of neosymbioimine (**36**) is composed of 15 carbons including three methyl groups, a sesquiterpene biosynthesis pathway can be considered. However, due to its inconsistency with the products in the normal sequential head-to-tail condensation of two molecules of isopentenyl diphosphate with dimethylallyl diphosphate, we proposed the polyketide synthesis pathway for construction of the characteristic C1–C12 moiety in **35** and **36**. From this point of view, the tricyclic ring in symbioimines can be considered to be biosynthesized by a cyclic imine formation, *endo*-transition state Diels–Alder reaction to provide *cis*-6,6,6-tricyclic ring stereospecifically, followed by epimerization. Studies of the biosynthetic pathways of symbioimines using isotope-labeled precursor incorporation studies are currently underway.

6.3 Biological Activities of Symbioimines

Symbioimine (**35**) inhibited osteoclastogenesis of the murine monocytic cell line RAW264, which can differentiate into osteoclasts following treatment with receptor activator of nuclear factor-κB ligand (RANKL) (EC_{50} = 44 μg/mL) [71]. RANKL induces the formation of osteoclast-like multinucleated cells in cultures of bone marrow cells [75]. Symbioimine (**35**) inhibited an increase in sRANKL-induced TRAP activity of preosteoclast cells. Meanwhile, it did not affect cell viability even at 100 μg/mL. Thus, symbioimine is a potential antiresorptive drug for the prevention and treatment of osteoporosis in postmenopausal women.

Symbioimine (**35**) also significantly inhibited cyclooxygenase-2 (COX-2) activity (32%) at 10 μM. Meanwhile, it had only weak inhibitory ability (5%) toward COX-1 at 10 μM [72]. The overexpression of COX-2 has been observed in many kinds of tumors, and its role in carcinogenesis and angiogenesis has been extensively investigated [76, 77]. Several COX-2-selective inhibitors, such as rofecoxib, celecoxib, and sulindac, have been developed. Because of its moderate subtype specificity, symbioimine (**35**) may be useful for the development of new nonsteroid anti-inflammatory drugs (NSAID) to treat COX-associated diseases, such as inflammatory diseases and cancer.

7 Conclusion

Along with the development of new analytical instruments and techniques over the past 30 years, a variety of marine alkaloids have been isolated and characterized from natural resources. Further chemical and biological studies on these marine alkaloids should contribute to a deeper understanding of their roles in nature. Also, intensive studies involving the comprehensive evaluation of these molecules may lead to the creation of a new field in bioscience.

References

1. Kuramoto M, Arimoto H, Uemura D (2003) Studies in bioactive marine alkaloids. J Synth Org Chem Jpn 61:1099
2. Kuramoto M, Arimoto H, Uemura D (2004) Bioactive marine alkaloids. A review. Mar Drugs 1:39

3. Gerwick WH, Tan LT, Sitachitta N (2001) Alkaloids of marine cyanobacteria. In: Cordell GA (ed) The Alkaloids, vol 57. Academic Press, San Diego, p 75
4. Fusetani N, Matsunaga S (1993) Bioactive sponge peptides. Chem Rev 93:1793
5. Shimizu Y (1993) Microalgal metabolites. Chem Rev 93:1685
6. Shimizu Y (1996) Microalgal metabolites: A new perspective. Annu Rev Microbiol 50:431
7. Daranas AH, Norte M, Fernández JJ (2001) Toxic marine microalgae. Toxicon 39:1101
8. Rosewater J (1961) The family Pinnidae in the Indo-Pacific. Indo-Pacific Mollusca 1:53/501/632
9. Zheng SZ, Huang FL, Chen SC, Tan XF, Zuo JB, Peng J, Xie RW (1990) The Isolation and bioactivities of pinnatoxin. Zhongguo Haiyang Yaowu (Chinese Journal of Marine Drugs) 9:33
10. Uemura D, Chou T, Haino T, Nagatsu A, Fukuzawa S, Zheng SZ, Chen H (1995) Pinnatoxin A: a toxic amphoteric macrocycle from the Okinawan bivalve *Pinna muricata*. J Am Chem Soc 117:1155
11. Chou T, Kamo O, Uemura D (1996a) Relative stereochemistry of pinnatoxin A, a potent shellfish poison from *Pinna muricata*. Tetrahedron Lett 37:4023
12. Chou T, Haino T, Kuramoto M, Uemura D (1996c) Isolation and structure of pinnatoxin D, a new shellfish poison from the Okinawan bivalve *Pinna muricata*. Tetrahedron Lett 37:4027
13. Takada N, Uemura N, Suenaga K, Chou T, Nagatsu A, Haino T, Yamada K, Uemura D (2001a) Pinnatoxins B and C, the most toxic components in the pinnatoxin series from the Okinawan bivalve *Pinna muricata*. Tetrahedron Lett 42:3491
14. McCauley JA, Nagasawa K, Lander PA, Mischke SG, Semones MA, Kishi Y (1998) Total synthesis of pinnatoxin A. J Am Chem Soc 120:7647
15. Nagasawa K (2000) Total synthesis of pinnatoxin A. J Synth Org Chem Jpn 58:877
16. Takada N, Uemura N, Suenaga K, Uemura D (2001b) Structural determination of pteriatoxins A, B and C, extremely potent toxins from the bivalve *Pteria penguin*. Tetrahedron Lett 42:3495
17. Takada N, Iwatsuki M, Suenaga K, Uemura D (2000) Pinnamine, an alkaloidal marine toxin, isolated from *Pinna muricata*. Tetrahedron Lett 41:6425
18. Djerassi C, Records R, Bunnenberg E, Mislow K, Miscowitz A (1962) Inherently dissymetric chromophores. Optical rotatory dispersion of α,β-unsaturated ketones and conformational analysis of cyclohexenones. J Am Chem Soc 84:870
19. Carmichael WW, Biggs DF, Gorham PR (1975) Toxicology and pharmacological action of anabaena flos-aquae toxin. Science 187:542
20. Devlin JP, Edwards OE, Gorham PR, Hunter NR, Pike RK, Stavric B (1977) Anatoxin A, a toxic alkaloid from anabaena-flos-aquae NRC-44H. Can J Chem 55:1367
21. Leete E, Marion L, Spenser ID (1954) The biogenesis of alkaloids. 12. The mode of formation of the tropine base of hyoscyamine. Can J Chem 32:1116
22. Kigoshi H, Hayashi N, Uemura D (2001) Stereoselective synthesis of pinnamine, an alkaloidal marine toxin from *Pinna muricata*. Tetrahedron Lett 42:7469
23. Fukuzawa S, Hayashi Y, Uemura D, Nagastu A, Yamada K, Ijyuin Y (1995) The isolation and structures of five new alkaloids, norzoanthamine, oxyzoanthamine, norzoanthaminone, cyclozoanthamine and epinorzoanthamine. Heterocycl Commun 1:207
24. Rao CB, Anjaneyula ASR, Sarma NS, Venkatateswarlu Y, Rosser RM, Faulkner DJ, Chen MHM, Clardy J (1984) Zoanthamine; a novel alkaloid from a marine zoanthid. J Am Chem Soc 106:7983
25. Rao CB, Anjaneyula ASR, Sarma NS, Venkatateswarlu Y, Rosser RM, Faulkner DJ, Chen MHM, Clardy J (1985) Alkaloids from a marine zoanthid. J Org Chem 50:3757

26. Rahman AU, Alvi KA, Abbas SA, Choudhary MI, Clardy J (1989) Zoanthaminone, a new alkaloid from a marine zoanthid. Tetrahedron Lett 30:6825
27. Kuramoto M, Hayashi K, Fujitani Y, Yamaguchi K, Tsuji T, Yamada K, Ijyuin Y, Uemura D (1997) Absolute configuration of norzoanthamine, a promising candidate for an osteoporotic drug. Tetrahedron Lett 38:5683
28. Nakamura H, Kawase Y, Maruyama K, Murai A (1998) Studies on polyketide metabolites of a symbiotic dinoflagellate, *Symbiodinium* sp: A new C30 marine alkaloid, zooxanthellamine, a plausible precursor for zoanthid alkaloids. Bull Chem Soc Jpn 71:781
29. Kuramoto M, Hayashi K, Yamaguchi K, Yada M, Tsuji T, Uemura D (1998) Structure-activity relationship of norzoanthamine exhibiting significant inhibition of osteoporosis. Bull Chem Soc Jpn 71:771
30. Yamaguchi K, Hayama T, Makita T, Tsuji T (1997) Time course of bone biomechanical parameters in ovariectomized mice. J Bone Miner Metab 15:138
31. Kuramoto M, Yamaguchi K, Tsuji T, Uemura D (2000) Zoanthamines, antiosteoporotic alkaloids. In: Fusetani N (ed) Drugs from the Sea. Karger, Basel, p 98
32. Kuiper GGJM, Enmark E, Pelto-Huikko M, Nilsson S, Gustafsson J (1996) Cloning of a novel estrogen receptor expressed in rat prostate and ovary. Proc Natl Acad Sci USA 93:5925
33. Turner CH, Burr DB (1993) Basic biomechanical measurements of bone: A tutorial. Bone 14:595
34. Ducy P, Desbois C, Boyce B, Pinero G, Story B, Dunstan C, Smith E, Bonadio J, Goldstein S, Gundberg C, Bradley A, Karsenty G (1996) Increased bone formation in osteocalcin-deficient mice. Nature 382:448
35. Miyashita M, Sasaki M, Hattori I, Sakai M, Tanino K (2004) Total synthesis of norzoanthamine. Science 305:495
36. van den Bosch H (1980) Intracellular phospholipases A. Biochim Biophys Acta 604:191
37. Arita H, Nakano T, Hanasaki K (1989) Thromboxane A2: its generation and role in platelet activation. Prog Lipid Res 28:273
38. Scheuer PJ, de Silva ED (1980) Manoalide, an antibiotic sesterterpenoid from the marine sponge *Luffariella variabilis* (polejaeff). Tetrahedron Lett 21:1611
39. Albizati KF, Holman T, Faulkner DJ, Glaser KB, Jacobs RS (1987) Luffariellolide, an anti-infmammatory sesterterpene from the marine sponge *Luffariella sp.* Experientia 43:949
40. Potts BCM, Faulkner DJ, de Carvalho MS, Jacobs RS (1992a) Chemical mechanism of inactivation of bee venom phospholipase A2 by the marine natural products manoalide, luffariellolide, and scalaradial. J Am Chem Soc 114:5093
41. Potts BCM, Faulkner DJ, Jacobs RS (1992b) Phospholipase A2 inhibitors from marine organisms. J Nat Prod 66:1701
42. Kim DK, Kudo I, Fujimori Y, Mizushima H, Masuda M, Kikuchi R, Ikizawa K, Inoue K (1990) Detection and subcellular localization of rabbit platelet phospholipase A2 which preferentially hydrolyzes an arachidonoyl residue. J Biochem 108:903
43. Kramer RM, Johansen B, Hession C, Pepinsky RB (1990) Structure and properties of a secretable phospholipase A2 from human platelets. Adv Exp Med Biol 275:35
44. Kramer RM, Sharp JD (1995) Recent insights into the structure, function and biology of cPLA2. Agents Actions Suppl 46:65
45. Chou T, Kuramoto M, Otani Y, Shikano M, Yazawa K, Uemura D (1996b) Pinnaic acid and tauropinnaic acid: two novel fatty acids composing a 6-azaspiro[4.5]decane unit from the Okinawan bivalve *Pinna muricata*. Tetrahedron Lett 37:3871

46. Kock AE, Halloran MM, Haskell CJ, Sah MR, Polverini PJ (1995) Angiogenesis mediated by soluble forms of E-selectin and vascular cell adhesion molecule-1. Nature 376:517
47. Osborn L, Hession C, Tizard R, Vassallo C, Huhovoskyi S, Chi-Rosso G, Hobb R (1988) Direct expression cloning of vascular cell adhesion molecule 1: a cytokine-induced endothelial protein that binds to lymphocytes. Cell 59:1203
48. Kuramoto M, Chou T, Yamada K, Chiba T, Hayashi Y, Uemura D (1996) Halichlorine, an inhibitor of VCAM-1 induction from the marine sponge *Halichondria okadai* Kadota. Tetrahedron Lett 37:3867
49. Arimoto H, Hayakawa I, Kuramoto M, Uemura D (1998) Absolute stereochemistry of halichlorine; a potent inhibitor of VCAM-1 induction. Tetrahedron Lett 39:861
50. Clive DLJ, Yu M, Wang J, Yeh VSC, Kang S (2005) Synthetic chemistry of halichlorine and the pinnaic acids. Chem Rev 105:4483
51. Carson MW, Kim G, Hentemann MF, Trauner D, Danishefsky SJ (2001) Concise stereoselective routes to advanced intermediates related to natural and unnatural pinnaic acid. Angew Chem Int Ed 40:4450
52. Carson MW, Kim G, Danishefsky SJ (2001) Stereochemistry of natural and unnatural pinnaic acids: a remarkable long-range stereochemical effect in the reduction of 17-oxo precursors of the pinnaic acids. Angew Chem Int Ed 40:4453
53. Trauner D, Danishefsky SJ (1999) Studies towards the total synthesis of halichlorine: asymmetric synthesis of the spiroquinolizidine subunit. Tetrahedron Lett 40:6513
54. Trauner D, Schwarz JB, Danishefsky SJ (1999) Total synthesis of (+)-halichlorine: an inhibitor of VCAM-1 expression. Angew Chem Int Ed 38:3542
55. Hayakawa I, Arimoto H, Uemura D (2003) Synthesis of (±)-pinnaic acid. Heterocycles 59:441
56. Bhadury P, Wright PC (2004) Exploitation of marine algae: biogenic compounds for potential antifouling applications. Planta 219:561
57. Yamada A, Kitamura H, Yamaguchi K, Fukuzawa S, Kamijima C, Yazawa K, Kuramoto M, Wang GYS, Fujitani Y, Uemura D (1997) Development of chemical substances regulating biofilm formation. Bull Chem Soc Jpn 70:3061
58. Takada N, Watanabe R, Suenaga K, Yamada K, Ueda K, Kita M, Uemura D (2001) Zamamistatin, a significant antibacterial bromotyrosine derivative, from the Okinawan sponge *Pseudoceratina purpurea*. Tetrahedron Lett 42:5265
59. Moody K, Thomson RH, Fattorusso E, Minale L, Sodano G (1972) Aerothionin and homoaerothionin: two tetrabromo spirocyclohexadienylisoxazoles from *Verongia* sponges. J Chem Soc Perkin Trans 1, p 18
60. Ogamino T, Obata R, Nishiyama S (2006) Symmetric synthesis of aerothionin, a marine dimeric spiroisoxazoline natural product, employing optically active spiroisoxazoline derivative. Tetrahedron Lett 47:727
61. Hayakawa I, Teruya T, Kigoshi H (2006) Revised structure of zamamistatin. Tetrahedron Lett 47:155
62. Nishiyama S, Yamamura S (1985) Total synthesis of (+/–)-aerothionin, (+/–)-homoaerothionin, and (+)-aerophobin-1. Bull Chem Soc Jpn 58:3453
63. Uemura D (1991) Antitumor polyethers. In: Scheuer PJ (ed) Bioorganic Marine Chemistry. Springer, Berlin Heidelberg New York, p 1
64. Cha JK, Christ WJ, Finan JM, Fujioka H, Kishi Y, Klein LL, Ko SS, Leder J, McWhorter WW Jr, Pfaff KP, Yonaga M, Uemura D, Hirata Y (1982) Stereochemistry of palytoxin 4. Complete structure. J Am Chem Soc 104:7369

65. Nakamura H, Asari T, Murai A, Kan Y, Kondo T, Yoshida K, Ohizumi Y (1995) Zooxanthellatoxin-A, a potent vasoconstrictive 62-membered lactone from a symbiotic dinoflagellate. J Am Chem Soc 117:550
66. Satake M, Ofuji K, Naoki H, James KJ, Furey A, McMahon T, Silke J, Yasumoto T (1998) Azaspiracid, a new marine toxin having unique spiro ring assemblies, isolated from Irish mussels, *Mytilus edulis*. J Am Chem Soc 120:9967
67. Nicolaou KC, Koftis TV, Vyskocil S, Petrovic G, Ling T, Yamada YM, Tang W, Frederick MO (2004) Structural revision and total synthesis of azaspiracid-1, part 2: Definition of the ABCD domain and total synthesis. Angew Chem Int Ed 43:4318
68. Onodera K, Nakamura H, Oba Y, Ojika M (2003) Zooxanthellamide A, a novel polyhydroxy metabolite from a marine dinoflagellate of *Symbiodinium* sp. Tetrahedron 59:1067
69. Onodera K, Nakamura H, Oba Y, Ojika M (2004) Zooxanthellamide B, a novel large polyhydroxy metabolite from a marine dinoflagellate of *Symbiodinium* sp. Biosci Biotechnol Biochem 68:955
70. Onodera K, Nakamura H, Oba Y, Ohizumi Y, Ojika M (2005) Zooxanthellamide Cs: vasoconstrictive polyhydroxylated macrolides with the largest lactone ring size from a marine dinoflagellate of *Symbiodinium* sp. J Am Chem Soc 127:10406
71. Kita M, Kondo M, Koyama T, Yamada K, Matsumoto T, Lee KH, Woo JT, Uemura D (2004) Symbioimine exhibiting inhibitory effect of osteoclast differentiation, from the symbiotic marine dinoflagellate *Symbiodinium* sp. J Am Chem Soc 126:4794
72. Kita M, Ohishi N, Washida K, Kondo M, Koyama T, Yamada K, Uemura D (2005) Symbioimine and neosymbioimine, amphoteric iminium metabolites from the symbiotic dinoflagellate Symbiodinium sp. Bioorg Med Chem 13:5253
73. Kita M, Uemura D (2005) Iminium alkaloids from marine invertebrates: structure, biological activity, and biogenesis. Chem Lett 34:454
74. Kita M, Sakai E, Uemura D (2006) Pursuit of novel bioactive marine metabolites. J Synth Org Chem Jpn 64:471
75. Soda H, Shim N, Nakagawa N, Yamaguchi K, Kiosk M, Mochizuki S, Tomoyasu A, Yano K, Got M, Murakami A, Stud E, Morinaga T, Hibachi K, Fukazawa N, Takahashi N, Soda T (1998) Osteoclast differentiation factor is a ligand for osteoprotegerin/osteoclastogenesis-inhibitory factor and is identical to TRANCE/RANKL. Proc Natl Acad Sci USA 95:3597
76. Warner TD, Giuliano F, Vojnovic I, Nukasa A, Mitchell JA, Vane JR (1999) Nonsteroid drug selectivities for cyclo-oxygenase-1 rather than cyclo-oxygenase-2 are associated with human gastrointestinal toxicity: A full in vitro analysis. Proc Natl Acad Sci USA 96:7563
77. Reddy BS, Hirose Y, Lubet R, Steele V, Kelloff G, Paulson S, Seibert K, Rao CV (2000) Chemoprevention of colon cancer by specific cyclooxygenase-2 inhibitor, celecoxib, administered during different stages of carcinogenesis. Cancer Res 60:293

Top Heterocycl Chem (2006) 6: 181–214
DOI 10.1007/7081_032

Published online: 7 April 2006

Synthetic Studies on Heterocyclic Antibiotics Containing Nitrogen Atoms

Hiromasa Kiyota

Graduate School of Agricultural Science, Tohoku University,
1-1 Tsutsumidori-Amamiyamachi, 981-8555 Aoba-ku Sendai, Japan
kiyota@biochem.tohoku.ac.jp

Abstract A variety of heterocyclic antibiotics and phytotoxins have been found to date. Total and semi-synthesis of these natural products and related analogous compounds are essential for the elucidation of structure–activity relationships and the development of useful agrochemicals and medicines. Early and recent examples are described, including synthetic studies of glutarimide antibiotics (cycloheximide and related aliphatic and macrolactone derivatives), antimycins (antimycin A and related diolides), and tabtoxins (amino acids with a hydroxy β-lactam ring).

Keywords Antibiotics · Antimycins · Glutarimides · Tabtoxins · Total synthesis

Abbreviations

AA	Antimycins A
aq.	aqueous
DCM	Dichloromethane
cat.	Catalyst
CHX	Cycloheximide
DEPC	Diethylphosphoryl chloride
DHP	3,4-Dihydro-2*H*-pyran
$(DHQ)_2PHAL$	1,4-Bis(9-*O*-dihydrochininyl)phthalazine
DIAD	Diisopropyl azodicarboxylate
dil.	Diluted
EDCI	*N*-Ethyl-*N*′-(3′-dimethylaminopropyl)carbodiimide
DNB	3,5-Dinitrobenzoyl
HOBt	1-Hydroxybenzotriazole
MMPP	Magnesium monoperoxyphthalate
MPM	*p*-Methoxyphenylmethyl (PMB)
MS4A	Molecular sieves 4 Å
NMM	*N*-Methylmorpholine
oxi.	Oxidation
quant.	Quantitative
sat.	Saturated
SAR	Structure–activity relationships
TBS	*t*-Butyldimethylsilyl
TβL	Tabtoxinine-β-lactam
TEMPO	2,2,6,6-Tetramethylpiperidinyloxy radical
TES	Triethylsilyl

1 Introduction

A variety of heterocyclic antibiotics and phytotoxins have been found to date. These compounds or their derivatives are essential for daily life as useful agrochemicals and medicines. Total and semi-synthesis of these natural products and related analogous compounds are important for the elucidation of structure–activity relationships and for further development of novel and effective agents. This chapter reviews synthetic studies of biologically active heterocyclic natural products containing nitrogen atoms in their structures, such as glutarimides, antimycins, and tabtoxins.

2
Glutarimide Antibiotics, the Cycloheximide Family

Cycloheximide (CHX, naramycin A, actidione; **1**) is one of the most famous antibiotics because of its unique, strong and broad antimicrobial spectra, and has been used as a protein synthesis inhibitor or a fungicide [1–4]. To date, a number of related compounds having a β-substituted glutarimide moiety have been isolated [4]. In this chapter, recent works on synthesis of the glutarimides are reviewed.

2.1
Cycloheximide and Derivatives

A variety of cycloheximide derivatives with a modified cyclohexanone ring, an aromatic ring, an aliphatic chain, or a lactam ring, have been synthesized.

2.1.1
Stereoisomers of Cycloheximide

Analogous compounds related to cycloheximide (CHX, **1**) have been investigated in the effort to reduce its high toxicity. Oritani's group reported the synthesis and antimicrobial activities of ten out of 16 (2^4) possible stereoisomers of cycloheximide (**1**) [5–7], including natural naramycin B (**2**) [8] and isocycloheximide (**3**) [9] (Fig. 1). The synthetic scheme was similar to that of **1** [10–12]. They prepared optically active *anti*- and *syn*-2,4-dimethylcyclohexanones by using microbial resolution originally developed by Oritani [13]. As shown in Scheme 1, racemic 1,2-*cis*-2,4-*trans*-2,4-dimethylcyclohexanol [(±)-**4**] was resolved to enantiomerically pure acetate (–)-**5** and (+)-**30** using culture broth of *Bacillus subtilis* var. *niger* in high yields [14]. The alcohol (–)-**5** was oxidized to ketone **6** and this was coupled with aldehyde **7** [15]. Aldol reaction via Li-enolate afforded naramycin B (**2**) and (–)-(6*S*, α*S*)-**1** as major products. On the other hand, four possible isomers were formed almost equally using Ti-enolate, but in

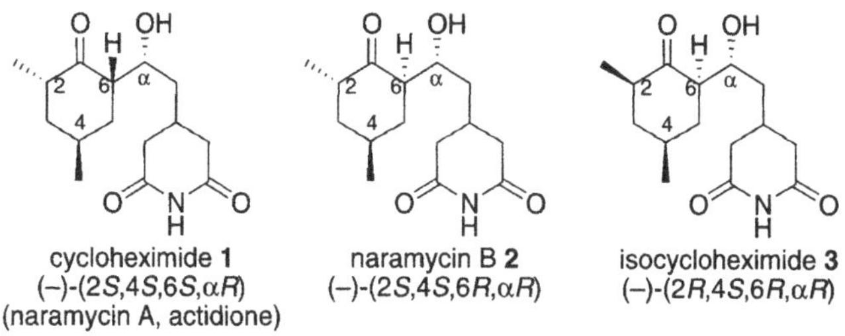

Fig. 1 Natural cycloheximide derivatives

low yields. Scheme 2 shows the preparation of isocycloheximide (**3**) and its isomers. Asymmetric hydrolysis of racemic acetate (±)-**8** gave (-)-acetate [(-)-**10**] with 92% ee [16]. Optical purity of (-)-**10** was enriched by recrystallization of the corresponding DNB ester, and oxidation gave pure ketone (+)-**11**. Aldol reaction provided natural **3**, neocycloheximide **12** [11] and α-epiisocycloheximide **13** [17]. The enantiomers of these three compounds were prepared similarly [7]. Antimicrobial assays against a yeast *Saccharomyces cerevisiae* and the rice blast disease fungus *Pyricularia oryzae* revealed that **1**, **2**, and *ent*-**13** (4-*epi*-**1**) showed strong growth inhibitory activity.

In addition, they prepared racemic steroisomers of des-methyl analogs of **1** (**14** to **17**, Fig. 2). As a result, only 4-desmethylcycloheximide (±)-**15** with the natural relative stereochemistry of CHX showed weak activity against *S. cerevisiae* and *Aspergillus niger* and *Cochliobolus miyabeanus* [18]. Consequently, $4S,\alpha R$-configuration is at least responsible for the CHX activity.

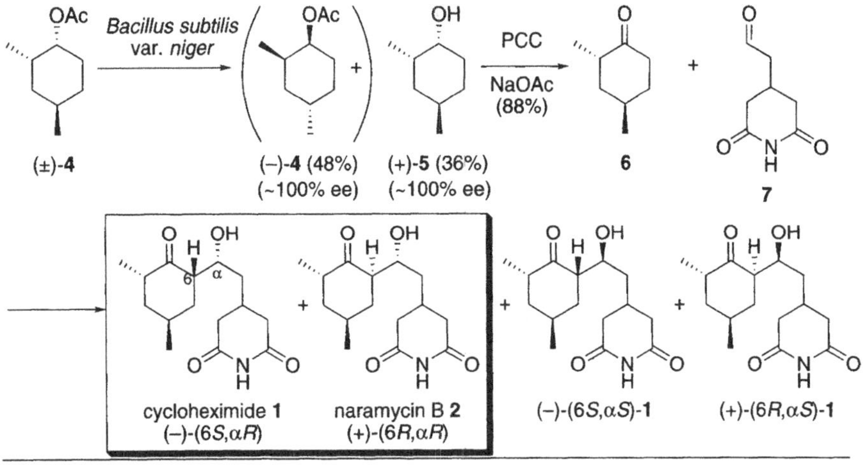

	cycloheximide 1	naramycin B 2	(–)-(6S,αS)-1	(+)-(6R,αS)-1
a) LDA (58%)	3[a]	38	50	9
b) LDA; TMSCl; $TiCl_3$ (19%)	29	21	32	18
c) LDA; TMSCl; $SnCl_4$ (8%)	10	70	0	21
Antimicrobial activity				
Saccharomyces cerevisiae	+++	++	–	++
Pyricularia oryzae	+++	+++	+	++

[a] The rate of diastereomers were determined by HPLC analysis.

Scheme 1 Oritani's synthesis of cycloheximide, naramycin B, and stereoisomers

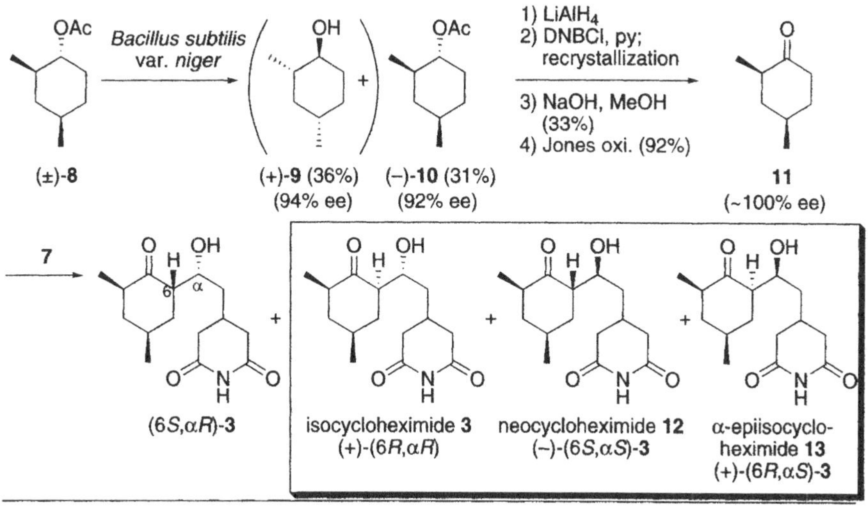

	(6S,αR)-3	isocycloheximide 3	neocycloheximide 12	α-epiisocycloheximide 13
a) LDA (64%)	0	21[a]	74	5
b) LDA; TMSCl; $TiCl_3$ (12%)	0	26	48	27
c) LDA; TMSCl; $SnCl_4$ (5%)	0	13	87	–
Antimicrobial activity				
Saccharomyces cerevisiae		+	–	++
Pyricularia oryzae		+	–	++

[a] The rate of diastereomers were determined by HPLC analysis.

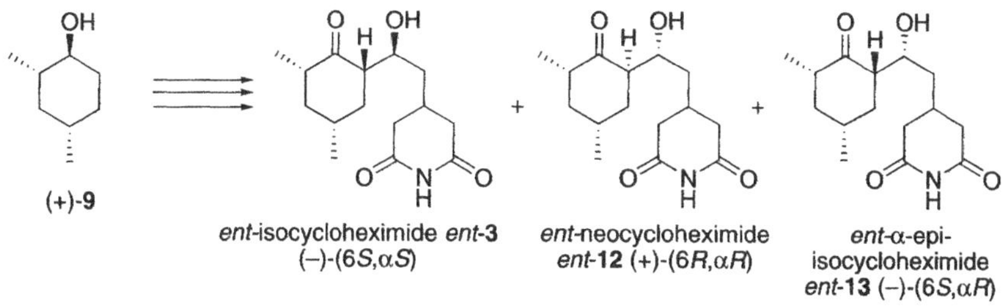

	ent-3	*ent*-12	*ent*-13
Antimicrobial activity			
Saccharomyces cerevisiae	+	–	+++
Pyricularia oryzae	+	+	+++

Scheme 2 Oritani's synthesis of isocycloheximide and stereoisomers

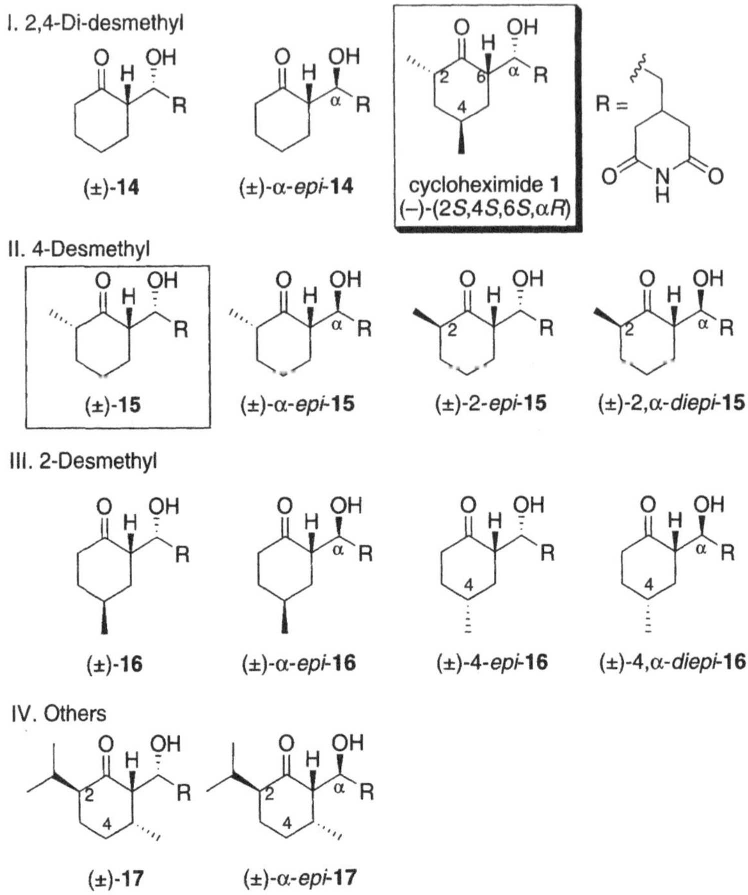

Fig. 2 Desmethyl cycloheximide analogs

2.1.2 Hydroxylated Derivatives, Streptovitacins A and C_2, and E-73

Streptovitacins A (STV-A, **18**) [9], B (**19**), and C_2 (**20**) [9,19] isolated from *Streptomyces griseus*, and E-73 (13-acetoxycycloheximide, **21**) [20] from *S. albulus* are oxygenated derivatives of **1** (Fig. 3). Oritani et al. synthesized **18**, **20**, and **21** to investigate the effect of oxygen functionalities on antimicrobial activities [21,22].

As shown in Scheme 3, they prepared two possible stereoisomers of **20** [21]. Introduction of a methyl group by Grignard reaction to (+)-(*R*)-pulegon (**22**) gave **23**, which was subjected to ozonolysis to afford **24**. The tertiary hydroxyl group was protected as TMS ether (**25**) and subsequent aldol reaction with aldehyde **7** provided (–)-(2*R*,4*R*,6*R*,α*S*)-STV-C_2 (**20a**) as a single diastereomer. On the other hand, racemic ketone (±)-**11** was hydroxylated via

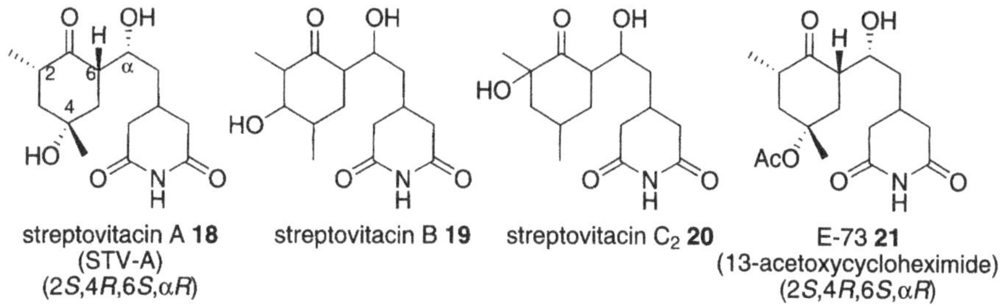

Fig. 3 Streptovitacins

enol acetate **26** using MMPP to give (±)-**24**, which afforded (±)-(2*R**,4*S**,6*S**, α*S**)-STV-C_2 (**20b**).

Scheme 4 indicates the synthesis of STV-A and E-73. Hydroxy ketone **27** was converted to **28** and the tertiary hydroxyl group was protected as TMS ether (**29**). Introduction of a 2-methyl group (**30**) and aldol reaction with **7** afforded a mixture of (±)-STV-A [(±)-**18**] and α-*epi*-**18** (15 : 85). Using Ti-

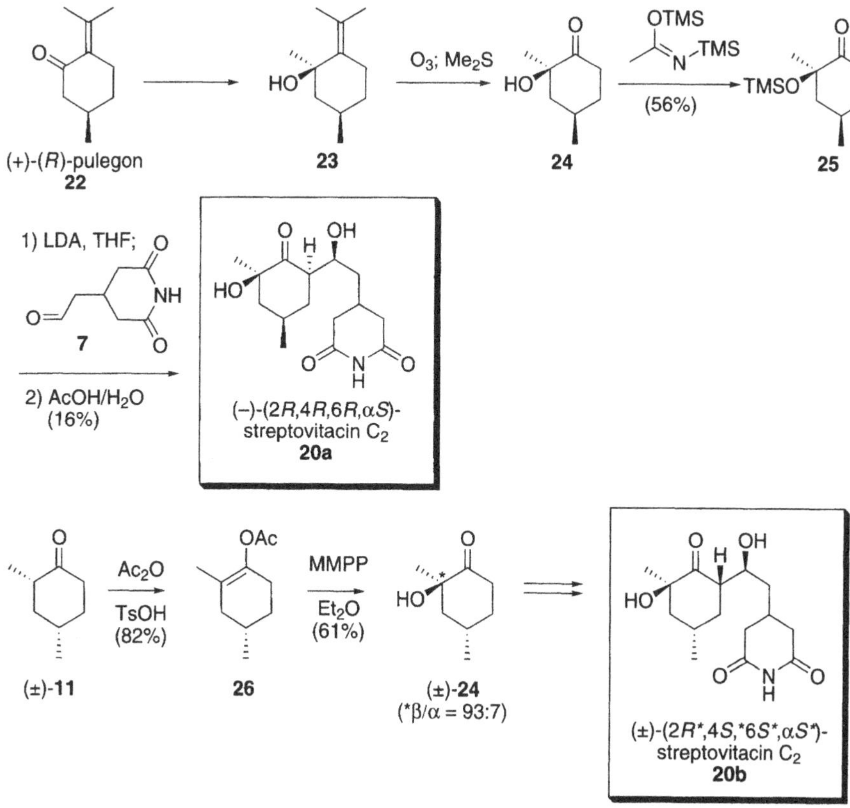

Scheme 3 Oritani's synthesis of streptovitacin C_2

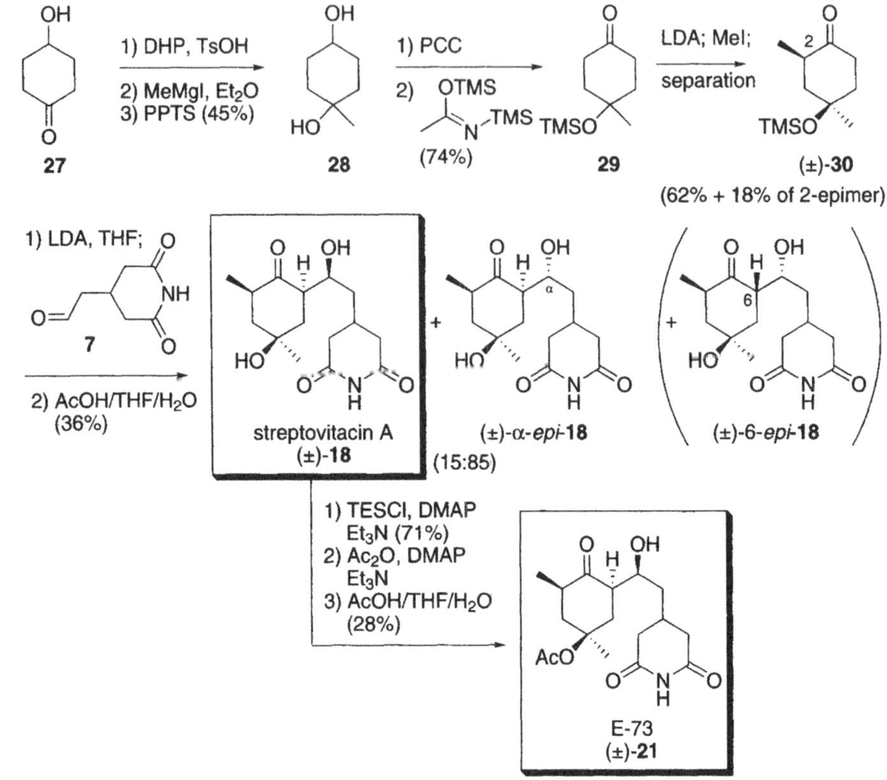

Scheme 4 Oritani's synthesis of streptovitacin A and E-73

enolate as shown in Scheme 1, the product ratio became **18**/*α-epi*-**18**+6-*epi*-**18** = 35 : 65. These isomers were hardly separable but pure samples were provided by HPLC and recrystallization. Compound (±)-**18** was converted to E-73 [(±)-**21**] in three steps [22]. Only (±)-**18** showed a weak growth inhibitory activity against *S. cerevisiae* [21, 22].

2.1.3 Aromatic Ring Derivatives, Actiphenol and Actiketal

The antimicrobial activity of cycloheximide derivatives with an unsaturated double bond in the cyclohexanone ring decreased as inactone (**31**) [23] and actiphenol (**32**) [24]. On the other hand, actiketal (RK-441S, **33**), a unique benzofuran-type derivative, was isolated from a culture extract of *Streptomyces pulveraceus* subsp. *epiderstagenes* [25]. This compound showed inhibitory activities towards Con A-induced blast formation in spleen cells (100% at 20 nM), and towards the incorporation of [^{3}H]thymidine into epidermal growth factor-stimulated Balb/MK cells (IC_{50} 14.5 μM). Kiyota–Oritani's group first tried a semi-synthesis from **32** (Scheme 5) [26], i.e.,

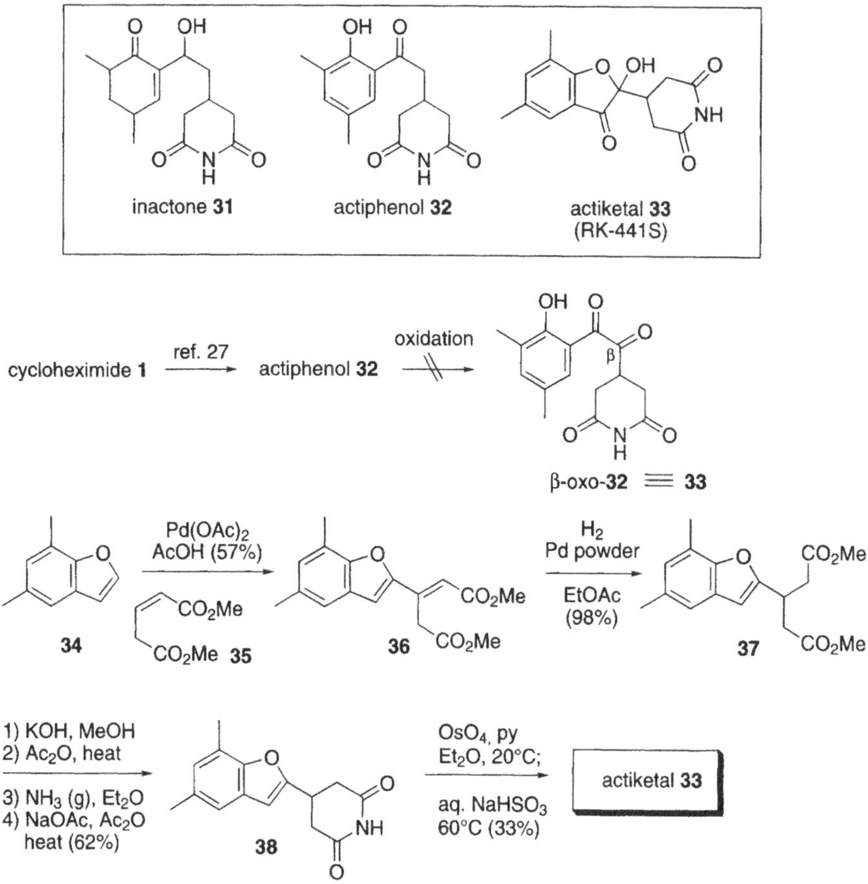

Scheme 5 Kiyota-Oritani's synthesis of actiketal

β-oxo-**32** corresponds to **33**. Preparation of **32** from cycloheximide **1** was according to Johnson [27]. However, all attempts to oxidize the β-position of **32** failed. A palladium-mediated oxidative coupling reaction [28] of **34** [29] with dimethyl glutaconate (**35**) gave **36** in a 57% yield. The double bond of the glutaconate residue was selectively hydrogenated (**37**), and the diester was transformed to glutarimide **38** [30]. Finally, osmium oxidation of the double bond of the furan part [31] afforded an over-oxidized product, actiketal (**33**) [26, 32].

2.1.4
Aliphatic Derivative, Streptimidone

Among a group of aliphatic derivatives **39–42** [33–36], streptimidone (**39**), isolated from various *Streptomyces* species [37, 38] and *Micromonospora coerulea* [39], showed strong antimicrobial activity against eukaryotic cells [38]. Oritani et al. synthesized all four diastereomers of streptimidone to investigate its antimicrobial activity (Scheme 6) [40]. Alcohol (*S*)-**43** [41] derived from methyl (S)-3-hydroxy-2-methylpropanoate (99.9% ee) was converted to aldehyde **44**, and this was coupled with phosphine oxide reagent **45** [42] to afford diene **46**. The TBS protecting group was removed, and the resulting hydroxyl group was oxidized to give ketone **47**. Finally, an aldol reaction of **47** with **7** [43], gave the unnatural enantiomer of streptimidone, (±)-(2′*S*,5′*R*)-**39**, and its 2′-epimer. The overall yields were 6.8%

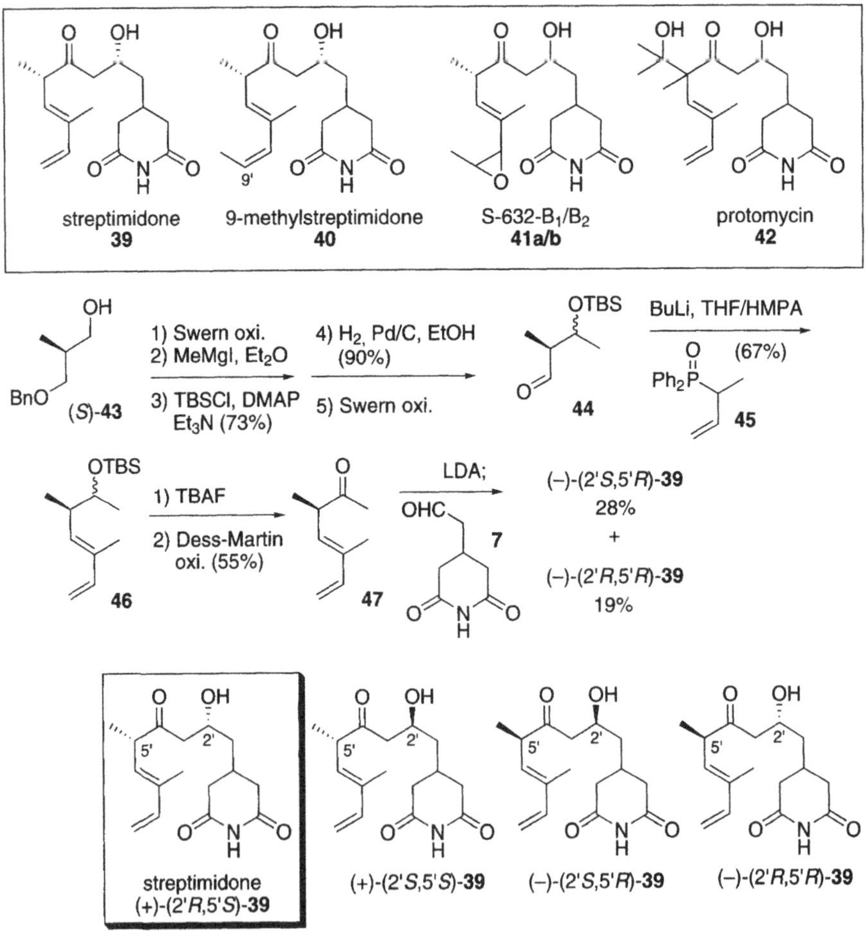

Antimicrobial activity	(+)-(2'R,5'S)-39	(+)-(2'S,5'S)-39	(−)-(2'S,5'R)-39	(−)-(2'R,5'R)-39
S. cerevisiae	++	+	+	−
C. miyabeanus	++	−	+	−

Scheme 6 Oritani's synthesis of streptimidone

and 4.4%, respectively. The natural (+)-streptimidone and its 2′-epimer were similarly prepared. Optical purity of these products were 90% ee. The antifungal activity of the four stereoisomers towards *Saccharomyces cereviciae* and *Cochliobolus miyabeanus* indicated that 2′*R*,5′*S*, especially in the 5′*S* configuration (the same orientation as that in CHX, **1**), was necessary for the activity [40].

2.1.5 Lactam Derivative, Epiderstatin

Epiderstatin (**48**), having a unique alkylidene lactam, was isolated as a potent inhibitor of the signal transduction induced by epidermal growth factor in quiescent Balb/MK cells [44–46]. However, the activity was later found to be due to contaminating 13-acetoxycycloheximide **21** [47]. The (3*S*, 5*S*)-absolute stereochemistry of **48** was established by X-ray analysis [48]. Dow's first synthesis also confirmed the relative stereochemistry [49] (Scheme 7).

Dimethyl glutarate (**49**) was converted to monothioglutarimides *trans*- and *cis*-**50** as a 1 : 1 mixture. Glutarimide fragment **53** was derived from **51** [15] via diazo ketone **52**. The mixture **50** was treated with NaH and condensed with 0.5 equivalents of **53**, the *S*-alkylated intermediate **54** being rearranged [50], to give epiderstatin **48** and *cis*-**48** as a 3 : 1 mixture. Treatment of the single isomer *cis*-**50** gave a 1 : 3 mixture. They determined the relative stereochemistry based on these results. Ubukata et al. followed this method and resolved four stereoisomers by chiral HPLC [47].

Ubukata's synthesis [51] started with ammonolysis followed by Swern oxidation of (±)-**55** [52] to give nitrile aldehyde **56**. Wittig–Horner reaction of **56** with phosphonate **58** derived from glutarimide acetic acid **57** [27] gave **59**. The nitrile was converted to amide **60**, which was subjected to selenolactamization [53] to afford **61**. Oxidation–elimination of the selenenide moiety gave **48** a diastereomeric mixture.

2.2 Macrolactones and Hydroxy Acids

Since the discovery of lactimidomycin (**62**) [54], which showed strong cytotoxicity against murine melanoma (B16-F10, IC50 0.03 μg/mL), several CHX derivatives with an unsaturated macrolactone ring or the corresponding open-chained hydroxy acid have been isolated to date (Fig. 4). 14-Membered lactone migrastatin (**63**), isolated from *Streptomyces* sp. MK929-43F1, is a potent inhibitor of human tumor cell migration [55–57]. The relative and absolute stereochemistry of **63** was determined by a modification of Mosher's method [58] and X-ray crystallographic analysis of the corresponding *N*-*p*-bromophenacyl derivative [59]. Later, 12-membered lactone congeners isomigrastatin (**64**) [60] and its L-cycteine conjugates NK30424A/B (**65a/b**) were

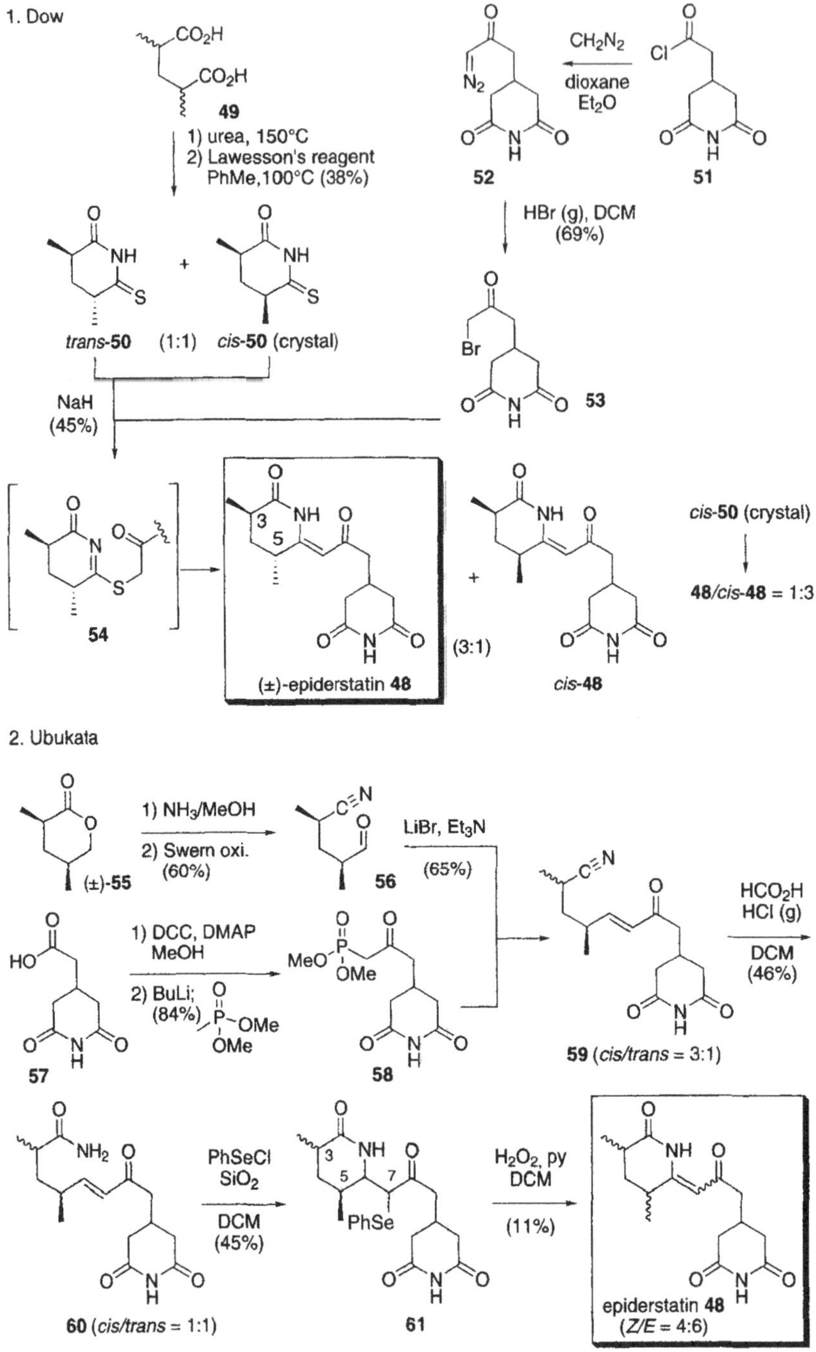

Scheme 7 Synthesis of epiderstatin

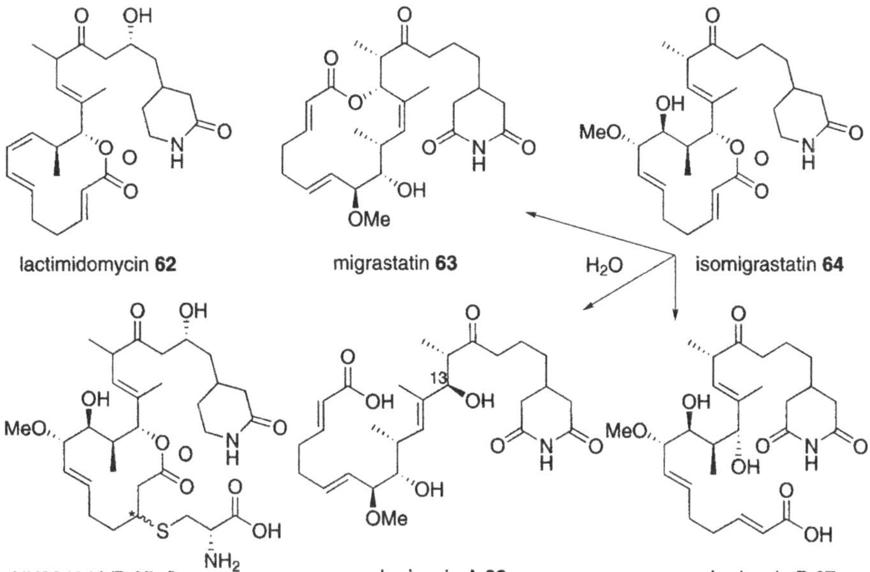

Fig. 4 Migrastatin and related compounds

isolated [61,62]. Their open-chained hydroxy acid derivatives dorrigocin A (**67**) and B (**68**) were also reported [60,63–65]. Very recently Shen et al. revealed that isomigrastatin (**64**) is the main product of *Streptomyces platensis* NRRL18993, and that the other derivatives **63**, **66**, and **67** could be derived from **64** via H_2O-mediated rearrangement [66]. They also reported other minor derivatives and determined the absolute stereochemistry of compounds **64–67** [67].

2.2.1 Total Synthesis of Migrastatin

Danishefsky et al. reported the total synthesis of migrastatin (**63**) (Scheme 8) [68–70]. Their synthesis started from tartrate derivative **68**. Partial reduction by DIBALH and the resulting formyl groups were vinylated to give diol **69**, which was then converted to aldehyde **70** [71]. Construction of two additional stereocenters was achieved by hetero Diels-Alder reaction mediated by titanium tetrachloride to afford γ-pyrone **71**. This was transformed to alcohol **73** via diol **72**. The other two stereocenters were constructed by Evans' *anti*-aldol reaction [72] (**74**), and removal of the chiral oxazolidinone moiety gave alcohol **75**. This was converted to phosphonate **76**, and subsequent Wittig-Horner reaction with aldehyde **7** provided **77**, which was further condensed with acid **78** to give ester **79**. Ring-closing

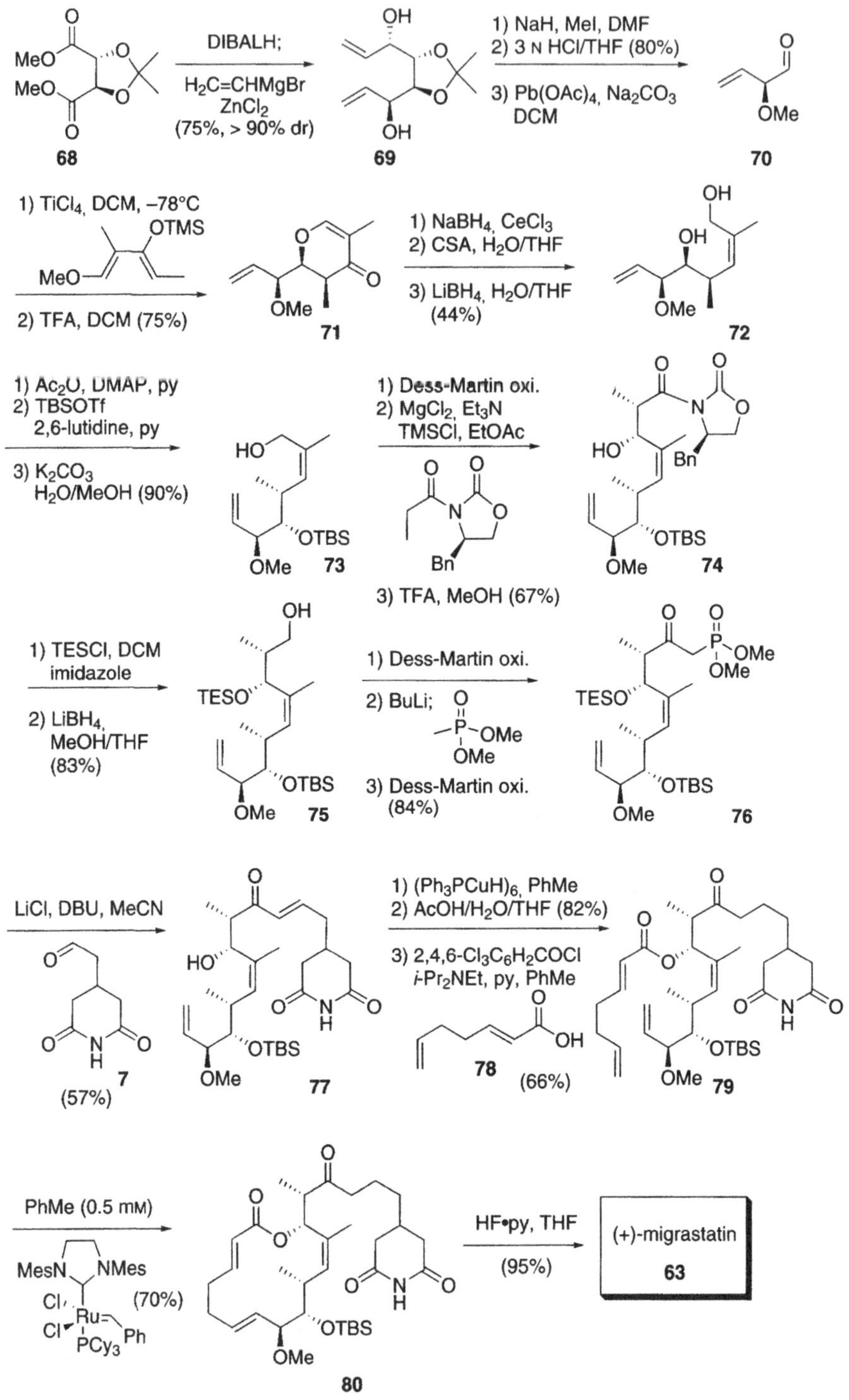

Scheme 8 Danishefsky's total synthesis of migrastatin

metathesis (RCM) was accomplished by using Grubbs' 2nd generation catalyst [73] to form macrolactone **80**. Deprotection of the TBS group gave (+)-migrastatin (**63**).

2.2.2 Migrastatin Analogs

Danishefsky et al. also prepared migrastatin analogs and investigated the structure–activity relationships (SAR) studies for chamber cell migration assay with human endothelial cells (HUVECs) [70, 74]. Scheme 9 shows the synthesis of two analogs. The intermediate **73** was converted to azide **81** and bromide **84**, which were respectively transformed to macrolactam **83** and macroketone **87** analogs lacking the glutarimide moiety. Among the analogs prepared, these two compounds in particular showed stronger activity than the parent compound **63**.

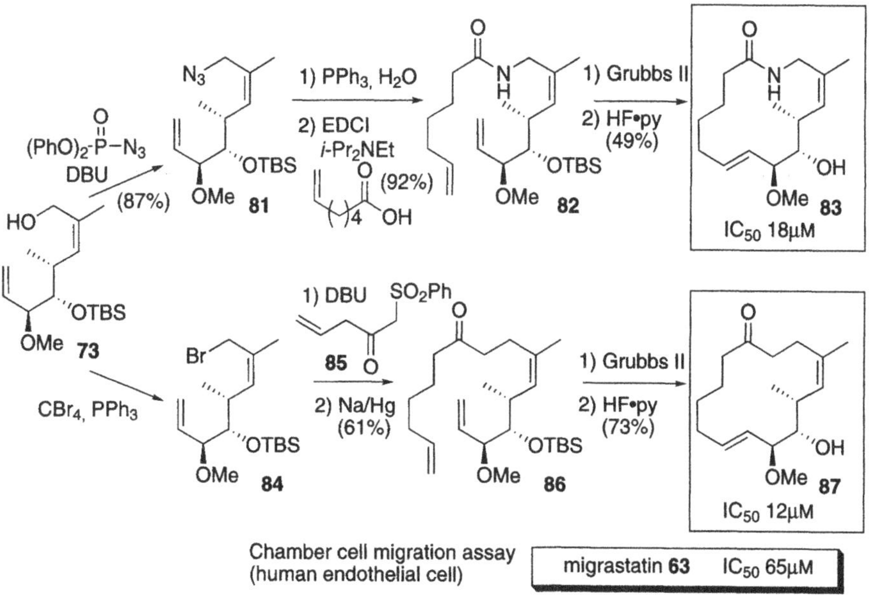

Scheme 9 Synthesis of migrastatin analogs

2.2.3 Synthetic Studies of Dorrigocin A

There has been only one report on the synthesis of these derivatives by other groups. Le Brazidec et al. reported the synthesis of C1–C13 fragment of unnatural 2,3-dihydro analog of dorrigocin A [75] (Scheme 10). Julia–Kocieński olefination [76] of **88** with **89** afforded **90**, which was subjected to Evans'

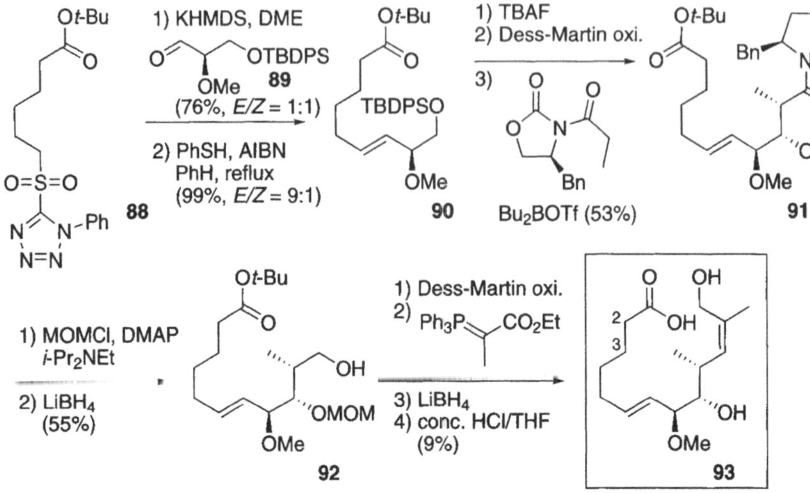

Scheme 10 Synthetic studies of dorrigocin A

syn-aldol reaction to give **91**. Removal of the chiral auxiliary (**92**) and chain elongation gave hydroxy acid **93**.

3
Antimycins and Related Antibiotics

Since 1946, a series of antibiotic antimycins A (AA, **94-1a**~**94-9**) have been isolated from various *Streptomyces* species [77–80] (Fig. 5). Antimycin A complex, a mixture of derivatives, has been widely used for biochemical studies. As it inhibits the electron transfer of ubiquinol–cytochrome *c* oxidoreductase [81], many scientists have investigated their structure-activity relationships and mechanism of action [82–89]. The related deacyl compounds, deisovalerylblastmycin (**95a**) [90], urauchimycins **95b** and **95c** [91], and kitamycins **95d** and **95e** [92] were isolated from *Streptomyces* species. In addition, the corresponding L-serine derivatives, UK-2A~2D (**96a–d**) [93] and UK-3A (**97**) [94], were added to this series.

3.1
Early Syntheses of Antimycin A_{3b}

3.1.1
Kinoshita's First Synthesis of Antimycin A_{3b}

The curious dilactone structure of AA, made up of L-threonine and the highly substituted hydroxy acid, has been an ideal target for synthetic chemists. The

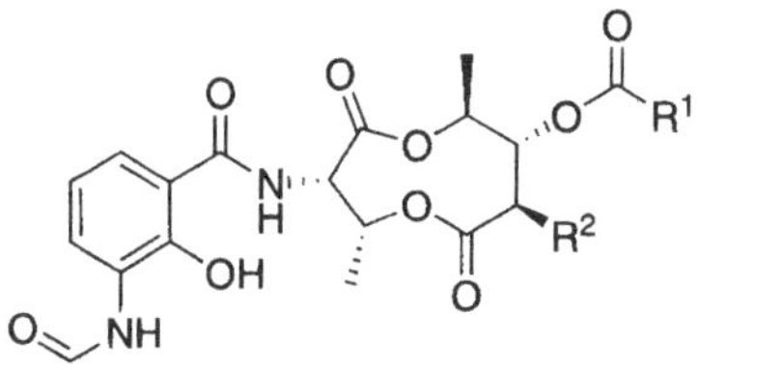

		R^1	R^2
antimycin A_{1a} (AA_{1a})	(+)-**94-1a**	*s*-Bu	hexyl
A_{1b}	(+)-**94-1b**	*i*-Bu	hexyl
A_{2a}	(+)-**94-2a**	*i*-Pr	hexyl
A_{2b}	(+)-**94-2b**	Pr	hexyl
A_{3a}	(+)-**94-3a**	(*S*)-*s*-Bu	Bu
A_{3b} (blastmycin)	(+)-**94-3b**	*i*-Bu	Bu
A_{4a}	(+)-**94-4a**	*i*-Pr	Bu
A_{4b}	(+)-**94-4b**	Pr	Bu
A_5	(+)-**94-5**	*i*-Bu	Et
A_6	(+)-**94-6**	Pr	Et
A_{7a}	(+)-**94-7a**	*i*-Pr	hexyl
A_{7b}	(+)-**94-7b**	Pr	Bu
A_{8a}	(+)-**94-8a**	*s*-Bu	Bu
A_{8b}	(+)-**94-8b**	*i*-Bu	Bu
antibiotic I-2743C	(+)-**94-9**	Me	Bu

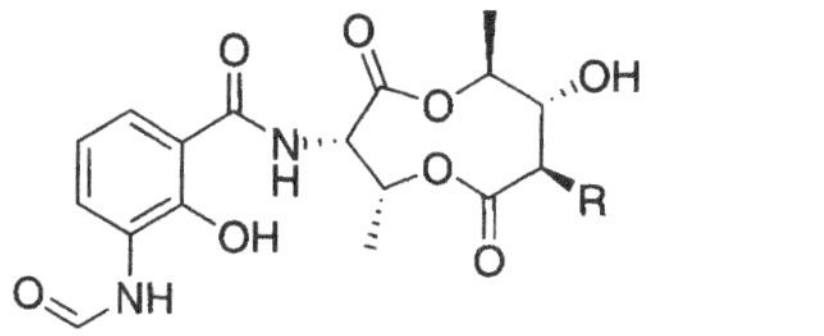

		R
deisovaleryl-blastmycin	(+)-**95a**	Bu
urauchimycin A	(+)-**95b**	$-CH_2CHMeEt$
B	(+)-**95c**	isopentyl
kitamycin A	(+)-**95d**	hexyl
B	(+)-**95e**	isopentyl

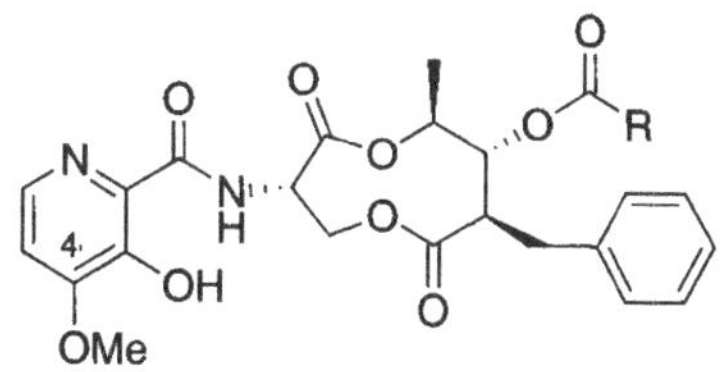

		R
UK-2A	(+)-**96a**	*i*-Pr
UK-2B	(+)-**96b**	*t*-Bu
UK-2C	(+)-**96c**	*i*-Bu
UK-2D	(+)-**96d**	*s*-Bu
UK-3A (4'-desmethoxy)	(+)-**97**	*i*-Pr

Fig. 5 Antimycins A and related compounds

first total synthesis of (+)-antimycin A_{3b} (blastmycin, **94-3b**) was achieved by Kinoshita et al. in 1970 [95, 96], however, separation of the diastereomers of **100** was necessary and the yield of macrolactonization to **101** was very low (0.8%). They also reported the improved synthesis of (+)-AA_{3a} and (+)-deisovalerylblastmycin (**95a**) in 1976 and 1979 as shown in Scheme 11 [97, 98]. Sugar derivative **102** was converted to L-threoninyl ester **104** via furanose **103**. Macrolactonization was accomplished by using Mukaiyama-Gerlach's method [99, 100] and the resulting diolide **105** was transformed to (+)-AA_{3b} (**94-3b**) and deisovalerylblastmycin (**95a**). They also prepared (−)-blastmycinolactol (**108**) and (+)-blastmycinone (**109**), degradation products of antimycin A_{3ab} [101].

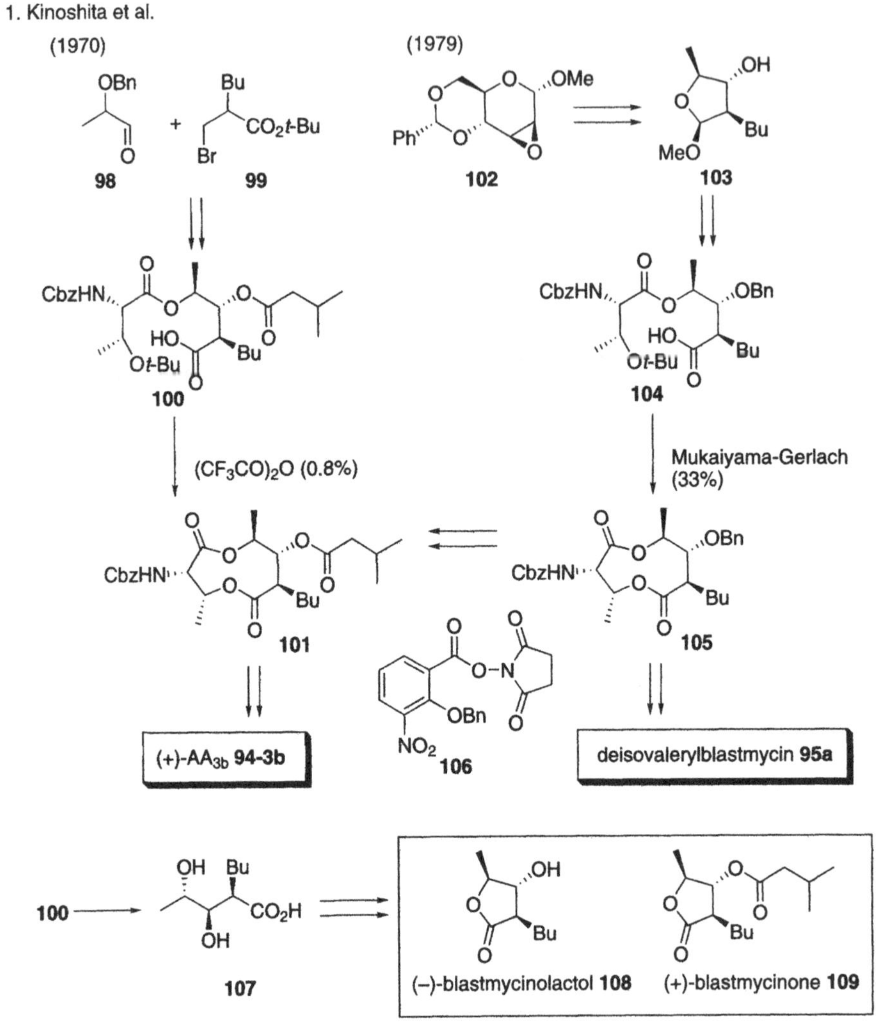

Scheme 11 Kinoshita's first synthesis of antimycin A

3.1.2
Synthetic Studies in the 1980s

Nakata–Oishi's group reported the synthesis of the dilactone intermediates (+)-**105** and (±)-**109** (Scheme 12) [102]. The key steps of Wasserman and Bambale's formal synthesis were alkylation of oxazole **112** and photooxygenation of the oxazole moiety of **115** [103]. Mulzer et al. [104] prepared all possible stereoisomers of blastimycinolactol **108** by using highly selective Nozaki–Hiyama–Kishi reaction [105]. Many of the synthetic studies of **108** and **109** reported to date are omitted in this review.

2. Nakata-Oishi et al. (1983)

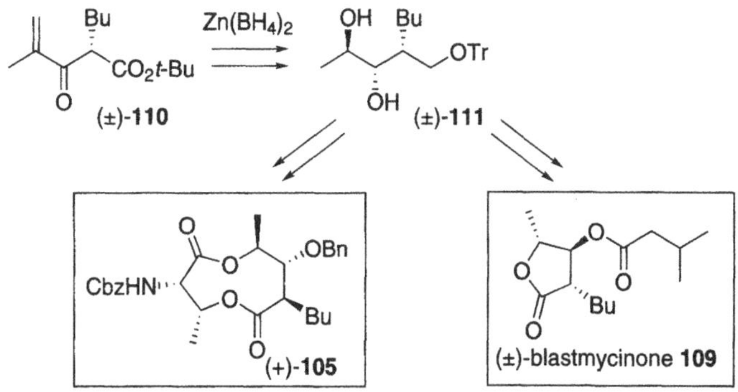

3. Wassermann & Gambale (1985)

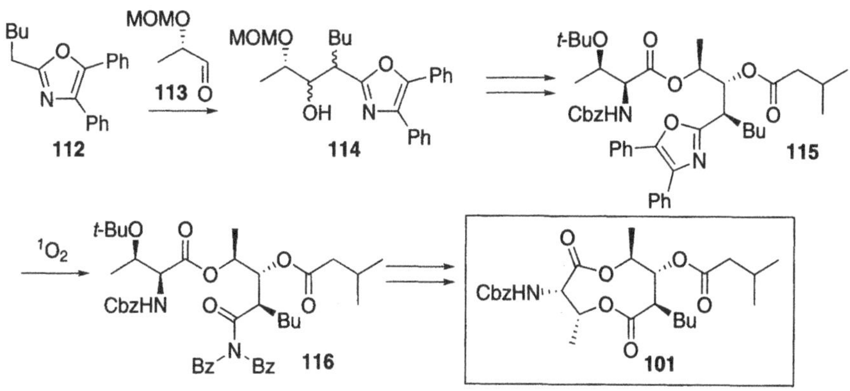

4. Mulzer et al. (1988)

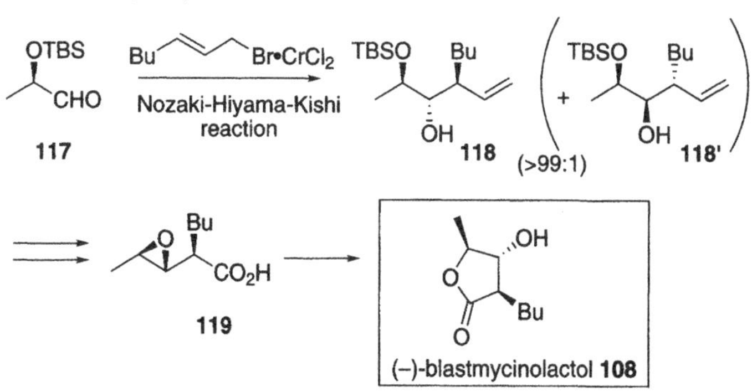

Scheme 12 Formal synthetic studies of antimycins A

3.2 Recent Syntheses of Antimycins

3.2.1 Oritani's Synthesis of (–)-Antimycin A_{3b}

Oritani et al. prepared unnatural enantiomer (–)-antimycin A_{3b} (*ent*-**94-3b**) and its deformylamidodehydroxy analog using chelation-controlled alkylation as the key step (Scheme 13) [106–108]. Aldehyde **113** [103] derived from ethyl (*S*)-lactate was treated with allylic stannyl compound under chelation control ($MgBr_2$) to give all-*syn* product **120** [104, 109]. The ratio of **120** to other diastereomers was 95 : 5. The hydroxyl group was protected as benzyl ether and the MOM ether was cleaved to afford **121**. The formed hydroxyl group was inverted by Mitsunobu reaction to give **122**. Then, the D-threonine residue **123** [110] was coupled with **122** to afford **124** in 87% yield. Direct introduction of **123** to **121** failed. The double bond of **124** was cleaved to give the enantiomer of Kinoshita's intermediate *ent*-**104** [98]. This compound was deprotected and condensed with acid **126** by using DEPC. In this step the 7-position was partially epimerized but the 7-epimer was not acylated in the next step, probably due to the steric hindrance. Finally, the unnatural enantiomer, (–)-AA_{3b} [(–)-*ent*-**94-3b**] and, in the same manner, deformylamidodehydroxyantimycin A_{3b} [(–)-*ent*-**127**] were synthesized. In addition, natural enantiomer (+)-**105** was prepared for the formal synthesis. These unnatural enantiomers (–)-*ent*-**94-3b** and (–)-*ent*-**127**, compared with natural antimycin mixture, scarcely inhibited the growth of *Saccharomyces cereviciae* nor the electron transport of rat liver mitochondria. Miyoshi et al. reported the importance of formylamido and hydroxyl substituents on the phenyl ring for the latter activity [87].

3.2.2 Tsunoda's Synthesis of Antimycin A_{3a} and A_{3b}

Tsunoda et al. reported the synthesis of both enantiomers of AA_{3b}, and determination of the absolute configuration of the AA_{3a} side chain [111, 112]. Their key step was the asymmetric aza-Claisen rearrangement reaction [113] of amide **129** derived from (*R*)-α-phenethylamine [(*R*)-**128**] (Scheme 14). The thermal 3,3-sigmatropic rearrangement proceeded to give **130** as a four diastereomeric mixture, from which (7*S*, 8*R*)-**130** and (7*R*, 8*S*)-**130** were isolated as an inseparable mixture (82 : 18). Iodolactonization and subsequent deiodination of **130** afforded lactone **131** of 64% ee. Hydrolysis of the lactone ring and prenylation gave **132**. The D-threonine moiety **133** was introduced by Mitsunobu reaction, followed by deprotection to give **134** (64% ee). Macrolactonization using Mukaiyama–Gerlach method [99, 100] and purification provided *ent*-**135** of $\sim$ 100% ee, which was condensed with acid **136** to afford

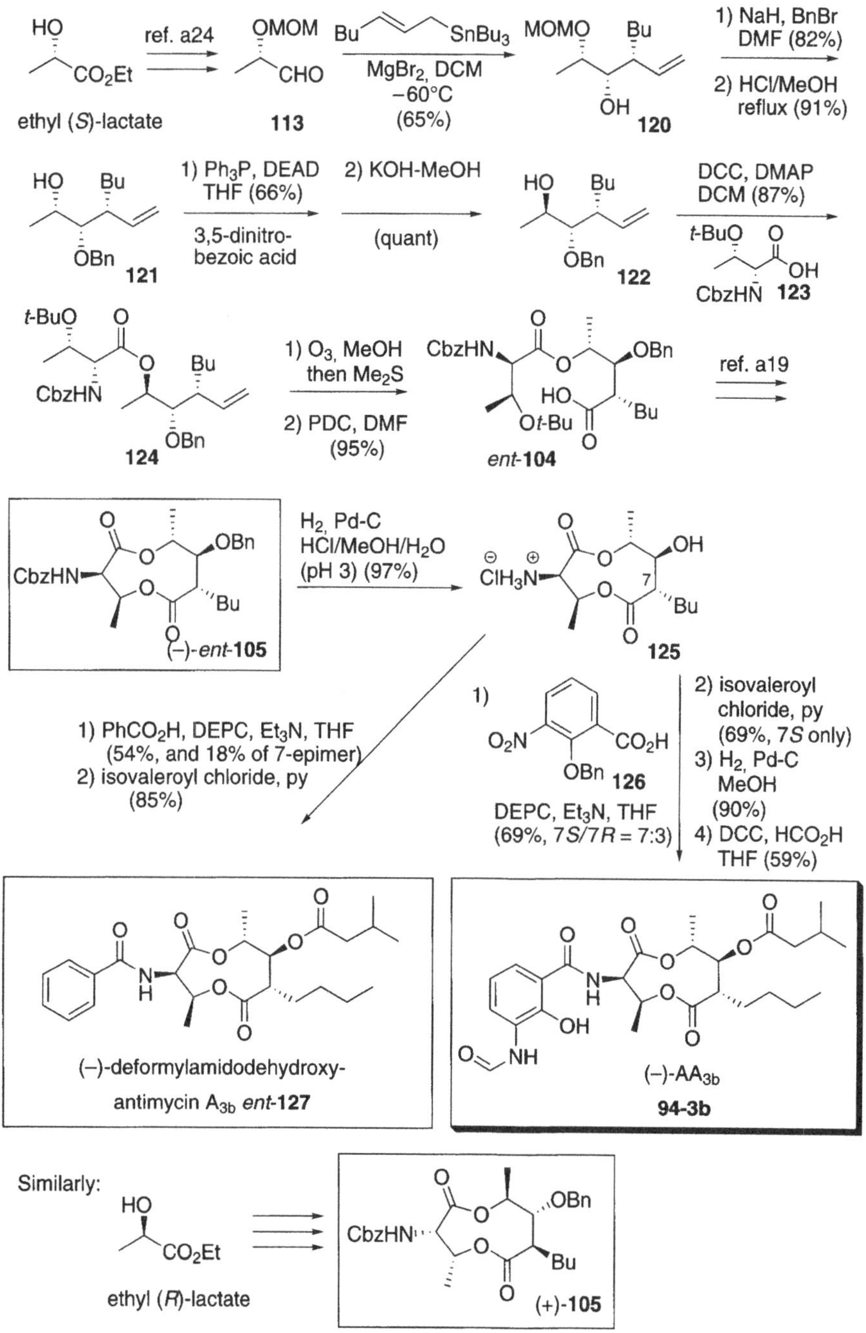

Scheme 13 Oritani's synthesis of antimycin A_{3b}

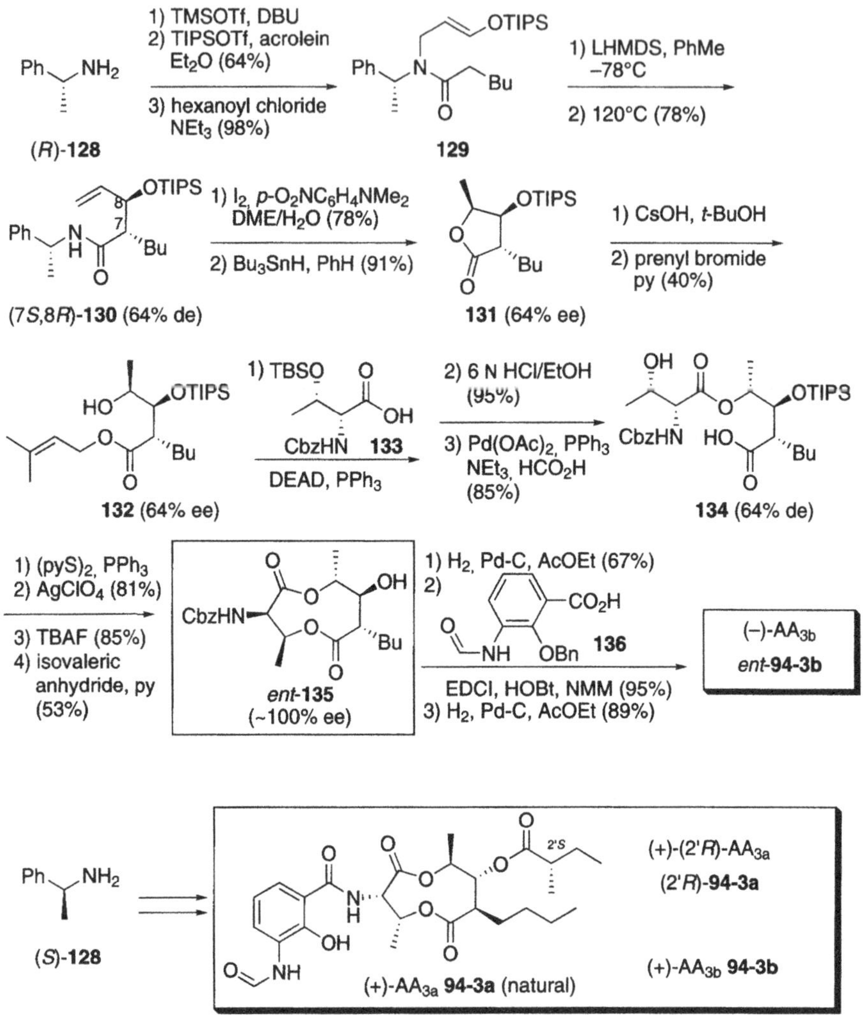

Scheme 14 Tsunoda's synthesis of antimycins A_{3a} and A_{3b}

(−)-AA_{3b}. In addition to natural (+)-AA_{3b}, the two possible stereoisomers of (+)-AA_{3a} were prepared and the absolute configuration of natural (+)-AA_{3a} was determined to be 2′*S* [112].

3.3
Synthesis of UKs

Shimano's group at Kaken Pharmaceutical Co. synthesized UK-2A, UK-3A, and their analogs and determine the stereochemistry (Scheme 15) [114, 115]. Preparation of the three successive asymmetric carbons were done by Evans'

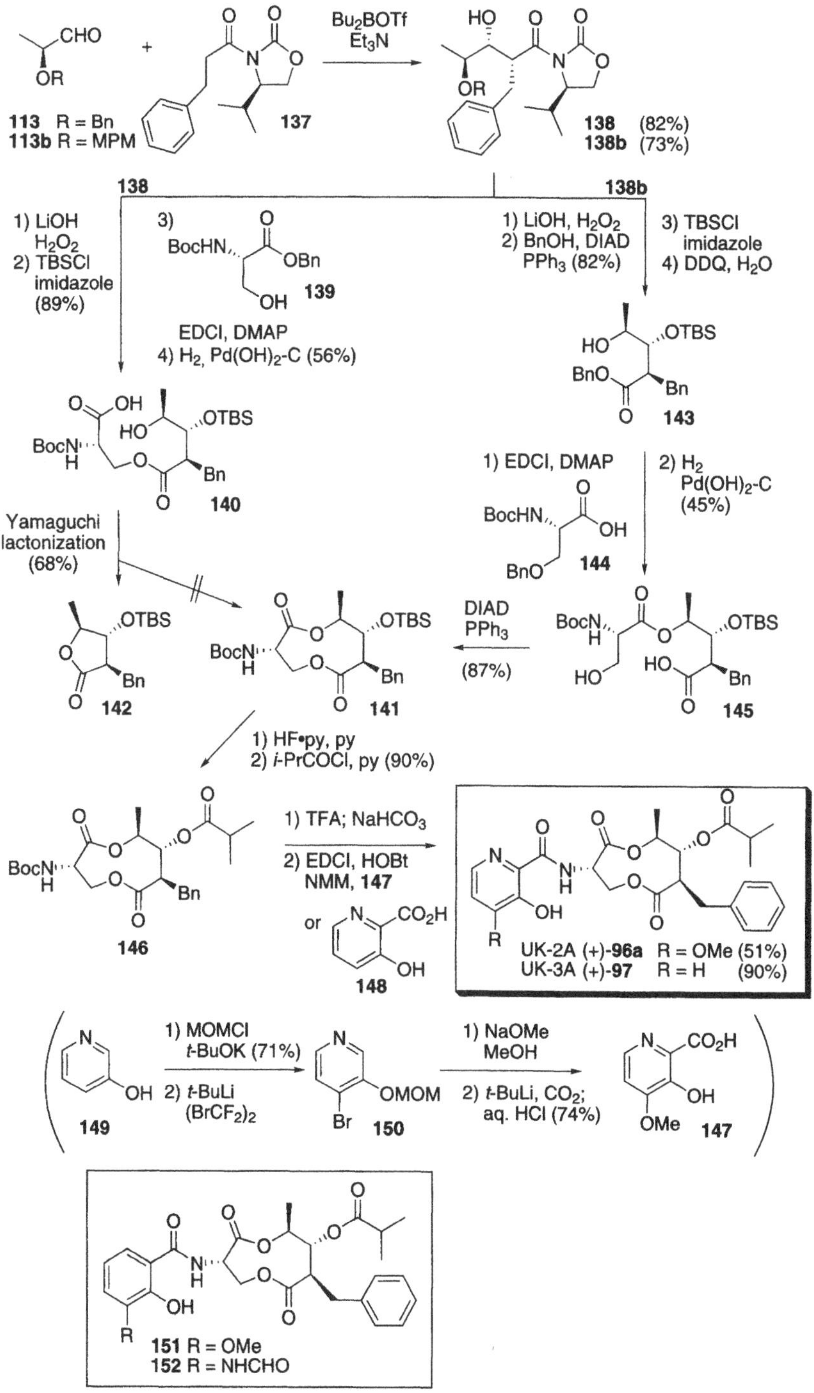

Scheme 15 Shimano's synthesis of UK-2A and UK-3A

syn-aldol reaction of **113** and **137** to afford **138**. Oxidative removal of the chiral oxazolidinone moiety, followed by condensation with the protected L-serine derivative **139**, gave ester **140**. However, macrolactonization of this hydroxy acid **140** failed and every attempt afforded not the desired dilactone **141** but γ-lactone **142** as a major product. In addition, **140** gradually decomposed to **142**. Next, they inversed the order of esterification. The aldol reaction using the corresponding MPM protected compound **113b** gave **138b**, which was converted to benzyl ester **143**. This was condensed with L-serine-derived acid **144**. Macrolactonization of this seco acid succeeded under Mitsunobu conditions to give **141** in 87% yield. Finally, deprotection and acylation provided UK-2A [(+)-**96a**] and UK-3A [(+)-**97**]. The pyridinecarboxylic acid derivative **147** was prepared from 3-hydroxypyridine **149** in four steps. They also synthesized antimycin A-UK hybridized analogs **151** and **152**, and the cytotoxic and antifungal activities of these compounds were tested. In comparison to antimycin A complex, all the synthetic compounds showed less activity against HEL, P-388, and EL-4 cells. On the other hand, UK-2A and **152** inhibited the growth of *Aspergillus* sp. and *Trichophyton* sp. more strongly than antimycin A complex.

4 Tabtoxins

4.1 Tabtoxinine-β-lactam (TβL), a Tobacco Wildfire Disease Toxin

Wildfire disease has been the most serious pest for tobacco since the early twentieth century [116]. Woolley et al. have isolated a phytotoxic compound tabtoxin from the phytopathogen, *Pseudomonas tabaci* [117], whose structure was later determined by Stewart as **153** (Fig. 6) [118]. Tabtoxin **153** and its serine-homolog **154** are inactive themselves. However, when they are hydrolyzed by host plant aminopeptidases [119], the resulting true phytotoxin [120] tabtoxinine-β-lactam (TβL, **155**) [121, 122] causes chlorosis by irreversible inactivation of the host plant glutamine synthetase (GS). Although **153** is available by fermentation, hydrolysis of the amide bond is complicated by isomerization to stable isotabtoxins (**156** and **157**) or tabtoxinine-δ-lactam (TδL, **158**) ($t_{1/2}$ = 24 h at pH 7.0 and 15 min at pH 4.5) [118]. The analogous chlorinated compound **159** was isolated from *Streptomyces* sp. [123]. The mechanism of action of TβL is postulated to be similar to that of GS inhibitor, L-methionine-*S*-sulfoximine (**160**) [124], whose in vivo phosphorylated product **161** mimics the true GS intermediate L-glutamic acid γ-phosphate **162**. *N*- or *O*-phosphorylated TβL (**163** or **164**) would mimic **162** similarly [118, 120].

TβL (**155**) is expected to be a selective pesticide because the tabtoxin resistance gene (*ttr*) has recently been cloned and a tabtoxin-resistant protein

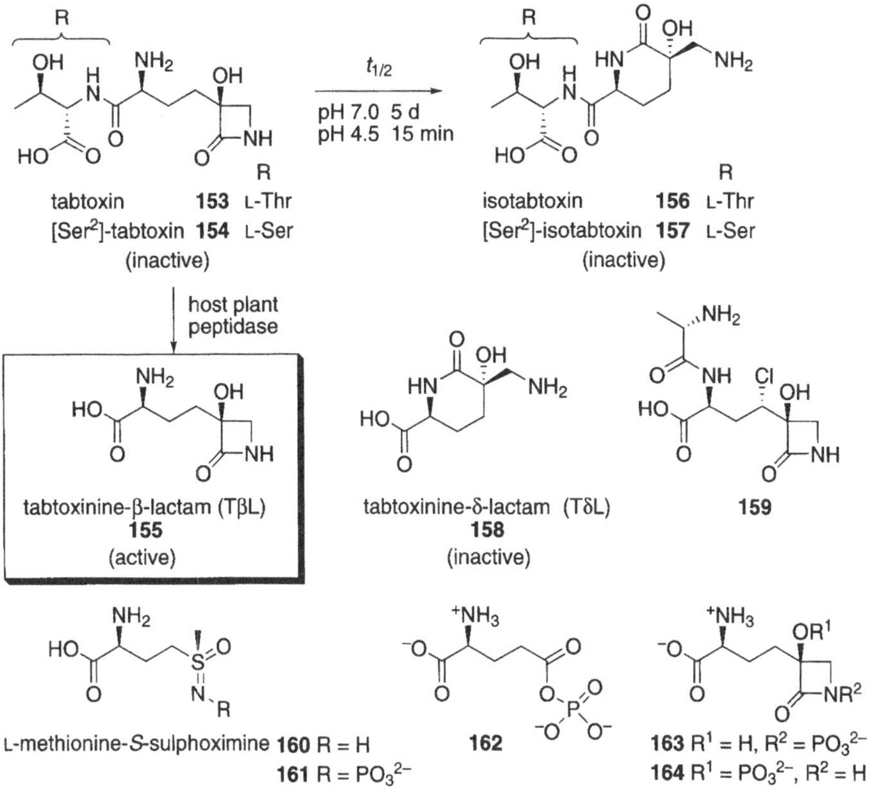

Fig. 6 Tabtoxinine-β-lactam and related compounds

(TTR) characterized [125]. Thus, an effective and careful synthetic procedure for TβL is desired.

4.2 Early Synthesis of Tabtoxin and Related Compounds

4.2.1 Rapoport's Synthesis of Tabtoxinine-δ-Lactam

In 1975, Rapoport et al. reported the synthesis of ($\pm$)-TδL (**158**) and confirmed the relative stereochemistry of tabtoxins (Scheme 16) [126]. Hydrogenation and methylation of isocinchomeronic acid (**165**) gave dimethyl pyperidine-2,5-dicarboxylate (**166**). After protection of the amino group with MPM group, the less hindered methoxycarbonyl group was hydrolyzed to afford **167**. The key rearrangement reaction [127] was accomplished in refluxing acetic anhydride, giving α-methylene lactam **168** in 75% yield. Finally, *m*-CPBA oxidation followed by ammonolysis provided ($\pm$)-TδL (**158**).

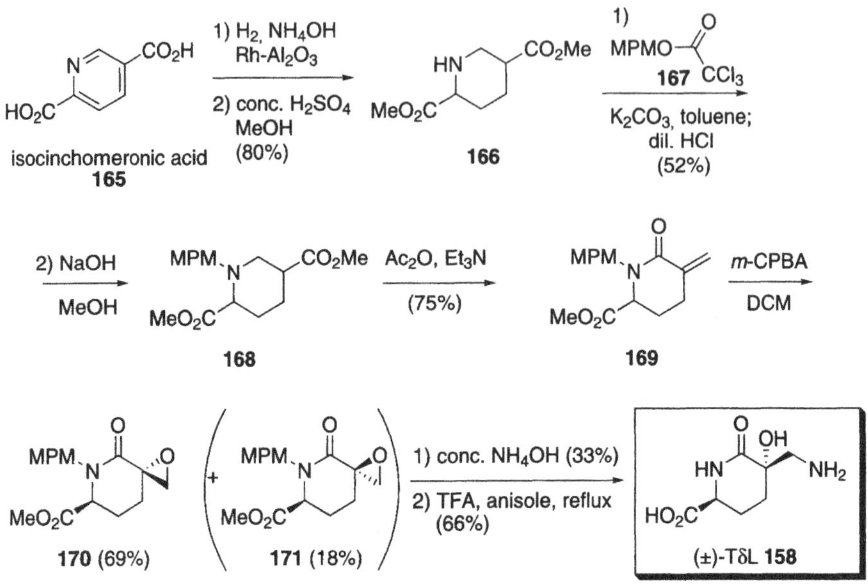

Scheme 16 Rapoport's synthesis of (±)-tabtoxinine-δ-lactam

4.2.2 Baldwin's Synthesis of Tabtoxin and (±)-TβL

Baldwin et al. used the hetero Diels–Alder reaction to simultaneously make two relative stereocenters, and achieved the first synthesis of racemic TβL (**155**) (Scheme 17) [128, 129] and optically active tabtoxin **153** [130, 131]. Benzyl nitrosoformate (**172**) and diene **173** reacted regioselectively, and the cyano group was reduced with $NaBH_3OTFA$ [132] to give bicyclic adduct **174**. After the amino group was protected with Boc group using Boc-ON (**175**) [133], the double bond of **176** was cleaved by the method of Starks [134]. The resulting dicarboxylic acid was mono-protected by decarboxylative esterification, and the Boc group was removed to give **177**. Then, the β-lactam was formed by Ohno's method [135] to give spirocyclic compound **178**. Finally, the N – O bond was cleaved by hydrogenolysis, giving (±)-TβL (**155**) in quantitative yield.

For the synthesis of tabtoxin (**153**), the less-hindered carboxyl group of **179**, derived from the amine **174**, was selectively condensed with di-*O*-benzyl-L-threonine (**180**) [136] via dipivaloyl ester, and the resulting two diastereomers were separated as benzhydryl esters **181** and **182** by recrystallization. Each diastereomer was converted to tabtoxin **153**, and the diastereomer **183**, which showed no biological activity on the tobacco plant.

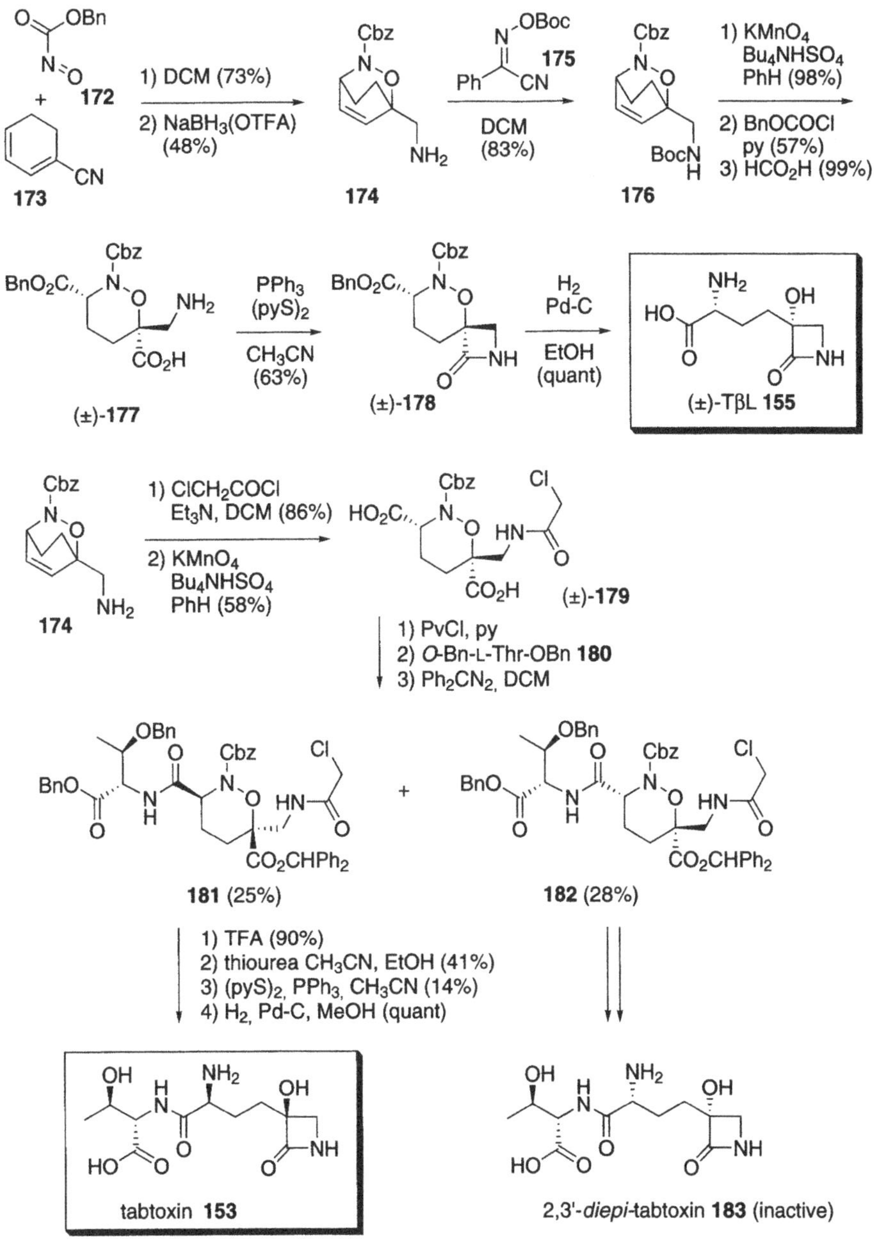

Scheme 17 Baldwin's synthesis of tabtoxin and (±)-tabtoxinine-β-lactam

4.3 Recent Synthesis of TβL

4.3.1 Dolle's Synthesis of (–)-TβL

In 1992, Dolle et al. synthesized (–)-TβL (**155**) and (3′*R*)-epimer from (*S*)-lactate (Scheme 18) [137]. Their strategy also involved the formation of spirocyclic compounds for separation of diastereomers. *E*-selective Wittig olefination of lactaldehyde **184** afforded alcohol **185**, whose hydroxyl group was

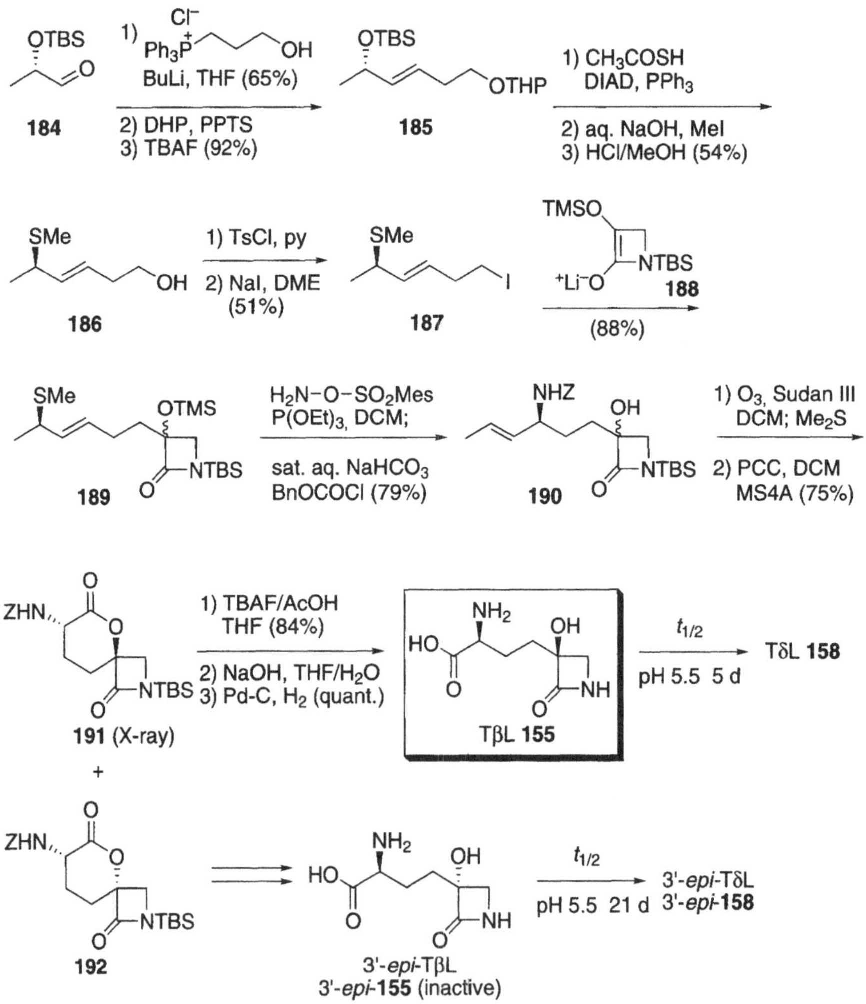

Scheme 18 Dolle's synthesis of (–)-tabtoxinine-β-lactam

inverted by Mitsunobu reaction using thioacetic acid, followed by saponification and methylation to give sulfide **186**. The corresponding iodide **187** was coupled with the β-lactam fragment **188** [138], then the key sulfilimine 2,3-sigmatropic rearrangement [139] provided amine **190** with > 85% retention of the stereochemistry. Ozonolysis of the double bond, followed by PCC oxidation, afforded bicyclic lactones **191** and **192**. Each diastereomer was converted to TβL (**155**) and 3′-*epi*-TβL (3′-*epi*-**155**). The epimer 3′-*epi*-**155** showed no inhibitory activity for GS.

4.3.2
Kiyota's Synthesis of (–)-TβL

Kiyota et al. applied asymmetric reactions to construct the stereocenter at the β-lactam ring of (–)-TβL (**155**) (Scheme 19) [140]. Their synthesis began with allylation of L-serine-derived iodide **193** [141]. Radical reaction with stannyl methacrylate **194** [142] gave α,β-unsaturated ester **195**. On the other hand, zinc-mediated coupling reaction [143] with bromide **196** successfully afforded allylic TIPS ether **199** in good yield. Construction of a quaternary chiral center was achieved by using Sharpless aminohydroxylation (81% de) [144], however, further conversion of the *N*-tosyl group of **198** failed. Sharpless asymmetric dihydroxylation [145] showed low diastereoselectivity for ester **195** (< 38% de) but good for ether **197** (95% de). Their preliminary study tried to convert diol ester **199** to the final compound. Thus, the corresponding amine **201** was cyclized by using Lynch's method [146, 147] via *N*,*O*-bis-protection (**202**) to give TβL (**155**, 24% de) [148].

Diol **200β** was used for the improved synthesis [138]. The primary hydroxyl group was converted to *N*-benzyloxy amide **203**, which was recrystallized as mesylate **204** to diastereomerically pure form. Then, intramolecular S_N2-type β-lactam formation was successful by Miller's procedure [149] after protection of the tertiary hydroxyl group, giving a mixture of **205** and carboxylic acid **206**. Finally, the mixture was deprotected to give (–)-TβL (**155**). The overall yield was 28% in 12 steps from **193**. **200α** was also transformed to 3′-*epi*-TβL (3′-*epi*-**155**) in a similar manner.

5
Conclusion

Examples of synthetic studies of heterocyclic antibiotics involving nitrogen atoms in their structure (glutarimides, antimycins, and tabtoxins) have been described. Further studies including total and semi-synthesis of analogous compounds as well as natural products are essential for elucidation of the SAR studies, and for the development of novel agrochemicals and medicines.

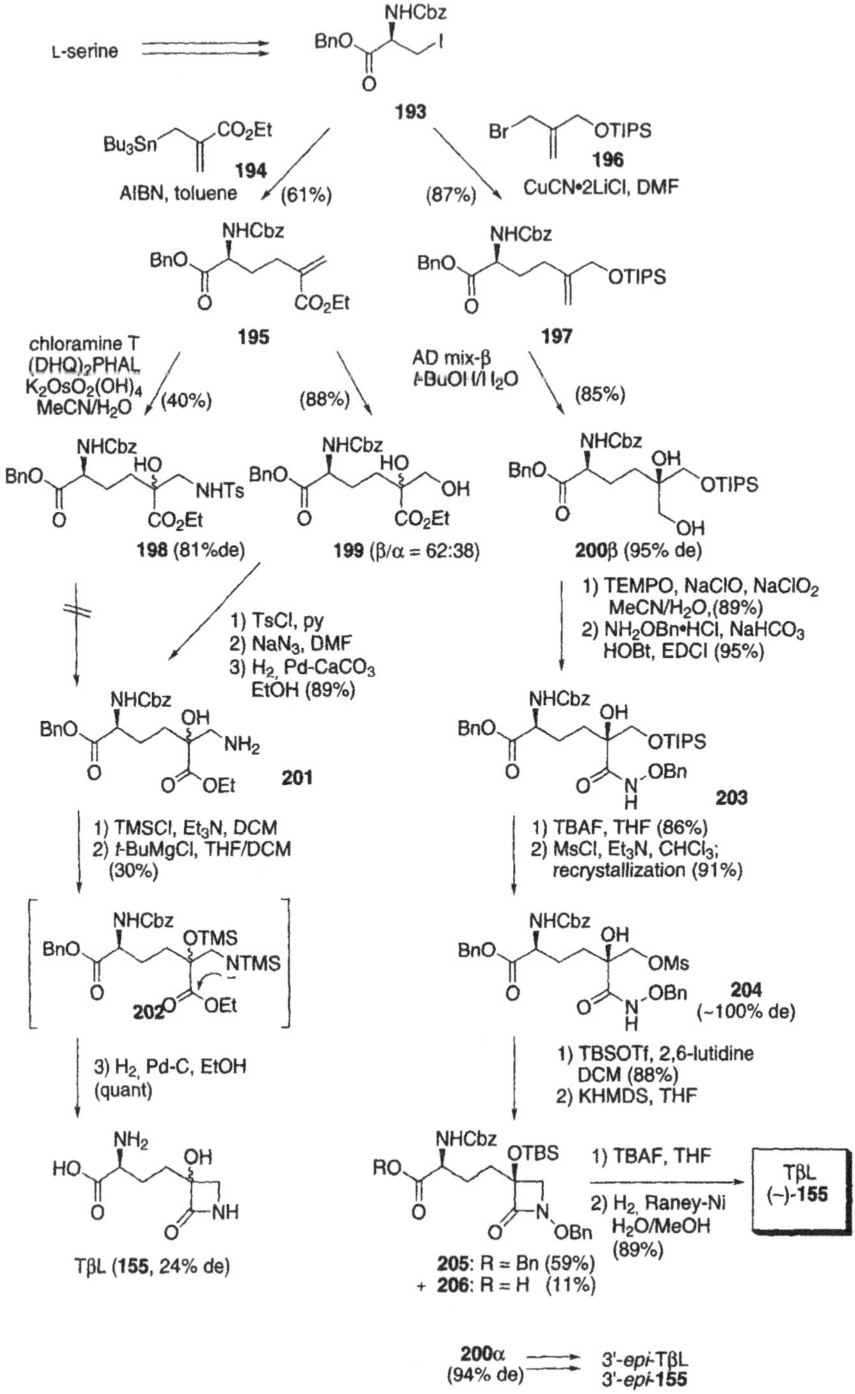

Scheme 19 Kiyota's synthesis of (–)-tabtoxinine-β-lactam

References

1. Whiffen AJ, Bohonons JN, Emerson RL (1946) J Bact 52:610
2. Kornfeld EC, Jones RG (1947) Science 108:437
3. Ford JH, Klomparens W, Hamner CL (1958) Plant Dis Rep 42:680
4. Johnson F (1971) Fortschr Org Naturst 21:140
5. Kudo S, Oritani T, Yamashita K (1984) Agric Biol Chem 48:2315
6. Kudo S, Oritani T, Yamashita K (1984) Agric Biol Chem 48:2739
7. Kudo S (1984) PhD thesis, Tohoku University
8. Okuda T, Suzuki M, Egawa Y, Ashino K (1959) Chem Pharm Bull 7:27
9. Eble TE, Bergy ME, Herr RR, Fox JA (1960) Antibiot Chemoth 10:479
10. Johnson F (1973) The total synthesis of antibiotics. In: ApSimon J (ed) The total synthesis of natural products, vol 1. Wiley, New York, p 451
11. Johnson F, Starkovsky NA, Gurowitz WD (1965) J Am Chem Soc 87:3492
12. Johnson F, Starkovsky NA, Paton AC, Carlson AA (1966) J Am Chem Soc 88:149
13. Oritani T, Yamashita K (1973) Agric Biol Chem 37:1687
14. Kudo S, Oritani T, Yamashita K (1987) Agric Biol Chem 51:263
15. Egawa Y, Suzuki M, Okuda T (1963) Chem Pharm Bull 11:589
16. Oritani T, Kudo S, Yamashita K (1982) Agric Biol Chem 46:757
17. Johnson F, Carlson AA, Starkovsky NA (1966) J Am Chem Soc 88:1327
18. Kondo H, Oritani T, Yamashita K (1988) Agric Biol Chem 52:129
19. Herr RR (1959) J Am Chem Soc 81:2595
20. Rao KV (1960) J Am Chem Soc 82:1129
21. Kudo S, Oritani T, Yamashita K (1985) Agric Biol Chem 49:2919
22. Kondo H, Oritani T, Yamashita K (1990) Agric Biol Chem 54:1531
23. Paul P, Tchelitcheff S (1955) Bull Chim Soc Fr:1316
24. Highet RJ, Prelog V (1959) Helv Chim Acta 42:1523
25. Sonoda T, Osada H, Uzawa J, Isono K (1991) J Antibiot 44:160
26. Kiyota H, Shimizu Y, Oritani T (2000) Tetrahedron Lett 41:5887
27. Johnson F (1962) J Org Chem 27:3658
28. Kasahara A, Izumi T, Yodono M, Saito R, Takeda T, Sugawara T (1973) Bull Chem Soc Jpn 46:1220
29. Stoermer R, Ghl F (1903) Chem Ber 36:2873
30. Matsuda F, Kawasaki M, Terashima S (1985) Tetrahedron Lett 26:4639
31. Ishii H, Ishikawa T, Takeda S, Ueki S, Suzuki M (1992) Chem Pharm Bull 40:1148
32. Kiyota H, Shimizu Y, Oritani T (2001) J Pesticide Sci 25:93
33. Saito N, Kitame F, Kikuchi M, Ishida N (1974) J Antibiot 27:206
34. Otani T, Sakaki T, Minami Y, Marunaka T, Yu QW (1989) J Antibiot 42:647
35. Otani T, Minami Y, Matsumoto H, Marunaka T, Low ZX, Yu QW (1989) J Antibiot 42:654
36. Sugawara R, Matsumae A, Hata T (1963) J Antibiot 16A:111
37. Frohardt RP, Dion HW, Jakubowski ZL, Rydeer A, French JC, Bartz QR (1959) J Am Chem Soc 81:5500
38. Allen MS, Becker AM, Rickards RW (1976) Aust J Chem 29:673
39. Kim BS, Moon SS, Hwang BK (1999) J Agric Food Chem 47:3372
40. Kondo H, Oritani T, Kiyota H (2000) Eur J Org Chem:3459
41. Nagaoka H, Kishi Y (1981) Tetrahedron 37:3873
42. Ukai J, Ikeda Y, Ikeda N, Yamamoto H (1983) Tetrahedron Lett 24:4029
43. Phillips DD, Acitelli MA, Meinwald J (1957) Helv Chim Acta 79:3517
44. Osada H, Sonoda T, Kusakabe H, Isono K (1989) J Antibiot 42:1599

45. Sonoda T, Osada H, Uramoto M, Uzawa J, Isono K (1989) J Antibiot 42:1607
46. Sonoda T, Osada H, Magae J, Isono K (1990) Agric Biol Chem 54:1259
47. Ubukata M, Radha Rani B, Cui CB, Osada H (1995) J Antibiot 48:1176
48. Sonoda T, Kobayashi K, Ubukata M, Osada H, Isono K (1992) J Antibiot 45:1963
49. Dow RL, Hain MA, Lowe III JA (1992) Tetrahedron Lett 33:309
50. Roth M, Dubs P, Gotschi E, Eschenmoser A (1971) Helv Chim Acta 54:710
51. Ubukata M, Sonoda T, Isono K (1992) Nat Prod Lett 1:149
52. Schierlein WK, Brufani M, Muntwyier R, Richle W (1971) Helv Chim Acta 54:44
53. Toshimitsu A, Terano K, Uemura S (1986) Tetrahedron Lett 25:5917
54. Sugawara K, Nishiyama Y, Toda S, Komiyama N, Hatori M, Moriyama T, Sawada Y, Kamei H, Konishi M, Oki T (1992) J Antibiot 45:1433
55. Nakae K, Yoshimoto Y, Sawa T, Homma Y, Hamada M, Takeuchi T, Imoto M (2000) J Antibiot 53:1130
56. Nakae K, Yoshimoto Y, Ueda M, Sawa T, Takahashi Y, Naganawa H, Takeuchi T, Imoto M (2000) J Antibiot 53:1228
57. Takemoto Y, Nakae K, Kawatani M, Takahashi Y, Naganawa H, Imoto M (2001) J Antibiot 54:1104
58. Ohtani I, Kusumi T, Kashman Y, Kakisawa H (1991) J Am Chem Soc 113:4092
59. Nakamura H, Takahashi Y, Naganawa H, Nakae K, Imoto M, Shiro M, Matsumura K, Watanabe H, Kitahara T (2002) J Antibiot 554:442
60. Woo EJ, Starks CM, Carney JR, Arslanian R, Cadapan L, Zavala S, Licari P (2002) J Antibiot 55:141
61. Takayasu Y, Tsuchiya K, Aoyama T, Sukenaga Y (2001) J Antibiot 54:1111
62. Takayasu Y, Tsuchiya K, Sukenaga Y (2002) J Antibiot 55:337
63. Karwowski JP, Jackson M, Sunga G, Shldon P, Poddig JB, Kohl WL, Adam S (1994) J Antibiot 47:862
64. Hochlowski JE, Whittern DN, Hill P, MacAlpine JB (1994) J Antibiot 47:870
65. Kadam S, McAlpine JB (1994) J Antibiot 47:875
66. Ju J, Lim SK, Jiang H, Shen B (2005) J Am Chem Soc 127:1622
67. Ju J, Lim SK, Jiang H, Seo JW, Shen B (2005) J Am Chem Soc 127:11930
68. Gaul C, Danishefsky SJ (2002) Tetrahedron Lett 43:9039
69. Gaul C, Njardarson JT, Danishefsky SJ (2003) J Am Chem Soc 125:6042
70. Gaul C, Njardarson JT, Shan D, Dorn DC, Wu KD, Tong WP, Huang XY, Moore MAS, Danishefsky SJ (2004) J Am Chem Soc 124:11326
71. Lee WW, Chang S (1999) Tetrahedron: Asymmetry 10:4473
72. Evans DA, Tedrow JS, Shar JT, Downey CW (2002) J Am Chem Soc 124:392
73. Scholl M, Ding S, Lee CW, Grubbs RH (1999) Org Lett 1:953
74. Shan D, Chen L, Njardarson JT, Gaul C, Ma XJ, Danishefsky SJ (2003) Proc Natl Acad Sci 102:3772
75. Le Brazidec JY, Gilson CA III, Boehm MF (2005) J Org Chem 70:8212
76. Blakemore PR, Cole JC, Kocienski PJ, Morley A (1998) Synlett:26
77. Dunshee BR, Leben C, Keitt GK, Strong FM (1946) J Am Chem Soc 71:2436
78. Bycroft DW (1988) Dictionary of antibiotics and related substances. Chapman and Hall, London
79. Barrow CJ, Oleynek JJ, Marinelli V, Sun HH, Kaplita P, Sedlock DM, Gillum AM, Chadwick CC, Cooper R (1997) J Antibiot 50:729
80. Omura S, Otani M, Hata T (1970) Kitasato Arch Exp Med 43:29
81. Hatefi Y, Haavik AG, Griffith DE (1962) J Biol Chem 237:1681
82. van Tamelen EE, Dickie JP, Loomans ME, Dewey RS, Strong FM (1961) J Am Chem Soc 83:1639

83. Tappel AL (1960) Biochem Pharmacol 3:289
84. Dickie JP, Loomans ME, Farley TM, Strong FM (1963) J Med Chem 6:424
85. Selwood DL, Livingstone DJ, Comley JC, O'Dowd AB, Hudson AT, Jackson P, Jandu KS, Rose VS, Stables JN (1990) J Med Chem 33:136
86. Kluepfel D, Sehgal SN, Vezina C (1970) J Antibiot 23:75
87. Tokutake N, Miyoshi H, Satoh T, Hatano T, Iwamura H (1994) Biochim Biophys Acta-Bioenerg 1185:271
88. Miyoshi H, Tokutake N, Imaede Y, Akagi T, Iwamura H (1995) Biochim Biophys Acta 1229:149
89. Usuki Y, Mitomo K, Adachi N, Ping X, Fujita KI, Sakanaka O, Iinuma K, Iio H, Taniguchi M (2005) Bioorg Med Chem Lett 15:2011
90. Ishiyama T, Endo T, Otake N, Yonehara H (1976) J Antibiot 29:804
91. Imamura N, Nishijuma M, Adachi K, Sano H (1993) J Antibiot 46:241
92. Hayashi K, Nozaki H (1999) J Antibiot 52:325
93. Hanafi M, Shibata K, Ueki M, Taniguchi M (1996) J Antibiot 49:1226
94. Ueki M, Kusumoto A, Hanafi M, Shibata K, Tanaka T, Taniguchi M (1997) J Antibiot 50:551
95. Kinoshita M, Umezawa S (1970) Bull Chem Soc Jpn 42:854
96. Kinoshita M, Aburaki S, Wada M, Umezawa S (1973) Bull Chem Soc Jpn 46:1279
97. Aburaki S, Kinoshita M (1976) Chem Lett:701
98. Aburaki S, Kinoshita M (1979) Bull Chem Soc Jpn 52:198
99. Mukaiyama T, Araki M, Takei H (1973) J Am Chem Soc 95:4763
100. Gerlach H, Thalmann A (1974) Helv Chim Acta 57:2661
101. Yonehara H, Takeuchi S (1958) J Antibiot Ser A 11:254
102. Nakata T, Fukui M, Oishi T (1983) Tetrahedron Lett 24:2657
103. Wassermann HH, Gambale RJ (1985) J Am Chem Soc 107:1423
104. Mulzer J, Schulze T, Strecker A, Denzer W (1988) J Org Chem 53:4098
105. Hiyama T, Okude Y, Kimura K, Nozaki H (1982) Bull Chem Soc Jpn 55:561
106. Miyoshi H, Kondo H, Oritani T, Saitoh I, Iwamura H (1991) FEBS Lett 292:61
107. Kondo H (1991) PhD thesis, Tohoku University
108. Kondo H, Oritani T, Kiyota H (2000) Heterocycl Commun 6:211
109. Keck GE, Boden EP (1984) Tetrahedron Lett 25:1879
110. Schröder E (1963) Liebigs Ann Chem 670:127
111. Tsunoda T, Nishii T, Yoshizuka M, Yamasaki C, Suzuki T, Itô S (2000) Tetrahedron Lett 41:7667
112. Nishii T, Suzuki S, Yoshida K, Arakaki K, Tsunoda T (2003) Tetrahedron Lett 44:7829
113. Tunoda T, Sakai M, Sasaki O, Sako Y, Hondo Y (1992) Tetrahedron Lett 33:1651
114. Shimano M, Shibata T, Kamei N (1998) Tetrahedron Lett 39:4363
115. Shimano M, Kamei N, Shibata T, Inoguchi K, Itoh N, Ikari T, Senda H (1998) Tetrahedron 54:12745
116. Wolf FA, Foster AC (1917) Science 46:361
117. Woolley DW, Shaffner G, Braun AC (1955) J Biol Chem 215:485
118. Stewart WW (1971) Nature 229:174
119. Sinden SL, Dubbin RD (1968) Nature 219:379
120. Uchytil TF, Durbin RD (1980) Experientia 36:301
121. Taylor PA, Schnoes HK (1972) Biochim Biophys Acta 286:107
122. Durbin RD, Uchytil TF, Steele JA, Ribeiro RLD (1978) Phytochemistry 17:147
123. Scannell JP, Pruess DL, Blount JF, Ax HA, Kellett M, Weiss F, Demny TC, Williams TH, Stempel A (1975) J Antibiot 28:1
124. Rowe WB, Ronzio RA, Meister A (1969) Biochemistry 8:2674

125. He H, Ding Y, Bartlam M, Sun F, Le Y, Qin X, Tang H, Zhang R, Joachimiak A, Liu J, Zhao N, Rao Z (2003) J Mol Biol 325:1019
126. Lee DL, Rapoport H (1975) J Org Chem 40:3491
127. Lee DL, Morrow CJ, Rapoport H (1974) J Org Chem 39:893
128. Baldwin JE, Otsuka M, Wallace PM (1985) Chem Commun:1549
129. Baldwin JE, Otsuka M, Wallace PM (1986) Tetrahedron 42:3097
130. Baldwin JE, Baily PD, Gallacher G, Singleton KA, Wallace PM (1983) Chem Commun:1049
131. Baldwin JE, Bailey PD, Gallacher G, Otsuka M, Singleton KA, Wallace PM (1984) Tetrahedron 40:3695
132. Umino N, Iwakuma T, Itoh N (1976) Tetrahedron Lett:2875
133. Itoh M, Hagiyama D, Kamiya T (1977) Bull Chem Soc Jpn 50:718
134. Starks CM (1971) J Am Chem Soc 93:195
135. Kobayashi S, Iimori T, Izawa T, Ohno M (1981) J Am Chem Soc 103:2406
136. Mizoguchi MT, Levin G, Woolley DW, Stewart JM (1968) J Org Chem 33:903
137. Doll RE, Li CS, Novelli R, Kruse LI, Eggleston D (1992) J Org Chem 57:128
138. Doll RE, Li CS, Shaw AN (1989) Tetrahedron Lett 30:4723
139. Doll RE, Hughes MJ, Li CS, Kruse LI (1989) J Chem Soc Chem Commun:1448
140. Kiyota H, Takai T, Saitoh M, Nakayama O, Oritani T, Kuwahara S (2004) Tetrahedron Lett 45:8191
141. Adlington RM, Baldwin JE, Basak A, Kozyrod RP (1983) J Chem Soc Chem Commun:944
142. Baldwin JE, Adlington RM, Birch DJ, Crawford JA, Sweeney JB (1986) J Chem Soc Chem Commun:1339
143. Knochel P, Singer RD (1993) Chem Rev 93:2117–2188
144. Li G, Sharpless KB (1996) Angew Chem Int Ed Engl 35:451
145. Kolb HC, Van Nioeuwenhze MS, Sharpless KB (1994) Chem Rev 94:2483–2547
146. Lynch JK, Hoolladay MW, Ryther KB, Bai H, Hsiao CN, Morton HE, Dickmann DA, Arnold W, King SA (1998) Tetrahedron: Asymmetry 9:2971
147. Baldwin JE, Adlingston RM, Gollins DW, Schoield CJ (1990) Tetrahedron 46:4733
148. Nakayama O (1999) MSc thesis, Tohoku University
149. Miller MJ, Mattingly PG, Morrison MA, Kerwin JF Jr (1980) J Am Chem Soc 102:7026

Author Index Volumes 1–6

The volume numbers are printed in italics

Subject Index

The manufacturer's authorised representative in the EU is Springer Nature Customer Service Centre GmbH, Europaplatz 3, 69115 Heidelberg, Germany. If you have any concerns regarding our products, please contact ProductSafety@springernature.com

Printed and bound by CPI Group (UK) Ltd, Croydon, CR0 4YY
22/09/2026
02221064-0007